Interdisciplinary Applied Mathematics

Volume 30

Problems in engineering, computational science, and the physical and biological sciences are using increasingly sophisticated mathematical techniques. Thus, the bridge between the mathematical sciences and other disciplines is heavily traveled. The correspondingly increased dialog between the disciplines has led to the establishment of the series: *Interdisciplinary Applied Mathematics*.

The purpose of this series is to meet the current and future needs for the interaction between various science and technology areas on the one hand and mathematics on the other. This is done, firstly, by encouraging the ways that that mathematics may be applied in traditional areas, as well as point towards new and innovative areas of applications; and, secondly, by encouraging other scientific disciplines to engage in a dialog with mathematicians outlining their problems to both access new methods and suggest innovative developments within mathematics itself.

The series will consist of monographs and high-level texts from researchers working on the interplay between mathematics and other fields of science and technology.

Interdisciplinary Applied Mathematics

Volumes published are listed at the end of the book

Panos Macheras Athanassios Iliadis

Modeling in Biopharmaceutics, Pharmacokinetics, and Pharmacodynamics

Homogeneous and Heterogeneous Approaches

With 131 Illustrations

Springer

Panos Macheras
School of Pharmacy
Zographou 15771
Greece
Macheras@pharm.uoa.gr

Athanassios Iliadis
Faculty of Pharmacy
Marseilles 13385 CX 0713284
France
Iliadis@pharmacie.univ-mrs.fr

Cover illustration: *Left panel: Stochastic description of the kinetics of a population of particles, Fig 9.15. Middle panel: Dissolution in topologically restricted media, Fig. 6.8B (reprinted with permission from Springer). Right panel: A pseudophase space for a chaotic model of cortisol kinetics, Fig.11.11.*

Mathematics Subject Classification (2000): 92C 45 (main n°), 62P10, 74H65, 60K20.

Library of Congress Control Number: 2005934524

ISBN-10: 0-387-28178-9
ISBN-13: 978-0387-28178-0

Printed in the United States of America. (MVY)

9 8 7 6 5 4 3 2 1

springeronline.com

- ♦ To our ancestors who inspired us
- ♦ To those teachers who guided us
- ♦ To our families

Preface

Η μεγάλη τέχνη βρίσκεται οπουδήποτε ο άνθρωπος κατορθώνει ν'αναγνωρίζει τον εαυτόν του και να τον εκφράζει με πληρότητα μες στο ελάχιστο.

Great art is found wherever man achieves an understanding of self and is able to express himself fully in the simplest manner.

Odysseas Elytis (1911-1996)
1979 Nobel Laureate in Literature
The magic of Papadiamantis

Biopharmaceutics, pharmacokinetics, and pharmacodynamics are the most important parts of pharmaceutical sciences because they bridge the gap between the basic sciences and the clinical application of drugs. The modeling approaches in all three disciplines attempt to:

- describe the functional relationships among the variables of the system under study and
- provide adequate information for the underlying mechanisms.

Due to the complexity of the biopharmaceutic, pharmacokinetic, and pharmacodynamic phenomena, novel physically physiologically based modeling approaches are sought. In this context, it has been more than ten years since we started contemplating the proper answer to the following complexity-relevant questions: Is a solid drug particle an ideal sphere? Is drug diffusion in a well-stirred dissolution medium similar to its diffusion in the gastrointestinal fluids? Why should peripheral compartments, each with homogeneous concentrations, be considered in a pharmacokinetic model? Can the complexity of arterial and venular trees be described quantitatively? Why is the pulsatility of hormone plasma levels ignored in pharmacokinetic-dynamic models? Over time we realized that questions of this kind can be properly answered only with an intuition about the underlying heterogeneity of the phenomena and the dynamics of the processes. Accordingly, we borrowed geometric, diffusional, and dynamic concepts and tools from physics and mathematics and applied them to the analysis of complex biopharmaceutic, pharmacokinetic, and pharmacodynamic phenomena. Thus, this book grew out of our conversations with fellow colleagues, correspondence, and joint publications. It is intended to introduce the concepts of fractals, anomalous diffusion, and the associated nonclassical kinetics, and stochastic modeling, within nonlinear dynamics and illuminate with their use the intrinsic complexity of drug processes in homogeneous and heterogeneous media. In parallel fashion, we also cover in this book all classical models that have direct relevance and application to the biopharmaceutics, pharmacokinetics, and pharmacodynamics.

The book is divided into four sections, with Part I, Chapters 1–3, presenting the basic new concepts: fractals, nonclassical diffusion-kinetics, and nonlinear dynamics; Part II, Chapters 4–6, presenting the classical and nonclassical models used in drug dissolution, release, and absorption; Part III, Chapters 7–9, presenting empirical, compartmental, and stochastic pharmacokinetic models; and Part IV, Chapters 10 and 11, presenting classical and nonclassical pharmacodynamic models. The level of mathematics required for understanding each chapter varies. Chapters 1 and 2 require undergraduate-level algebra and calculus. Chapters 3–8, 10, and 11 require knowledge of upper undergraduate to graduate-level linear analysis, calculus, differential equations, and statistics. Chapter 9 requires knowledge of probability theory.

We would like now to provide some explanations in regard to the use of some terms written in italics below, which are used extensively in this book starting with *homogeneous* vs. *heterogeneous* processes. The former term refers to kinetic processes taking place in well-stirred, Euclidean media where the classical laws of diffusion and kinetics apply. The term *heterogeneous* is used for processes taking place in disordered media or under topological constraints where classical diffusion-kinetic laws are not applicable. The word *nonlinear* is associated with either the kinetic or the dynamic aspects of the phenomena. When the kinetic features of the processes are nonlinear, we basically refer to Michaelis–Menten-type kinetics. When the dynamic features of the phenomena are studied, we refer to nonlinear dynamics as delineated in Chapter 3.

A *process* is a real entity evolving, in relation to time, in a given environment under the influence of internal mechanisms and external stimuli. A *model* is an image or abstraction of reality: a mental, physical, or mathematical representation or description of an actual process, suitable for a certain purpose. The model need not be a true and accurate description of the process, nor need the user have to believe so, in order to serve its purpose. Herein, only mathematical models are used. Either processes or models can be conceived as boxes receiving inputs and producing outputs. The boxes may be characterized as gray or black, when the internal mechanisms and parameters are associated or not with a physical interpretation, respectively. The *system* is a complex entity formed of many, often diverse, interrelated elements serving a common goal. All these elements are considered as *dynamic processes* and *models*. Here, deterministic, random, or chaotic real processes and the mathematical models describing them will be referenced as *systems*. Whenever the word "system" has a specific meaning like *process* or *model*, it will be addressed as such.

For certain processes, it is appropriate to describe globally their properties using numerical techniques that extract the basic information from measured data. In the domain of linear processes, such techniques are correlation analysis, spectral analysis, etc., and in the domain of nonlinear processes, the correlation dimension, the Lyapunov exponent, etc. These techniques are usually called *nonparametric models* or, simply, indices. For more advanced applications, it may be necessary to use models that describe the functional relationships among the system variables in terms of mathematical expressions like difference or differential equations. These models assume a prespecified parametrized structure. Such models are called *parametric models.*

Usually, a mathematical model simulates a process behavior, in what can be termed a *forward problem.* The *inverse problem* is, given the experimental measurements of behavior, what is the structure? A difficult problem, but an important one for the sciences. The inverse problem may be partitioned into the following stages: hypothesis formulation, i.e., model specification, definition of the experiments, identifiability, parameter estimation, experiment, and analysis and model checking. Typically, from measured data, nonparametric indices are evaluated in order to reveal the basic features and mechanisms of the underlying processes. Then, based on this information, several structures are assayed for candidate parametric models. Nevertheless, in this book we look only into various aspects of the forward problem: given the structure and the parameter values, how does the system behave?

Here, the use of the term "model" follows Kac's remark, "models are caricatures of reality, but if they are good they portray some of the features of the real world" [1]. As caricatures, models may acquire different forms to describe the same process. Also, Fourier remarked, "nature is indifferent toward the difficulties it causes a mathematician," in other words the mathematics should be dictated by the biology and not vice versa. For choosing among such competing models, the "parsimony rule," Occam's "razor rule," or Mach's "economy of thought" may be the determining criteria. Moreover, modeling should be dependent on the purposes of its use. So, for the same process, one may develop models for process identification, simulation, control, etc. In this vein, the tourist map of Athens or the system controlling the urban traffic in Marseilles are both tools associated with the real life in these cities. The first is an identification model, the second, a control model.

Over the years we have benefited enormously from discussions and collaborations with students and colleagues. In particular we thank P. Argyrakis, D. Barbolosi, A. Dokoumetzidis, A. Kalampokis, E. Karalis, K. Kosmidis, C. Meille, E. Rinaki, and G. Valsami. We wish to thank J. Lukas whose suggestions and criticisms greatly improved the manuscript.

A. Iliadis
Marseilles, France
August 2005

P. Macheras
Piraeus, Greece
August 2005

Contents

List of Figures

Part I

BASIC CONCEPTS

Biopharmaceutics, pharmacokinetics, and pharmacodynamics have been developed and expanded, as have many other scientific subjects, based on the concepts of *homogeneity* and *linearity*.

Homogeneity is a presupposition in almost all research in these fields. Some typical examples can be quoted. The shape of a drug particle is assumed to be an ideal sphere and its surface smooth. The permeability of the intestinal complex membrane is considered constant along the gastrointestinal tract while the concentration of drug is postulated to be homogeneous in the gastrointestinal fluids. Homogeneous conditions are assumed for each one of the hypothetical compartments of pharmacokinetic models and the effect compartment of pharmacokinetic-dynamic models. However, common intuition and scientific knowledge tell us that the drug particle is not an ideal sphere, and its surface is not smooth: the permeability of the gastrointestinal wall is position- and time-dependent. Moreover, the assumed concentration homogeneity of drug in (1°) the gastrointestinal tract, (2°) the peripheral compartments of compartmental systems, and (3°) the biophase at receptor's level, is synonymous with a well mixed system.

Linearity is the basic assumption behind classical biopharmaceutics, pharmacokinetics, and pharmacodynamics. Formally, a system is linear if the output of an operation is proportional to the input. This property of proportionality along with the property of independence, i.e., the response of the system to an input is equal to the sum of the outcomes, is the fundamental feature of linear systems. In contrast, nonlinear kinetic phenomena are frequently observed in pharmacokinetic and pharmacodynamic studies, when the output (e.g., the area under the time–concentration curve) is nonlinearly related to the input (e.g., the dose). However, we deal with complicated biological systems consisting of a large number of interrelated components and processes. Linear approaches neglect these intrinsic relations and therefore are not complete.

In various fields of research, e.g., physics, chemistry, and physiology, scientists are increasingly finding that at the research level it is the nonlinear phenomena that control the game; physical or physiological heterogeneity is everywhere, while heterogeneous conditions prevail in numerous physical, physiological, and biochemical processes. Today, science shows that the real world is relentlessly nonlinear, and therefore the techniques of nonlinear dynamics are required to analyze the nonlinear phenomena. In parallel, structural and functional heterogeneities can be described and understood with the concept of fractals.

Kinetic processes in various scientific fields are traditionally treated with classical kinetics. The latter is quite satisfactory for reactions and processes in well-stirred media, i.e., under homogeneous conditions. In fact, the kinetics of diffusion-controlled processes and reactions in 3-dimensional homogeneous systems obey the classical laws of diffusion where the rate constant of the process is linearly proportional to the diffusion coefficient. However, this proportionality is not valid for systems with smaller dimensions, fractal spaces, or disordered systems since the laws of transport are different in these media. Accordingly, fractal kinetics has been developed since classical kinetics was found to be un-

Table 1: Classical and nonclassical considerations of the in vitro and in vivo drug processes.

Fields	**Classical**	**Nonclassical**
Geometry	Euclidean	Fractal
Topology	Ordered media	Disordered media
Diffusion	Regular	Anomalous
Kinetics	Deterministic	Stochastic
Dynamics	Linear	Nonlinear

satisfactory under dimensional constraints, e.g., phase boundaries, understirred media, or membrane reactions.

The concepts delineated above are quoted in Table 1 in a comparative manner. The clear message of the middle column of Table 1 is that classical behavior is expected when the phenomena are taking place in Euclidean spaces and/or ordered media (e.g., well-stirred systems) since the classical laws of diffusion and kinetics are valid. According to the third column of Table 1, deviations from the normal behavior appear when the processes take place in fractal spaces and/or under topological constraints since neither diffusion nor kinetics follow the classical pattern in that case. Finally, the dynamics of the systems are linear when the variable of interest is considered to be detached from the remainder of the system. On the contrary, nonlinear dynamic behavior exists when the interaction of the studied variable with the other variables constituting the system is taken into account.

We do believe that it is only through the understanding of underlying principles that adequate exploration in the fields of biopharmaceutics, pharmacokinetics, and pharmacodynamics can be carried out. To this end, in this first, introductory part of the book we deal with the basic nonclassical concepts. Chapter 1 gives the essentials of fractals such as self-similarity, scaling laws, and fractal dimension. Chapter 2 deals with the basics of diffusion-kinetics in either Euclidean or disordered media. The concepts of nonlinear dynamics are discussed in Chapter 3, where reference is made to a number of characteristics of dynamic systems, e.g., phase space, attractor, bifurcation.

1

The Geometry of Nature

The proper route to an understanding of the world is an examination of our errors about it.

Euclid (325-265 BC)

Our understanding of nature has been based on the classical geometric figures of smooth line, triangle, circle, cube, sphere, etc. Each of these regular forms can be determined by a characteristic scale. For example, the length of a straight line can be measured with a ruler that has a finer resolution than the entire length of the line. In general, each Euclidean object has a unique value for its characteristics (length, area, or volume). It is also known that when these objects are viewed at higher magnification they do not reveal any new features.

In the real world, however, the objects we see in nature and the traditional geometric shapes do not bear much resemblance to one another. Mandelbrot [2] was the first to model this irregularity mathematically: clouds are not spheres, mountains are not cones, coastlines are not circles, and bark is not smooth, nor does lightning travel in a straight line. Mandelbrot coined the word *fractal* for structures in space and processes in time that cannot be characterized by a single spatial or temporal scale. In fact, the fractal objects and processes in time have multiscale properties, i.e., they continue to exhibit detailed structure over a large range of scales. Consequently, the value of a property of a fractal object or process depends on the spatial or temporal characteristic scale measurement (ruler size) used.

The physiological implications of the fractal concepts are serious since fractal structures and processes are ubiquitous in living things, e.g., the lung, the vascular system, neural networks, the convoluted surface of the brain, ion channel kinetics, and the distribution of blood flow through the blood vessels. Besides, many applications of fractals exist for the morphology of surfaces, e.g., the surface area of a drug particle, surface reactions on proteins. Thus, fractal geometry allows scientists to formulate alternative hypotheses for experimental observations, which lead to more realistic explanations compared to the traditional approaches. These fractal hypotheses can be expressed in terms quantifying the

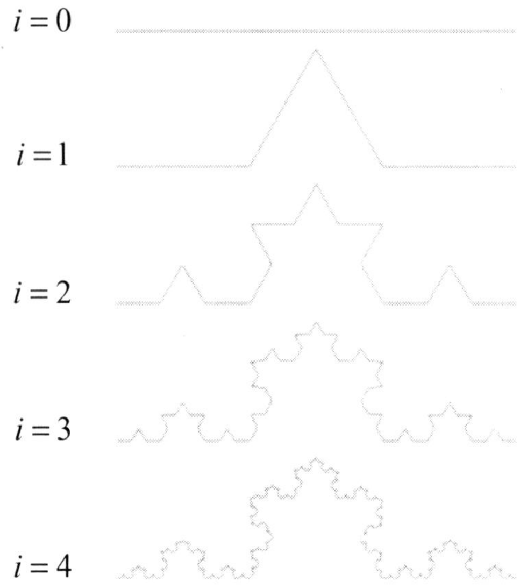

Figure 1.1: The first four iterations of the Koch curve.

fractal properties of the system under study as delineated below.

1.1 Geometric and Statistical Self-Similarity

The most interesting property of fractals is geometric self-similarity, which means that the parts of a fractal object are smaller exact copies of the whole object. Replacement algorithms are used to generate geometric fractals. For example, the Koch curve shown in Figure 1.1 can be produced after four successive replacements according to the following replacement rule: two lines of the same length replace the middle third of the length of the line at each step. Infinite recursions can be applied resulting in a continuous increase of the "line" length by a factor of 4/3 at each successive step. This continuous ramification of the Koch curve leads to a surprising result if one attempts to measure the length of its perimeter: the length is dependent on the ruler size used for its measurement. In fact, the smaller the ruler size used, the longer the perimeter. Accordingly, when we deal with fractal objects or processes we say that their characteristics (length in this case) "scale" with the measurement resolution.

Similar algorithms for area and volume replacement can be used to create fractals from 2- or 3-dimensional objects. The fractals shown in Figure 1.2 are called the Sierpinski triangle (gasket) and Menger sponge. They have been generated from an equilateral triangle and a cube, respectively, applying the following replacement algorithms:

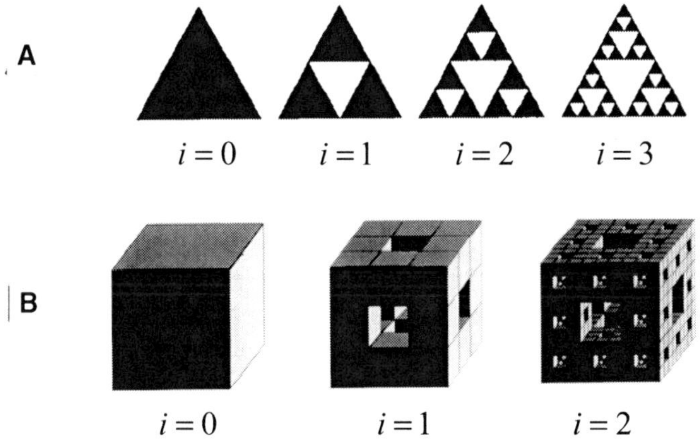

Figure 1.2: Generation of the (**A**) Sierpinski triangle (gasket) (the first three iterations are shown), (**B**) Menger sponge (the first two iterations are shown) from their Euclidean counterparts.

- Sierpinski triangle: At each step an equilateral triangle with area equal to one-quarter of the remaining triangle is removed.
- Menger sponge: At each step one-third of the length of the side of each cube is removed taking care to apply this rule in 3 dimensions and avoiding removal of corner cubes. This means that if the original cube has been constructed from $3 \times 3 \times 3 = 27$ small cubes, after the first iteration 20 small cubes are remaining (6 are removed from the center of the faces and one is removed from the center of the cube).

These line, area, and volume replacement rules give fractal structures (Figures 1.1 and 1.2), which are quite different from the original Euclidean objects. This obvious difference in shape has implications when one considers physical measurements or (bio)chemical processes taking place in Euclidean vs. fractal spaces. For example, surface and/or surface/volume ratios are extremely important for reactions or transport processes taking place at interfaces of different phases like liquid–solid boundaries, e.g., drug dissolution, drug uptake from the gastrointestinal mucosa. In general, objects with fractal surfaces are very efficient for surface reactions.

Replacement rules are expressed mathematically by difference equations, which can be used to generate the fractal structures. These equations are usually called maps and have the form

$$z_{i+1} = g(z_i), \tag{1.1}$$

where z_i and z_{i+1} are the input and output, respectively, at two successive steps, while the functional form of g in (1.1) depends on the exact features of the recursion process. The discrete nature of (1.1) allows for a recursive creation of the fractal object utilizing the output z_{i+1} as the next input z_i. In this respect, (1.1) operates like a copy machine, which produces the self-similar object in accord with the rule imposed on g.

The replacement rules used for the generation of fractal objects ensure the geometric self-similarity discussed above. However, the fractal objects or processes we encounter in nature are not generated by exact mathematical rules. For example, some biological objects with fractal structure like the venular and arterial tree cannot be characterized by geometric self-similarity; rather they possess statistical self-similarity. The fractal is statistically self-similar since the characteristics (such as the average value or the variance or higher moments) of the statistical distribution for each small piece are proportional to the characteristics that concern the whole object. For example, the average rate at which new vessels branch off from their parent vessels in a physiological structure can be the same for large and small vessels. This is due to the fact that portions of fractal biological objects resemble the whole object instead of being exact copies of the whole. The term *random fractal* is used for these fractal structures to underline their statistical character. Also, statistical self-similarity can be observed when time series data are recorded for physiological processes, e.g., the electroencephalogram or the electrocardiogram. In this case, we speak of statistical self-similarity in time and not in space.

At this point, a distinction should be made between geometrically and statistically self-similar fractals. The pure mathematical basis of geometric fractals does not impose any restriction on the range of application of their scaling laws. In contrast, scaling laws for statistically self-similar fractals adhering to biological objects or processes are subject to the limitations imposed by the physiology and/or the resolution of the measurement technique. In other words, experimental data usually obey scaling laws over a finite range of resolution measurements. This important aspect of scaling laws, with regard to the range of their application, should be carefully considered when one is applying scaling principles for the analysis of experimental data.

1.2 Scaling

The issue of scaling was touched upon briefly in the previous section. Here, the quantitative features of scaling expressed as scaling laws for fractal objects or processes are discussed. Self-similarity has an important effect on the characteristics of fractal objects measured either on a part of the object or on the entire object. Thus, if one measures the value of a characteristic $\theta(\omega)$ on the entire object at resolution ω, the corresponding value measured on a piece of the object at finer resolution $\theta(r\omega)$ with $r < 1$ will be proportional to $\theta(\omega)$:

$$\theta(r\omega) = k\theta(\omega), \tag{1.2}$$

where k is a proportionality constant that may depend on r. When statistical self-similarity in time for recordings of an observable is examined, the scale $r\omega$

is a finer time resolution than scale ω. Relation (1.2) reveals that there is a constant ratio k between the characteristic $\theta(\omega)$ measured at scale ω and the same characteristic $\theta(r\omega)$ measured at scale $r\omega$.

The above-delineated dependence of the values of the measurements on the resolution applied suggests that there is no true value of a measured characteristic. Instead, a scaling relationship exists between the values measured and the corresponding resolutions utilized, which mathematically may have the form of a scaling power law:

$$\theta(\omega) = \beta\omega^{\alpha}, \tag{1.3}$$

where β and a are constants for the given fractal object or process studied. Equation (1.3) can be written

$$\ln\theta(\omega) = \ln\beta + \alpha\ln\omega.$$

This equation reveals that when measurements for fractal objects or processes are carried out at various resolutions, the log-log plot of the measured characteristic $\theta(\omega)$ against the scale ω is linear. Such simple power laws, which abound in nature, are in fact self-similar: if ω is rescaled (multiplied by a constant), then $\theta(\omega)$ is still proportional to ω^{a}, albeit with a different constant of proportionality. As we will see in the rest of this book, power laws, with integer or fractional exponents, are one of the most abundant sources of self-similarity characterizing heterogeneous media or behaviors.

1.3 Fractal Dimension

The objects considered are sets of points embedded in a Euclidean space. The dimension of the Euclidean space that contains the object under study is called the *embedding dimension*, d_e, e.g., the embedding dimension of the plane is $d_e = 2$ and of 3-dimensional space is $d_e = 3$.

One is accustomed to associating topological dimensions with special objects: dimension 1 with a curve, dimension 2 with a square, and dimension 3 with a cube. Because there are severe difficulties for the definition of the topological dimension d_t, it is convenient to associate the topological dimension of an object with its *cover dimension* d_o.

A curve in the plane is covered with three different arrangements of disks (Figure 1.3 center). In the right part of the figure there are only pairs of disks with nonempty intersections, while in the center part there are triplets and in the left part even quadruplets. Thus, one can arrange coverings of the curve by only one intersection of each disk with another, and the cover dimension of a line is defined as $d_o = d_t = 1$.

A set of points (Figure 1.3 top) can be covered with disks of sufficiently small radius so that there is no intersection between them. Their covering dimension is $d_o = d_t = 0$. A surface (Figure 1.3 bottom) has covering dimension $d_o = d_t = 2$, because one needs at least two overlapping spheres to cover the surface. The same ideas generalize to higher dimensions.

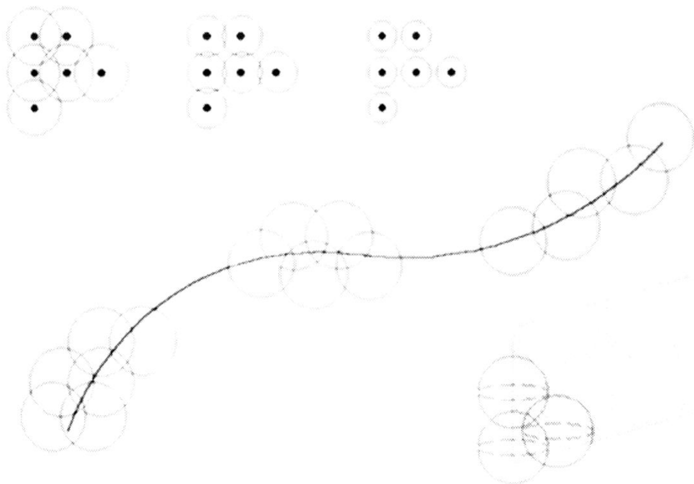

Figure 1.3: The cover dimension.

Similarly, the degree of irregularity of a fractal object is quantified with the *fractal dimension*, d_f. This term is used to show that apart from the Euclidean integer dimensions (1 or 2 or 3) for the usual geometric forms, fractal objects have noninteger dimensions. The calculation of d_f using the concept of self-similarity requires in essence the knowledge of the replacement rule, which dictates how many similar pieces m are found when the scale is reduced by a given factor r at each step. Thus, if we count the number m of the exact copies of the entire geometric fractal that are observed when the resolution of scale is changed by a factor of r, the value of d_f can be derived from

$$d_f = \frac{\ln m}{\ln r} \tag{1.4}$$

after logarithmic transformation of

$$m = r^{d_f}. \tag{1.5}$$

For example, the fractal dimension of the Koch curve is 1.2619 since four ($m = 4$) identical objects are observed (cf. levels $i = 0$ and $i = 1$ in Figure 1.1) when the length scale is reduced by a factor $r = 3$, i.e., $d_f = \ln 4 / \ln 3 \approx 1.2619$. What does this noninteger value mean? The Koch curve is neither a line nor an area since its (fractal) dimension lies between the Euclidean dimensions, 1 for lines and 2 for areas. Due to the extremely ramified structure of the Koch curve, it covers a portion of a 2-dimensional plane and not all of it and therefore its "dimension" is higher than 1 but smaller than 2.

Similarly, the first iteration in the generation of the Sierpinski gasket, (Figure 1.2 A) involves the reduction of the scale by a factor $r = 2$ and results in 3 identical black equilateral triangles ($m = 3$); thus, $d_f = \ln 3/\ln 2 \approx 1.5815$. For the Menger sponge, (Figure 1.2 B), the reduction of the scale by a factor $r = 3$ results in $m = 20$ identical cubes, i.e., $d_f = \ln 20/\ln 3 \approx 2.727$. Both values of d_f are consistent with their dimensions since the Sierpinski gasket lies between 1 and 2, while the Menger sponge lies between 2 and 3.

Equations (1.4) and (1.5) are also valid for Euclidean objects. For example, if one creates $m = 16$ identical small squares in a large square by reducing the length scale by one-fourth, $r = 4$, the value of d_f is $\ln 16/\ln 4 = 2$, which is the anticipated result, i.e., the topological dimension $d_t = 2$ for a plane.

1.4 Estimation of Fractal Dimension

Irrespective of the origin of fractals or fractal-like behavior in experimental studies, the investigator has to derive an estimate for d_f from the data. Since strict self-similarity principles cannot be applied to experimental data extracted from irregularly shaped objects, the estimation of d_f is accomplished with methods that unveil either the underlying replacement rule using self-similarity principles or the power-law scaling. Both approaches give identical results and they will be described briefly.

1.4.1 Self-Similarity Considerations

In principle, the object under study is covered with circles for 1- and 2-dimensional objects or spheres for 3-dimensional objects. This process is repeated using various sizes ω for circles or spheres, while overlapping may be observed. Then, the minimum number of "balls" (circles or spheres) $m(\omega)$ of size ω needed to cover the object are calculated. Finally, the fractal dimension, which in this case is called the *capacity dimension*, d_c is calculated from the relationship

$$d_c = \lim_{\omega \to 0} \frac{\ln m(\omega)}{\ln(1/\omega)}. \tag{1.6}$$

Note that (1.6) relies on the self-similarity concept since the number of identical objects m and the scale factor r in (1.5) have been replaced by the number of "balls" $m(\omega)$ and the reciprocal of the size $1/\omega$, respectively. The limit ($\omega \to 0$) is being used to indicate the estimation of d_c at the highest possible resolution, i.e., as the "ball" size ω decreases continuously.

The reference situation implied in this definition is that at $\omega = 1$, one "ball" covers the object. A clearer definition of d_c is

$$d_c = \frac{\ln\left[m(\omega)/m(1)\right]}{\ln(1/\omega)},$$

or in general, if at $\omega = 1$, k "balls" cover the object,

$$d_c = \frac{\ln\left[m(k\omega)/m(k)\right]}{\ln(k/k\omega)}$$

and

$$d_c = -\frac{\mathrm{d}\ln\left[m\left(\omega\right)\right]}{\mathrm{d}\ln\omega}. \tag{1.7}$$

The capacity dimension tells us how the number of "balls" changes as the size of the "balls" is decreased. This method is usually called *box counting* since the method is implemented in computers with a variety of algorithms utilizing rectangular grids instead of "balls". Dimensions d_f and d_c are quite similar, and the differences between them are usually much smaller than the error of estimates [3].

1.4.2 Power-Law Scaling

When the scaling law (1.3) of the measured characteristic θ can be derived from the experimental data (ω, θ), an estimate of the fractal dimension d_f of the object or process can be obtained as well. In order to apply this method one has first to derive the relationship between the measured characteristic θ and the function of the dimension $g(d_f)$, which satisfies

$$\theta \propto \omega^{g(d_f)}, \tag{1.8}$$

where ω represents the various resolutions used. Then, the exponents of (1.3) and (1.8) are equated,

$$g(d_f) = \alpha, \tag{1.9}$$

and (1.9) is solved in terms of d_f to derive an estimate for d_f.

The form of the function $g(d_f)$ in (1.9) depends on the measured characteristic θ [4]. For instance:

- When the characteristic is the mass of the fractal object, the exponent of (1.8) corresponds to the value of d_f, $d_f = \alpha$.
- When the characteristic is the average density of a fractal object, $d_f = d_e + \alpha$, where d_e is the embedding dimension.
- For measurements regarding lengths, areas, or volumes of objects, a simple equation can be derived using scaling arguments, $d_f = d_e - \alpha$.

Apart from the estimation of d_f from experimental data for mass, density, and purely geometric characteristics, the calculation of d_f for a plethora of studies dealing with various characteristics like frequency, electrical conductivity, and intensity of light is also based on the exact relationship that is applicable in each case between d_f and the scaling exponent α, (1.9).

1.5 Self-Affine Fractals

The replacement rule we have used so far to generate geometric fractals creates isotropic fractals. In other words, the property of geometric self-similarity is the same in all directions. Thus, a unique value for the fractal dimension d_f is being used to quantify an irregular structure. When either the replacement algorithm or the actual physical object exhibits an asymmetry in different directions, then the anisotropic fractal is characterized as a *self-affine fractal*. For example, if one divides a large square into 6 identical small parallelograms and discards 3 of them in an alternate series at each iteration, the result is a disconnected self-affine fractal. Obviously, the unequal horizontal and vertical sides of the parallelograms produced with the successive replacements follow different scaling laws in accord with the dimensions of the sides. The basic difference between self-similarity and self-affinity lies in the fact that self-similar fractals become identical upon simple magnification (classical scaling), while to become identical, self-affine fractals should be scaled by different amounts of the spatial directions. Accordingly, there is no single value of d_f for self-affine fractals; it varies with the ruler size used for measurements. Usually, the box-counting method is applied in conjunction with (1.6) with limits $\omega \to 0$ and $\omega \to \infty$; two estimates for d_f are derived, namely, $d_{f,local}$ and $d_{f,global}$, respectively, and used to characterize a self-affine fractal. Both values indicate limiting values of the fractal dimension: the former is relevant when the size of the boxes decreases infinitely, while the latter corresponds to the largest length scale used for measurements.

1.6 More About Dimensionality

The concept of fractals has helped us to enrich the notion of dimensionality. Apart from the classical systems with dimensions 1, 2 and 3 there are disordered systems with noninteger dimensions.

In the simplest case, a system is called Euclidean or nonfractal if its topological dimension d_t is identical to the fractal dimension d_f. This means $d_t = d_f = 1$ for a curve, $d_t = d_f = 2$ for a surface, and $d_t = d_f = 3$ for a solid. The following relationship holds for the three expressions of dimensionality

$$d_t \leq d_f \leq d_e.$$

Although we have used the value of the fractal dimension d_f as a means to quantify the degree of disorderliness, it is the magnitude of the difference $d_f - d_t$ that in essence reflects how irregular (disordered) the system is. Geometrically speaking, this difference $d_f - d_t$ allows the disordered system to accommodate structure within structure, and the larger this difference is, the more disordered the system.

The above-defined d_f and d_t are *structural parameters* characterizing only the geometry of a given medium. However, when we are interested in processes like diffusion or reactions in disordered media, we need *functional parameters*, which are associated with the notion of time in order to characterize the dynamic behavior of the species in these media. The spectral or fracton dimension d_s and random-walk dimension d_w are two such parameters, and they will be defined in Section 2.2.

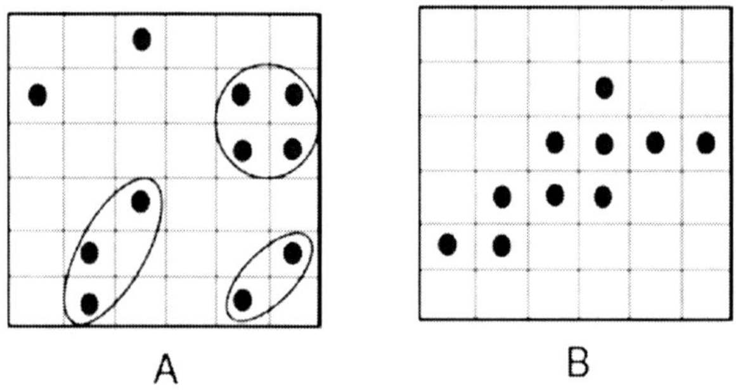

Figure 1.4: A 6×6 square lattice site model. The dots correspond to multifunctional monomers. (**A**) The encircled neighboring occupied sites are clusters (branched intermediate polymers). (**B**) The entire network of the polymer is shown as a cluster that percolates through the lattice from left to right.

1.7 Percolation

The origins of percolation theory are usually attributed to Flory and Stockmayer [5–8], who published the first studies of polymerization of multifunctional units (monomers). The polymerization process of the multifunctional monomers leads to a continuous formation of bonds between the monomers, and the final ensemble of the branched polymer is a network of chemical bonds. The polymerization reaction is usually considered in terms of a lattice, where each site (square) represents a monomer and the branched intermediate polymers represent clusters (neighboring occupied sites), Figure 1.4 A. When the entire network of the polymer, i.e., the cluster, spans two opposite sides of the lattice, it is called a *percolating cluster*, Figure 1.4 B.

In the model of bond percolation on the square lattice, the elements are the bonds formed between the monomers and not the sites, i.e., the elements of the clusters are the connected bonds. The extent of a polymerization reaction corresponds to the fraction of reacted bonds. Mathematically, this is expressed by the probability p for the presence of bonds. These concepts can allow someone to create randomly connected bonds (clusters) assigning different values for the probability p. Accordingly, the size of the clusters of connected bonds increases as the probability p increases. It has been found that above a critical value of $p_c = 0.5$ the various bond configurations that can be formed randomly share a common characteristic: a cluster percolates through the lattice. A more realistic case of a percolating cluster can be obtained if the site model of a square lattice is used with probability $p = 0.6$, Figure 1.5. Notice that the critical value of p_c is 0.593 for the 2-dimensional site model. Also, the percolation thresholds vary according to the type of model (site or bond) as well as with the dimensionality of the lattice (2 or 3).

Figure 1.5: A percolation cluster derived from computer simulation in a 300×300 square site model with $p = 0.6$. Only the occupied sites that belong to the percolating cluster are shown.

The most remarkable properties of percolation clusters arise from their sudden inception when the bond concentration (probability) reaches the critical threshold value $p = p_c$. At this specific value the emerged cluster spans two opposite sides of the lattice and if one conceives of the bonds as channels, the cluster allows a fluid to flow through the medium from edge to edge. Accordingly, the terms *percolation* and *percolation transition* have been coined in an attempt to capture the sudden change in the geometry and the phase transition. In the same vein, the probability p_∞ that a bond belongs to the percolating cluster undergoes a sharp transition, i.e., $p_\infty = p = 0$ for $p_\infty = p < p_c$, while p_∞ becomes finite following a power law when $p > p_c$:

$$p_\infty \propto (p - p_c)^\lambda ,$$

where λ is an exponent usually called the *critical exponent.* According to the findings in this field of research the critical exponent λ depends exclusively on the dimensionality of the system. This independence from other factors is characterized as *universality.*

Important characteristics of the clusters like the mass q and the typical length ξ of the clusters, usually called the *correlation length,* obey power laws too:

$$q \propto |p - p_c|^{-\mu} , \qquad \xi \propto |p - p_c|^{-\nu} ,$$

where μ and ν are also critical exponents. These laws allow reconsideration of the fractal properties of the clusters. According to the last equation the clusters are self-similar as long as the length scale used for measurements is shorter than ξ. For example, the giant cluster shown in Figure 1.5 is a random fractal and as such has a characteristic value for its fractal dimension d_f. However, the calculation of the fractal dimension for the percolating cluster of Figure 1.5 should be performed with radii ρ shorter than ξ. In other words, when $\rho < \xi$ the self-similar character of the cluster is kept and the scaling law holds. Indeed, when the box-counting method is applied, the scaling law $q \propto \rho^{1.89}$ between the mass q (calculated from the mass of ink or equivalently from the number of dots) and the radius ρ of the box is obtained. This means that $d_f = 1.89$ for the percolating cluster of Figure 1.5 since the characteristic measured is the mass for various radii ρ, and no further calculations are required in accord with (1.8). On the contrary, for measurements with $\rho > \xi$, self-similarity no longer exists.

2

Diffusion and Kinetics

Everything changes.

Heraclitus of Ephesus (544-483 BC)

The principles of physical and chemical laws are essential for the understanding of drug kinetics in mammalian species. This also applies to pharmacodynamics since the interaction of drug with the receptor(s) relies on the physicochemical principles of the law of mass action. In reality one can consider the entire course of drug in the body as consecutive and/or concurrent processes of diffusion and convection. For example, the oral administration of a drug may include, among many others, the following processes:

- dissolution in the gastrointestinal fluids (diffusion),
- transport in the chyme by intestinal peristalsis (convection),
- transcellular uptake (diffusion),
- transport with the blood to organs (convection),
- transfer from the bloodstream into the interstitial and intracellular spaces (diffusion),
- interaction with receptors at the effect site (diffusion),
- transfer from tissues back into blood (diffusion),
- glomerular filtration (convection),
- transport with the urine into the efferent urinary tract (convection),
- reabsorption from the tubular lumen to the peritubular capillary (diffusion).

The above convection processes are the result of the movement of a liquid in bulk, i.e., the flow of the biological fluid. Consequently, convection processes are particularly dependent on physiology. For example, the glomerular filtration of a drug is extremely important from a therapeutic point of view, but it is solely determined by the physiological condition of the patient, e.g., the glomerular filtration rate. This is so, since a common translational velocity is superposed on the thermal motions of all drug molecules in any element of volume. On the other hand, convection processes for the dissolved and undissolved drug in the gastrointestinal tract are much more complicated. Here, physiology still plays a major role but dietary conditions and the type of formulation are important too. The picture becomes even more complicated if one takes into account the oscillatory nature of intestinal motility, which is related to the food intake. Despite the complexity involved, the term convection implies that both dissolved drug molecules and undissolved drug particles along with the gastrointestinal fluid molecules are transported together without separation of individual components of the solution/suspension.

On the other hand, diffusion is the random migration of molecules or small particles arising from motion due to thermal energy. Here, drug diffusive fluxes are produced by differences in drug concentrations in different regions. Thus, diffusion is one of the most significant process in all fields of pharmaceutical research either in vitro or in vivo. This is justified by the fact that everything is subject to thermal fluctuations, and drug molecules or particles immersed in aqueous environments are in continuous riotous motion. Therefore, understanding of these random motions is crucial for a sound interpretation of drug processes.

2.1 Random Walks and Regular Diffusion

Particles under the microscope exhibiting Brownian motion demonstrate clearly that they possess kinetic energy. We are also familiar with the diffusional spreading of molecules from the classical experiment in which a drop of dye is carefully placed in an aqueous solution. Fick's laws of diffusion describe the spatial and temporal variation of the dye molecules in the aqueous solution. However, before presenting Fick's differential equation, attention will be given to a proper answer for the fundamental question: How much do the molecules move on average during diffusional spreading?

The correct answer to the above question is a law of physics: "the mean square displacement is proportional to time." We can intuitively reach this conclusion with particles executing an imaginary 1-dimensional random walk. A simple model is presented in Figure 2.1, ignoring the detailed structure of the liquid and temperature effects and assuming no interaction between particles. The particles are placed at $z = 0$ and start their random walk at $t = 0$ moving at a distance δ either to the right or to the left once every $t_\circ$ units of time; thus, the particles execute i steps in time $t = it_\circ$. Equal probabilities (1/2) are assigned for each movement of the particles (either to the right or to the left).

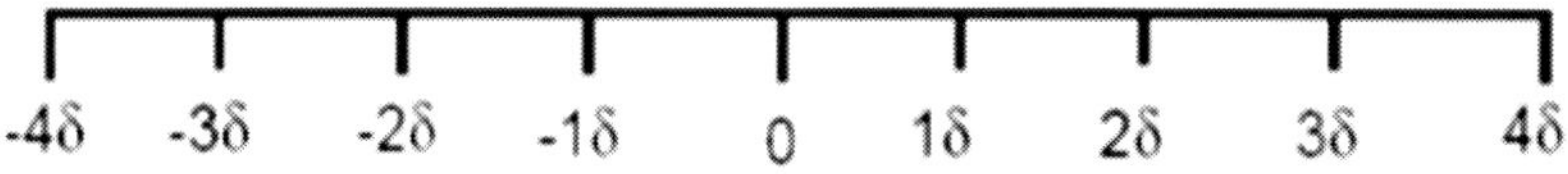

Figure 2.1: A one-dimensional random walk of particles placed at $z = 0$ at $t = 0$. The particles occupy only the positions 0, $\pm\delta$, $\pm 2\delta$, $\pm 3\delta$, $\pm 4\delta$.

This means that the successive jumps of particles are statistically independent and therefore the walk is unbiased. We say that the particles are blind since they have no "memory" of their previous movement(s).

The question arises: How far will a particle travel in a given time interval? The average distance a particle travels is given by mean square displacement evaluated as follows: The position of a particle along the z axis after i steps z_i is

$$z_i = z_{i-1} \pm \delta, \tag{2.1}$$

where z_{i-1} is the position of the particle at the previous $(i-1)$th step. Taking the square of (2.1) we get the square displacement

$$z_i^2 = z_{i-1}^2 \pm 2\delta z_{i-1} + \delta^2,$$

which if averaged for the total number of particles, provides their mean square displacement $\left\langle z_i^2 \right\rangle$:

$$\left\langle z_i^2 \right\rangle = \left\langle z_{i-1}^2 \right\rangle \pm 2\delta \left\langle z_{i-1} \right\rangle + \delta^2 = \left\langle z_{i-1}^2 \right\rangle + \delta^2. \tag{2.2}$$

The second term in the brackets vanishes since the plus sign corresponds to half of the particles and the minus sign to the other half. Given that $z_0 = 0$ and applying (2.2) for the successive steps $1, 2, \ldots, i$, we get

$$\left\langle z_1^2 \right\rangle = \delta^2, \quad \left\langle z_2^2 \right\rangle = 2\delta^2, \quad \ldots, \quad \left\langle z_i^2 \right\rangle = i\delta^2. \tag{2.3}$$

Since as previously mentioned the number of steps is proportional to time $(i = t/t_\circ)$, we can express the positioning of particles as a function of time t using (2.3):

$$\left\langle z^2 (t) \right\rangle = \left(\delta^2 / 2t_\circ \right) t. \tag{2.4}$$

The use of 2 in the denominator of the previous equation will be explained in Section 2.4. The last expression shows that the mean square displacement of the particles is proportional to time, t:

$$\left\langle z^2 (t) \right\rangle \propto t. \tag{2.5}$$

The same result is obtained if one considers a simple random walk in two dimensions, i.e., the walk is performed on a 2-dimensional lattice. Here, the walker (particle) moves either vertically or horizontally at each time step ($t_\circ$ units of time) with equal probabilities. Two configurations for eight-time-step

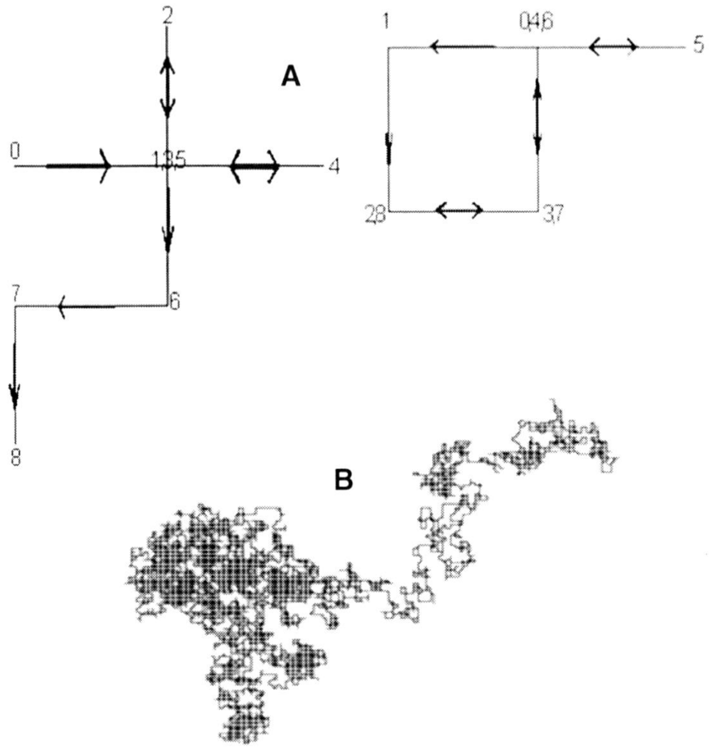

Figure 2.2: (**A**) Two configurations of eight-step random walks in two dimensions. The numbers correspond to the successive eight steps and the arrows indicate the direction of movement. (**B**) A random walk of 10, 000 steps.

random walks are shown in Figure 2.2 A, along with the trail of a random walk of 10, 000 steps, Figure 2.2 B. In the general case and assuming that the lattice spacing is δ, the position of the walker on the plane after i steps z_i is

$$z_i = \delta \sum_{j=1}^{i} u_j,$$

where u_j is a (unit) vector pointing to a nearest-neighbor site; it represents the jth step of the walk on the two dimensional lattice. The mean displacement $\langle z_i \rangle$ of the walker can be obtained if z_i is averaged for the total number of walkers, $\langle z_i \rangle = 0$. This equation is obtained from the previous one since $\langle u_j \rangle = 0$. Moreover, the mean square displacement can be obtained from the previous equation if one takes into account that $\langle u_j u_j \rangle = 1$, and $\langle u_j u_k \rangle = 0$:

$$\begin{aligned} \langle z_i^2 \rangle &= \left\langle \left[\delta \sum_{j=1}^{i} u_j \right]^2 \right\rangle \\ &= \delta^2 \langle (u_1 + u_2 + \cdots + u_i)(u_1 + u_2 + \cdots + u_i) \rangle \\ &= \delta^2 \sum_{j=1}^{i} \langle u_j u_j \rangle + \delta^2 \sum_{\substack{j=1 \\ k \neq j}}^{i} \langle u_j u_k \rangle = i\delta^2. \end{aligned} \tag{2.6}$$

Substituting $i = t/t_\circ$ in the last equation, (2.4) is recovered using the factor $\frac{1}{2}$ for the derivation once again.

The theory for motion in three dimensions results in the same law if the same assumptions are applied and motions in the three directions are statistically independent. The important result for regular diffusion is that its time dependence is universal regardless of the dimension of the medium. This square root relation (2.5) has striking consequences for the distance covered by diffusing molecules. It takes *four* times as long to get *twice* as far while a particle can cover *half* the distance in a *quarter* of the time. Thus, transport by diffusion is very slow if there is far to go, but very rapid over very short distances. For example, the exchange and transport of solutes within cells and between cells and capillaries can be effectively maintained by diffusion due to the small size and close spacing of cells and capillaries in the body of mammals. On the contrary, the slowness of diffusion over large distances points to the necessity for a circulatory system to bring oxygen, for example, from the lungs to the brain or glucose from the liver to the muscles of the arms. To permit these exchanges, the bulk flow of blood carries a large number of solutes around the body in the vascular system by convection.

Equation (2.4) will help us to define and understand the meaning of the diffusion coefficient $\mathcal{D}$. This term corresponds to the proportionality constant of (2.4),

$$\mathcal{D} \triangleq \frac{\delta^2}{2t_\circ}, \tag{2.7}$$

has dimensions of area$\times$time^{-1} and takes different values for different solutes in a given medium at a given temperature. Hence, the value of $\mathcal{D}$ is characteristic for a given solvent (or better, medium structure) at a given temperature of the diffusing tendency of the solute. For example, a small drug molecule in water at 25 °C has $\mathcal{D} \approx 10^{-5}\,\text{cm}^2/\,\text{s}$, while a protein molecule like insulin has $\mathcal{D} \approx 10^{-7}\,\text{cm}^2/\,\text{s}$. Using these values one can roughly calculate the time required for the drug and protein molecules to travel a distance of 1 mm; it takes $(0.1)^2/10^{-5} \approx 1000\,\text{s} \approx 16.6\,\text{min}$ for the drug and 1666.6 min for insulin. Hence, the value of $\mathcal{D}$ is heavily dependent on the size of the solute molecules. These numerical calculations are very useful in obtaining insight into the rapidity or slowness of a solute migration, e.g., drug release from controlled release formulations when regular diffusion is the operating mechanism.

2.2 Anomalous Diffusion

In the previous section we analyzed the random walk of molecules in Euclidean space and found that their mean square displacement is proportional to time, (2.5). Interestingly, this important finding is not true when diffusion is studied in fractals and disordered media. The difference arises from the fact that the nearest-neighbor sites visited by the walker are equivalent in spaces with integer dimensions but are not equivalent in fractals and disordered media. In these media the mean correlations between different steps $\langle u_j u_k \rangle$ are not equal to zero, in contrast to what happens in Euclidean space; cf. derivation of (2.6). In reality, the anisotropic structure of fractals and disordered media makes the value of each of the correlations $u_j u_k$ structurally and temporally dependent. In other words, the value of each pair $u_j u_k$ depends on where the walker is at the successive times j and k, and the Brownian path on a fractal may be a "fractal of a fractal" [9]. Since the correlations $u_j u_k$ do not average out, the final important result is $\langle u_j u_k \rangle \neq 0$, which is the underlying cause of anomalous diffusion. In reality, the mean square displacement does not increase linearly with time in anomalous diffusion and (2.5) is no longer exact.

To characterize the dynamic movement of particles on a fractal object, one needs two additional parameters: the spectral or fracton dimension d_s and the random-walk dimension d_w. Both terms are quite important when diffusion phenomena are studied in disordered systems. This is so since the path of a particle or a molecule undergoing Brownian motion is a random fractal. A typical example of a random fractal is the percolation cluster shown in Figure 1.5.

The definition of spectral dimension d_s refers to the probability $p(t)$ of a random walker returning to its origin after time t:

$$p(t) \propto t^{-d_s/2}. \tag{2.8}$$

According to (2.8), the value of d_s governs the decrease of the probability $p(t)$ with time. When diffusion is considered in Euclidean spaces the various dimensionality terms become identical: $d_t = d_s = d_f$. However, in fractal spaces the following inequalities hold: $d_t < d_s < d_f < d_e$, where d_e is the embedding dimension. For example, we found for the Sierpinski gasket (Figure 1.2 A) $d_f = 1.5815$, while $d_s = 1.3652$ and the embedding dimension in this case is $d_e = 2$. The meaning of d_s can be understood if one considers a walker executing a random walk on a ramified system, like the Sierpinski gasket with $d_f = 1.5815$, Figure 1.2 A. Due to the system's ramification, the walker has many alternatives of movement in the branched system, and therefore the probability of the walker being back at the origin is small. Hence, the value of d_s goes up in accord with (2.8) and is higher than one ($d_s > 1$), i.e., the topological dimension of a curve. In actual practice, the calculation of d_s is accomplished numerically. Analytical solutions for d_s are available when the recursion algorithm of the system is known, e.g., Sierpinski gasket.

Finally, a stochastic viewpoint may be associated with the relation (2.8) since the spectral dimension also characterizes the number $n(t)$ of distinct sites visited by the random walker up to time t:

$$n(t) \propto t^{d_s/2}. \tag{2.9}$$

The random-walk dimension d_w is useful whenever one has a specific interest in the fractal dimension of the trajectory of the random walk. The value of d_w is exclusively dependent on the values of d_f and d_s:

$$d_w = \min\left[2\frac{d_f}{d_s}, d_f\right].$$

The type of the random walk (recurrent or nonrecurrent) determines the minimum value of the two terms in the brackets of the previous equation. If the walker does not visit the same sites (nonrecurrent) then $d_w = 2d_f/d_s$. If the walk is of recurrent type then the walker visits the same sites again and again and therefore the walker covers the available space (space-filling walk). Consequently, the meaning of d_w coincides with d_f ($d_w = d_f$). The mean square displacement in anomalous diffusion follows the pattern

$$\left\langle z^2(t)\right\rangle \propto t^{2/d_w}, \tag{2.10}$$

where d_w is the fractal dimension of the walk and its value is usually $d_w > 2$. The exponent d_w arises from the obstacles of the structure such as holes, bottlenecks, and dangling ends, i.e., the diffusional propagation is hindered by geometric heterogeneity. The previous equation is the fundamental relation linking the propagation of the diffusion front to the structure of the medium, and it recovers also the classical law of regular diffusion when $d_w = 2$.

In conclusion, the dynamic movement of particles on a fractal object may be described by functional characteristics such as the spectral dimension d_s and the random-walk dimension d_w. This anomalous movement of the molecules induces heterogeneous transport and heterogeneous reactions. Such phenomena present a challenge to several branches of science: chemical kinetics, surface and solid state physics, etc. Consequently, one may argue that all mechanisms involved in drug absorption, metabolism, enzymatic reactions, and cell microscopic reactions can be analyzed in the new heterogeneous context since these processes are taking place under topological constraints.

2.3 Fick's Laws of Diffusion

Apart from the above considerations of diffusion in terms of the distance traveled in time, the amount of substance transported per unit time is useful too. This approach brings us to the concept of the rate of diffusion. The two considerations are complementary to each other since the diffusion of molecules at the microscopic level results in the observed "flux" at the macroscopic level. Fick's laws of diffusion describe the flux of solutes undergoing classical diffusion.

The simplest system to consider is a solution of a solute with two regions of different concentrations c_l and c_r to the left and right, respectively, of a

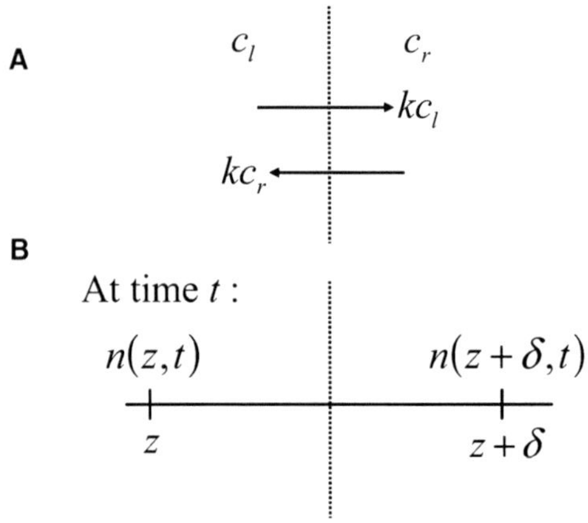

Figure 2.3: A solute diffuses across a plane. (**A**) Solute diffusion from two regions of different concentrations c_l and c_r; the plane indicates the boundary of the regions. The transfer rate of material is proportional to concentrations c_l and c_r. (**B**) At a given time t there are $n(z,t)$ and $n(z+\delta,t)$ molecules at positions z and $z+\delta$, respectively.

boundary separating the two regions, Figure 2.3. In reality, the rate of diffusion is the net flux, i.e., the difference between the two opposite unidirectional fluxes. There will be a net movement of solute molecules to the right if $c_l > c_r$ or to the left if $c_l < c_r$. When $c_l = c_r$, the unidirectional fluxes are equal and the net flux is zero. Since the two fluxes across the boundary from left to right and vice versa are proportional to c_l and c_r, respectively, the net flux is proportional to the concentration difference across the boundary.

The derivation of Fick's first law of diffusion requires a reconsideration of Figure 2.3 A in terms of the 1-dimensional random walk as shown in Figure 2.3 B. Let us suppose that at time t, there are $n(z,t)$ molecules at the left position z and $n(z+\delta,t)$ molecules at the right position $z+\delta$, Figure 2.3 B. Since equal probabilities (1/2) are assigned for the movement of the molecules (either to the right or to the left), half of the $n(z,t)$ and $n(z+\delta,t)$ molecules will cross the plane at the next instant of time $t+t_\circ$, moving in opposing directions. The net number of molecules crossing the plane to the right is $-\frac{1}{2}\left[n\left(z+\delta,t\right)-n\left(z,t\right)\right]$ and the corresponding net flux J of the diffusate is

$$J\left(z,t\right)=-\frac{1}{2\mathcal{A}t_\circ}\left[n\left(z+\delta,t\right)-n\left(z,t\right)\right],$$

where $\mathcal{A}$ is the area of the plane and $t_\circ$ is the time interval. Multiplying and dividing the right part by δ^2 and rearranging, we get

$$J\left(z,t\right)=-\frac{\delta^2}{2t_\circ}\frac{1}{\delta}\left[\frac{n\left(z+\delta,t\right)}{\mathcal{A}\delta}-\frac{n\left(z,t\right)}{\mathcal{A}\delta}\right].$$

The terms in the brackets express the concentration of molecules per unit volume $\mathcal{A}\delta$, i.e., $c(z+\delta,t) \equiv c_r(t)$ and $c(z,t) \equiv c_l(t)$ at positions $z+\delta$ and z, respectively, while the term $\delta^2/2t_\circ$ is the diffusion coefficient $\mathcal{D}$; the presence of 2 in the denominator explains its use in (2.4). We thus obtain

$$J(z,t) = -\mathcal{D}\frac{c(z+\delta,t) - c(z,t)}{\delta}.$$

Since the term in the brackets in the limit $\delta \to 0$ is the partial derivative of $c(z,t)$ with respect to z, one can write

$$J(z,t) = -\mathcal{D}\frac{\partial c(z,t)}{\partial z}. \tag{2.11}$$

The minus sign indicates that the flow occurs from the concentrated to the dilute region of the solution. Equation (2.11) is Fick's first law, which states that the net flux is proportional to the gradient of the concentration function (at z and t). Flux has dimensions of mass$\times$area$^{-1}\times$time^{-1}.

Since the flux J is the flow of material $\dot{q}(z,t)$ from the left to the right through the surface $\mathcal{A}$, (2.11) is rewritten as follows:

$$\dot{q}(z,t) = -\mathcal{D}\mathcal{A}\frac{\partial c(z,t)}{\partial z}. \tag{2.12}$$

From this relationship it is clear that the force acting to diffuse the material q through the surface is the concentration gradient $\partial c/\partial z$. This gradient may be approximated by differences

$$\frac{\partial c(z,t)}{\partial z} \approx \frac{\Delta c(z,t)}{\Delta z} = \frac{c(z+\delta,t) - c(z,t)}{\delta} = \frac{c_r(t) - c_l(t)}{\delta}, \tag{2.13}$$

and the previous expression becomes

$$\dot{q}(t) \triangleq R_{lr} = -\frac{\mathcal{D}\mathcal{A}}{\delta}\left[c_r(t) - c_l(t)\right], \tag{2.14}$$

where R_{lr} is the transfer rate of material. This equation usually takes one of two similar forms:

$$\dot{q}(t) = -CL_{lr}\left[c_r(t) - c_l(t)\right] \qquad \text{or} \qquad \dot{q}(t) = -P\mathcal{A}\left[c_r(t) - c_l(t)\right]. \tag{2.15}$$

The new introduced parameter $CL_{lr} \triangleq \mathcal{D}\mathcal{A}/\delta$ is called *clearance*, and it has dimensions of flow, volume$\times$time^{-1}. The clearance has a bidirectional use and indicates the volume of the solution that is cleared from drug per unit of time because of the drug movement across the plane. For an isotropic membrane, structural and functional characteristics are identical at both sides of the membrane, $CL_{lr} = CL_{rl}$. In practice, the term "clearance" is rarely used except for the irreversible removal of a material from a compartment by unidirectional path ways of metabolism, storage, or excretion. The other new parameter $P \triangleq \mathcal{D}/\delta$ characterizes the diffusing ability of a given solute for a given membrane, and it is called *permeability*. Permeability has dimensions of length$\times$time^{-1}.

We now write a general mass conservation equation stating that the rate of change of the amount of material in a region of space is equal to the rate of flow across the boundary plus any that is created within the boundary. If the region is $z_1 < z < z_2$ and no material is created

$$\frac{\partial}{\partial t}\int_{z_1}^{z_2} \mathrm{d}q\,(z,t) = \frac{\partial}{\partial t}\int_{z_1}^{z_2} c\,(z,t)\,\mathrm{d}z = J\,(z_1,t) - J\,(z_2,t)\,.$$

Here, if we assume $\mathcal{D}$ constant in (2.11) and $z_2 = z_1 + \Delta z$, at the limit $\Delta z \to 0$, this relation leads to

$$\frac{\partial c\,(z,t)}{\partial t} = \mathcal{D}\frac{\partial^2 c\,(z,t)}{\partial z^2}. \tag{2.16}$$

This is the second Fick's law stating that the time rate of change in concentration (at z and t) is proportional to the curvature of the concentration function (at z and t). There is a clear link between the two laws (2.11) and (2.16).

In order to examine the relevance of the two laws, let us consider that the layer separating the two regions in Figure 2.3 A is not thin but has an appreciable thickness δ, while z is the spatial coordinate along it. According to (2.11), if $\partial c/\partial z$ is constant then the flux J is constant. Consequently, $\partial^2 c/\partial z^2 = 0$ in (2.16). This means that the concentration is stationary. This happens when c is a linear function of z ($\partial c/\partial z$ is constant). Under these conditions, as many drug molecules diffuse in from the side of higher concentration as diffuse out from the side of lower concentration. This can be accomplished experimentally if the concentration gradient for the two regions of Figure 2.3 A is maintained constant, e.g., c_l and c_r are kept fixed. Under these conditions Fick's first law of diffusion (2.11) dictates a linear $c\,(z,t)$ profile and a constant flux, Figure 2.4 A. However, in the general case $\partial c/\partial z$ is not constant (Figure 2.4 B). In reality, plot A is the asymptotic behavior of the general case B as t goes to infinity. The solution of (2.11) for an initial distribution $c(z,0) = 0$ (there is no solute inside the layer initially) and boundary conditions $c(0,t) = c_l\,(t)$ and $c(\delta,t) = c_r\,(t)$ yields [10]

$$\begin{aligned} c\,(z,t) &= c_l\,(t) + [c_r\,(t) - c_l\,(t)]\,\frac{z}{\delta} \\ &+\frac{2}{\pi}\sum_{i=1}^{\infty}\frac{c_r\,(t)\cos i\pi - c_l\,(t)}{i}\sin\left(i\pi\frac{z}{\delta}\right)\exp\left(-\frac{\mathcal{D}}{\delta^2}i^2\pi^2 t\right). \end{aligned} \tag{2.17}$$

Fick's second law (2.16) dictates a nonlinear $c\,(z,t)$ profile (Figure 2.4 B) in accord with (2.17).

If we postulate that molecules move independently, the concentration $c\,(z,t)$ at some point z is proportional to the probability density $p\,(z,t)$ of finding a molecule there. Thus, the diffusion partial differential equation (2.16) holds when probability densities are substituted for concentrations:

$$\frac{\partial p\,(z,t)}{\partial t} = \mathcal{D}\frac{\partial^2 p\,(z,t)}{\partial z^2}. \tag{2.18}$$

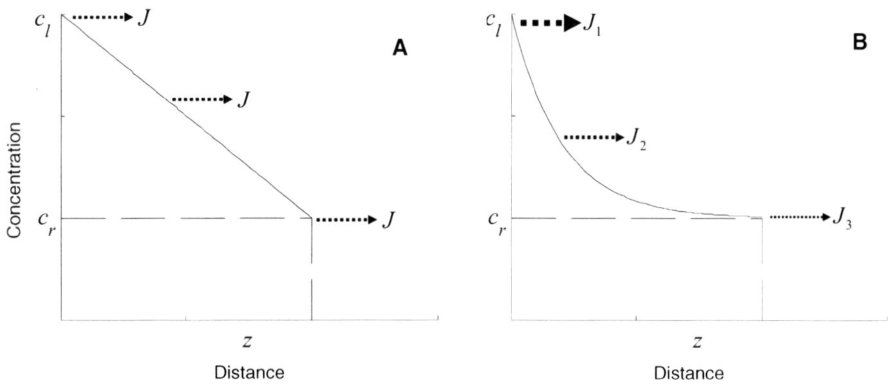

Figure 2.4. Schematic of concentration distance profiles derived from Fick's laws when (**A**) $\partial c / \partial z$ is constant, thus J is constant and (**B**) $\partial c / \partial z$ is not constant, thus $J_1 > J_2 > J_3$.

If a molecule is initially placed at $z = 0$, then the solution of the previous equation is

$$p(z,t) = (4\pi \mathcal{D} t)^{-1/2} \exp\left(-\frac{z^2}{4\mathcal{D}t}\right).$$

For $t \gg 1$ at any z, we obtain $p(z,t) \propto t^{-1/2}$. This behavior in a homogeneous medium corresponds to (2.8), giving the probability density in a fractal medium with spectral dimension d_s.

2.4 Classical Kinetics

Pharmacy, like biology and physiology, is wet and dynamic. Drug molecules immersed in the aqueous environment of intravascular, extravascular, and intracellular fluids participate in reactions, such as reversible binding to membrane or plasma proteins; biotransformation or transport processes, e.g., drug release from a sustained release formulation; drug uptake from the gastrointestinal membrane; and drug permeation through the blood–brain barrier. This classification is very rough since some of these processes are more complex. For example, drug release is basically a mass transport phenomenon but may involve reaction(s) too, e.g., polymer dissolution and/or polymer transition from the rubbery to the glassy state. However, irrespective of the detailed characteristics, the common and principal component of the underlying mechanism of numerous drug processes is diffusion. This is the case for the ubiquitous passive transport processes that rely on diffusion exclusively. The value of $\mathcal{D}$ depends on the nature of the environment of the diffusing species. If the environment changes from one point to another, the value of $\mathcal{D}$ may depend on position. Usually, we deal with systems in which the environment of the diffusing species is the same everywhere, so that $\mathcal{D}$ is a constant. The diffusion coefficient is constant for diffusion of dilute solute in a uniform solvent. This case takes in a large number of important situations, and if the dilute solute is chemically the same as the solvent but is isotopically tagged, then the diffusion is termed self-diffusion. In contrast, chemical reactions can be either reaction-limited or diffusion-limited. In the following sections we will discuss them separately.

2.4.1 Passive Transport Processes

There appear to be two main ways for solutes to pass through cell membranes, namely, transcellular and paracellular. The most important is the transcellular route, whereby compounds cross the cells by traversing the cell membrane following either passive diffusion or carrier-mediated transport. Undoubtedly, the transcellular passive diffusion is the basic mechanism of solute permeation through cell membranes. According to this mechanism the solute leaves the fluid bathing the membrane, dissolves in the substance of the membrane, diffuses across in solution, and then emerges into the intracellular fluid. Accordingly, the mathematical treatment of drug diffusion across a membrane can be based on (2.12), which is a very useful expression of Fick's first law of diffusion. This equation is used extensively in the pharmaceutical sciences. It describes the mass (number of molecules, or moles, or amount) transported per unit time, $\dot{q}$, across an area $\mathcal{A}$ with a concentration gradient $\partial c / \partial z$ at right angles to the area. According to this definition, the numerical value of the diffusion coefficient $\mathcal{D}$, expressed in mass units, corresponds to the amount of solute that diffuses per unit time across a unit area under the influence of a unit concentration gradient.

For a passive transport process, the concentration gradient across the membrane can be considered constant and therefore the gradient can be approximated by differences as in (2.13) to obtain

$$\dot{q}(t) = \frac{\mathcal{D}' \mathcal{A}}{\delta} \left[c_l(t) - c_r(t) \right],$$

where $\mathcal{D}'$ is a modified diffusion coefficient, for restricted diffusion inside the membrane. The value of $\mathcal{D}'$ is much smaller than the diffusion coefficient $\mathcal{D}$ in free solution. The minus sign is not used in the previous equation since the rate of transport corresponds to the solute transfer from the external to the internal site ($c_l > c_r$). Furthermore, if sink conditions prevail ($c_l \gg c_r$), the previous equation can be simplified to

$$\dot{q}(t) = CLc(t) = P\mathcal{A}c(t). \tag{2.19}$$

The last equation reveals that estimates for P can be obtained in an experimental setup if the permeation rate $\dot{q}(t)$ and the total membrane area $\mathcal{A}$ available for transport are measured and the drug concentration $c(t)$ in the donor compartment remains practically constant. What is implicit from all the above is that the diffusion coefficient $\mathcal{D}'$ is at the origin of the definition of the clearance CL and permeability P, and these parameters are incorporated into the global rate constant of the rate equations used in pharmacokinetics. For example, the first-order absorption rate constant k_a in the following equation is proportional to the diffusion coefficient $\mathcal{D}'$ of drug in the gastrointestinal membrane:

$$\dot{c}_b(t) = k_a c_{GI}(t),$$

where $c_b(t)$ and $c_{GI}(t)$ denote drug concentration (amount absorbed/volume of distribution) in blood and in the gastrointestinal lumen (amount dissolved in the gastrointestinal fluids/volume of gastrointestinal fluids), respectively. In other words, $\mathcal{D}'$ controls the rate of drug absorption from the gastrointestinal tract.

2.4.2 Reaction Processes: Diffusion- or Reaction-Limited?

Pharmacokinetics has been based on the concepts of classical chemical kinetics. However, the applicability of the rate equations used in chemical kinetics presupposes that the reactions are really reaction-limited. In other words, the typical time for the two chemical species to react when placed in close proximity (reaction time t_{reac}) is larger than the typical time needed for the two species to reach each other (diffusion time t_{diff}) in the reaction space. When the condition $t_{\text{reac}} > t_{\text{diff}}$ is met, then one can use the global concentrations of the reactant species in the medium to obtain the classical rate equations of chemical kinetics. This is so since the rate of the reaction is proportional to the global concentrations of the reactant species (law of mass action). The inequality $t_{\text{reac}} > t_{\text{diff}}$ underlines the fact that the two reactant species have encountered each other more than one time previously in order to react effectively.

The opposite case, $t_{\text{reac}} < t_{\text{diff}}$, indicates that the two reactant species actually react upon their first encounter. The diffusion characteristics of the species control the rate of the reaction, and therefore these reactions are called *diffusion limited*. Consider for example a system consisting of species A and B with n_A and n_B molecules of A and B, respectively. The problem of the reaction rate between A and B is in essence reduced to the rate at which A and B molecules will encounter one another. The principal parameters governing the reaction rate are the diffusion coefficients $\mathcal{D}_A$ and $\mathcal{D}_B$ of the reactant species since they determine the diffusing tendency of the species. Focusing on B molecules, it can be proven that the rate of B molecules diffusing to an A molecule is proportional to the diffusion coefficient of B, the number of B molecules, and the distance between A and B, namely, $4\pi\mathcal{D}_B(\rho_A + \rho_B)n_B$, where $\rho_A + \rho_B$ is the distance between the centers of A and B molecules; accordingly, the total rate of A and B encounters is $4\pi\mathcal{D}_B(\rho_A + \rho_B)n_B n_A$. In an analogous manner the total rate of A and B encounters, viewed in terms of the A molecules, is $4\pi\mathcal{D}_A(\rho_A + \rho_B)n_B n_A$. The mean of these separate rates provides a reasonable expression for the rate per unit volume for A and B molecules separately:

$$\text{Rate of } A \text{ and } B \text{ encounters} = 2\pi(\mathcal{D}_A + \mathcal{D}_B)(\rho_A + \rho_B)n_A n_B.$$

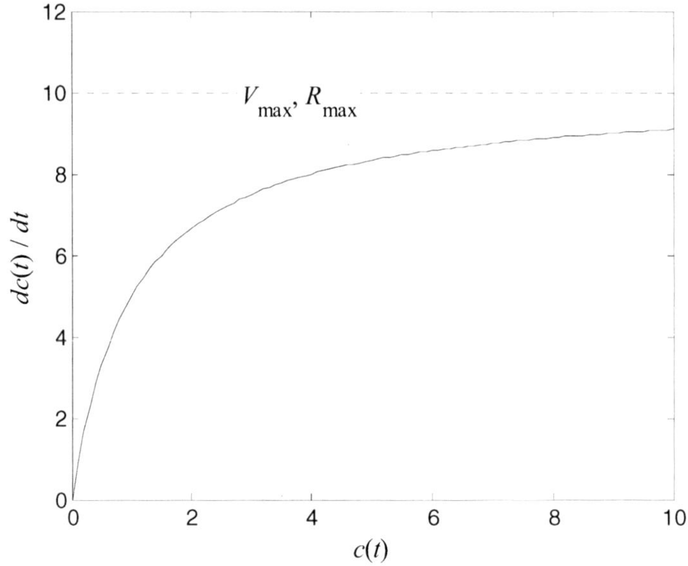

Figure 2.5: The rate of biotranformation or carrier-mediated transport vs. solute concentration. The plateau value corresponds to $V_{\max}$ or $R_{\max}$. k_M and $V_{\max}$ were set to 1 and 10, respectively, with arbitrary units.

Although the previous equation signifies the importance of the diffusion characteristics of the reactant species, it cannot be used to describe adequately the rate of the reaction. The reason is that the concept of global concentrations for the n_A and n_B molecules is meaningless, since a unit volume cannot be conceived due to the local fluctuations of concentrations. Hence, the local concentrations of the reactants determine the rate of the reaction for diffusion-limited reactions. Accordingly, local density functions with different diffusion coefficients for the reactant species are used to describe the diffusion component of reaction–diffusion equations describing the kinetics of diffusion-limited reactions.

2.4.3 Carrier-Mediated Transport

The transport of some solutes across membranes does not resemble diffusion and suggests a temporary, specific interaction of the solute with some component (protein) of the membrane characterized as "carrier," e.g., the small-peptide carrier of the intestinal epithelium. The rate of transport increases in proportion to concentration only when this is small, and it attains a maximal rate that cannot be exceeded even with a large further increase in concentration. The kinetics of carrier-mediated transport is theoretically treated by considering carrier–solute complexes in the same manner as enzyme-substrate complexes following the principles of enzyme–catalyzed reactions in Michaelis–Menten kinetics. In both biotransformation and carrier-mediated transport, unrestricted diffusion is considered for the reactant species. Due to the analogous formulation of the two processes, the equations describing the rates of biotransformation,

$$\dot{c}(t) = \frac{V_{\max} c(t)}{k_M + c(t)}, \tag{2.20}$$

and carrier-mediated transport,

$$\dot{c}(t) = \frac{R_{\max} c(t)}{k_M + c(t)}, \tag{2.21}$$

are similar. In these expressions, $c(t)$ is the solute (substrate) concentration, k_M is the Michaelis constant, $V_{\max}$ is the maximum biotransformation rate, and $R_{\max}$ is the maximum transport rate. Both equations indicate that the rate of biotransformation or carrier-mediated transport become independent of substrate (solute) concentration when this is large. In this case, the rate of biotransformation or carrier-mediated transport is said to exhibit *saturation kinetics*. The graphical representation of the previous equations is shown in Figure 2.5.

2.5 Fractal-like Kinetics

The undisputable dogma of chemistry whether in chemical synthesis or classical chemical kinetics, is to "stir well the system." The external stirring re-randomizes the positioning of the reactant species, and therefore the rate of the reaction follows the classical pattern imposed by the order of the reaction. However, many reactions and processes take place under dimensional or topological constraints that introduce spatial heterogeneity. A diffusion process under such conditions is highly influenced, drastically changing its properties. A general well-known result is that in such constrained spaces, diffusion is slowed down and diffusion follows an anomalous pattern. Obviously, the kinetics of the diffusion-limited reactions (processes) are then sensitive to the peculiarities of the diffusion process. In other words, the transport properties of the diffusing species or the reactants largely determine the kinetics of the diffusion-limited processes. Under these circumstances one can no longer rely on classical rate equations and a different approach is necessary. The drastic and unexpected consequences of nonclassical kinetics of diffusion-limited reactions are called *fractal-like kinetics*; the essentials for this "understirred" type of kinetics are delineated below.

2.5.1 Segregation of Reactants

Classical homogeneous kinetics assumes that the reactants are located in a 3-dimensional vessel, and that during the reaction process the system is constantly stirred, thus causing the positions (locations) of the reactants to be constantly re-randomized as a function of time. However, there are important chemical reactions, which are called "heterogeneous," in which the reactants are spatially constrained by either walls or phase boundaries, e.g., liquid–solid boundaries. This is the case for in vivo drug dissolution as well as for many bioenzymatic and membrane reactions. Due to dimensional or topological constraints these heterogeneous reactions take place under understirred conditions. The most dramatic manifestation of such highly inefficient stirring is the spontaneous segregation of reactants in $A+B$ reactions [11–13]. This means that correlations begin to develop between the reactants' positions, which subsequently have a profound effect on the rate of a diffusion-controlled reaction. The build-up of such correlations is strongly dependent on the dimensionality, being more pronounced the further one goes below 3-dimensional spaces. This is so because quantitatively the parameter values in the diffusion laws are very different in different dimensions. In addition, if the space where the reaction takes place is not smooth, but highly irregular, this has an added effect on the building of such correlations. This happens if the space is a fractal structure characterized by its own dimensionality, which as discussed in Chapter 1 could be different from the integer 1, 2, or 3.

An important segregation effect is related to the violation of Wenzel's old law for heterogeneous reactions; this law states that the larger the interface, the higher the reaction rate [14]. Thus, the most classical way to speed up a heterogeneous process, e.g., drug dissolution, is to grind the material in order to increase the surface area. At the macroscopic level, this law has been verified in numerous physicochemical studies [15] as well as in in vitro drug dissolution studies and in vivo bioavailability studies using micro instead of macro drug particles. However, violation of Wenzel's law has been observed in simulation studies [16, 17] at the microscopic level. Simulations for the catalytic reaction $A+B \rightarrow AB \uparrow$, which takes place only on the rims of surfaces, indicate that the steady-state rate per unit surface area is not constant but rather depends on the size of the sample. In reality, lower reaction rates were observed for a connected catalyst compared to a disjointed one despite the fact that equal lengths for both designs were used. This is due to the lower segregation of the reactants on the rims of the disjointed catalyst, which results in a higher rate coefficient for the catalytic reaction. The clear message taken from these results is that shredding a sample not only increases the surface area but can also increase the reactivity per unit area. The latter observation violates Wenzel's law.

2.5.2 Time-Dependent Rate Coefficients

The spatial reactant correlations result in building a depletion zone around each reactant, which grows steadily with time. This means that in the close neighborhood of each reactant there is a void, a space that is empty of reactants. The net result is that the reactant distribution for the two-reactant case ($A + B \rightarrow C$) shows clear segregation of unlike species (A from B) and aggregation of like species (either A or B). Naturally, the diffusion-controlled reaction slows down, since as reactants get further apart, they must travel longer distances to find another reactant to react with (cf. equation 2.9). A curious effect now is that the rate constant k of the reaction is no longer "constant", but depends on the growth of this depletion zone and consequently is time-dependent:

$$k(t) = k_{\circ}t^{-\lambda} \qquad (t > t_{\circ}),$$

where $k(t)$ is the instantaneous rate coefficient since it depends on time t, and λ is the fractal kinetics exponent with $0 \leq \lambda < 1$. In fact, $k(t)$ crosses over from a constant regime at short times, $t < t_{\circ}$, to a power-law decrease at longer times, $t > t_{\circ}$. The switching time $t_{\circ}$ depends on the experimental conditions. This behavior is the hallmark of fractal kinetics [16].

Under homogeneous conditions (e.g., vigorous stirring), $\lambda = 0$ and therefore $k(t)$ is a constant giving back the classical kinetics result. The previous equation has been applied to the study of various reactions in fractals as well as in many other nonclassical situations. For instance, theory, simulations, and experiments have shown that the value of λ for $A + A$ reactions is related to the spectral dimension d_s of the walker (species) as follows [9, 18]:

$$\lambda = 1 - \frac{d_s}{2}.$$

From this relationship, we obtain $\lambda = 1/3$ since the value of d_s is $\approx 4/3$ for $A + A$ reactions taking place in random fractals in all embedded Euclidean dimensions [9, 19]. It is also interesting to note that $\lambda = 1/2$ for an $A + B$ reaction in a square lattice for very long times [12]. Thus, it is now clear from theory, computer simulation, and experiment that elementary chemical kinetics are quite different when reactions are diffusion limited, dimensionally restricted, or occur on fractal surfaces [9, 11, 20–22].

We emphasize that the fractal-like kinetic characteristics are not observed only under "bing-bang" type conditions (also called batch) as discussed above but also under quasi-steady-state conditions (cf. Section 8.5.1). Consider, for example, the homodi-meric reaction with two molecules of a single substrate reacting to form product ($A + A \rightarrow C$). Under homogeneous conditions the rate at quasi-steady state will be proportional to substrate concentration squared, $c^2(t)$, i.e., it is time-independent (by definition). However, the rate for the bimolecular $A + A$ diffusion-limited reaction under topological or dimensional constraints will be proportional to $c^{\gamma}(t)$. Surprisingly, the effective reaction order γ is higher than 2 and is related to the spectral dimension d_s and in turn to the fractal kinetics exponent λ [9]:

$$\gamma = 1 + \frac{2}{d_s} = 1 + (1 - \lambda)^{-1},$$

with $d_s \leq 2$. Typical values for the Sierpinski gasket and the percolation cluster are $\gamma = 2.46$ and $\gamma = 2.5$, respectively. If $d_s = 1$, so that diffusion is compact, then $\gamma = 3$ for the bimolecular $A + A$ reaction. In all these cases, the mechanism of diffusion is bimolecular. However, the increase in the effective reaction order arises from the distribution of the species, which as time goes by becomes "less random," i.e., it is actually more ordered.

Before we close this section some major, unique kinetic features and conclusions for diffusion-limited reactions that are confined to low dimensions or fractal dimensions or both can now be derived from our previous discussion. First, a reaction medium does not have to be a geometric fractal in order to exhibit fractal kinetics. Second, the fundamental linear proportionality $k \propto \mathcal{D}$ of classical kinetics between the rate constant k and the diffusion coefficient $\mathcal{D}$ does not hold in fractal kinetics simply because both parameters are time-dependent. Third, diffusion is compact in low dimensions and therefore fractal kinetics is also called *compact kinetics* [23,24] since the particles (species) sweep the available volume compactly. For dimensions $d_s > 2$, the volume swept by the diffusing species is no longer compact and species are constantly exploring mostly new territory. Finally, the initial conditions have no importance in classical kinetics due to the continuous re-randomization of species but they may be very important in fractal kinetics [16].

2.5.3 Effective Rate Equations

The dependence of the kinetics on dimensionality is due to the physics of diffusion. This modifies the kinetic differential equations for diffusion-limited reactions, dimensionally restricted reactions, and reactions on fractal surfaces. All these chemical kinetic patterns may be described by power-law equations with time-invariant parameters like

$$\dot{c}(t) = -\kappa c^{\gamma}(t), \qquad c(t_0) = c_0, \tag{2.22}$$

with $\gamma \geq 2$. Under these conditions, the traditional rate law for the $A + A$ reaction with concentration squared exhibits a characteristic reduction of the rate constant with time:

$$\dot{c}(t) = -k(t)\, c^{2}(t), \qquad c(t_0) = c_0, \tag{2.23}$$

where $k(t) = k_{\circ} t^{-\lambda}$. Conversely, (2.23) is equivalent to a time-invariant rate law (2.22) with an increased kinetic order γ. New parameters λ and $k_{\circ}$ are given by

$$\lambda = (\gamma - 2) / (\gamma - 1) \quad \text{and} \quad k_{\circ} = \kappa^{1/(\gamma-1)} (\gamma - 1)^{(2-\gamma)/(\gamma-1)}$$

with $0 \leq \lambda < 1$.

In traditional chemical kinetics $\lambda = 0$, the rate constant is time-invariant, and the effective kinetic order γ equals the molecularity 2. As the reaction becomes increasingly diffusion-limited or dimensionally restricted, λ increases, the rate constant decreases more quickly with time, and the kinetic order in the time-invariant rate law increases beyond the molecularity of the reaction. When the reaction is confined to a 1-dimensional channel, $\gamma = 3.0$, or it can be as large as 50 when isolated on finely dispersed clusters or islands [9,21]. The kinetic order is no longer equivalent to the molecularity of the reaction. The increase in kinetic order results in behavior with a higher effective cooperativity. The kinetic orders in some cases reflect the fractal dimension of the physical surface on which the reaction occurs.

This anomaly stems from the nonrandomness of the reactant distributions in low dimensions. Although in a classical reaction system the distribution of the reactants stays uniformly random, in a fractal-like reaction system the distribution tends to become "less random." Similar changes take place in other reactions and other spaces. Such findings are well established today, and they have been observed experimentally and theoretically. Also, results from Monte Carlo simulations (a powerful tool in this field) are in very good agreement with these findings.

The solution of the differential equations above is a power function of time, namely $c(t) = \beta t^{\alpha}$ with parameters β and α satisfying the initial condition $c(t_0) = c_0$. Usually β and α are estimated by curve fitting on experimental data, and the parameters of (2.22) and (2.23) are obtained by

$$\kappa = -\alpha\beta^{1/\alpha} \quad \text{and} \quad \gamma = 1 - 1/\alpha$$

and

$$k_{\circ} = -\alpha/\beta \quad \text{and} \quad \lambda = 1 + \alpha,$$

respectively. Since we have assumed $\gamma \geq 2$ or $0 \leq \lambda < 1$, the parameter α satisfies $-1 \leq \alpha < 0$.

2.5.4 Enzyme-Catalyzed Reactions

In the same vein and under dimensionally restricted conditions, the description of the Michaelis–Menten mechanism can be governed by power-law kinetics with kinetic orders with respect to substrate and enzyme given by noninteger powers. Under quasi-steady-state conditions, Savageau [25] defined a fractal Michaelis constant and introduced the fractal rate law. The behavior of this fractal rate law is decidedly different from the traditional Michaelis–Menten rate law:

- the effective k_M decreases as the concentration of enzyme increases, and
- the kinetic order of the overall reaction with respect to total enzyme is greater than unity.

These properties are likely to have an important influence on the behavior of intact biochemical systems, e.g., within the living cell, enzymes do not function in dilute homogeneous conditions isolated from one another. The postulates of the Michaelis–Menten formalism are violated in these processes and other formalisms must be considered for the analysis of kinetics in situ. The intracellular environment is very heterogeneous indeed. Many enzymes are now known to be localized within 2-dimensional membranes or quasi-1-dimensional channels, and studies of enzyme organization in situ [26] have shown that essentially all enzymes are found in highly organized states. The mechanisms are more complex, but they are still composed of elementary steps governed by fractal kinetics.

The power-law formalism was used by Savageau [27] to examine the implications of fractal kinetics in a simple pathway of reversible reactions. Starting with elementary chemical kinetics, that author proceeded to characterize the equilibrium behavior of a simple bimolecular reaction, then derived a generalized set of conditions for microscopic reversibility, and finally developed the fractal kinetic rate law for a reversible Michaelis–Menten mechanism. By means of this fractal kinetic framework, the results showed that the equilibrium ratio is a function of the amount of material in a closed system, and that the principle of microscopic reversibility has a more general manifestation that imposes new constraints on the set of fractal kinetic orders. So, Savageau concluded that fractal kinetics provide a novel means to achieve important features of pathway design.

2.5.5 Importance of the Power-Law Expressions

Power-law expressions are found at all hierarchical levels of organization from the molecular level of elementary chemical reactions to the organismal level of growth and allometric morphogenesis. This recurrence of the power law at different levels of organization is reminiscent of fractal phenomena. In the case of fractal phenomena, it has been shown that this self-similar property is intimately associated with the power-law expression [28]. The reverse is also true; if a power function of time describes the observed kinetic data or a reaction rate higher than 2 is revealed, the reaction takes place in fractal physical support.

The power-law formalism is a mathematical language or representation with a structure consisting of ordinary nonlinear differential equations whose elements are products of power-law functions. The power-law formalism meets two of the most important criteria for judging the appropriateness of a kinetic representation for complex biological systems: the degree to which the formalism is systematically structured, which is related to the issue of mathematical tractability, and the degree to which actual systems in nature conform to the formalism, which is related to the issue of accuracy.

2.6 Fractional Diffusion Equations

Before closing this chapter we would like to mention briefly a novel consideration of diffusion based on the recently developed concepts of fractional kinetics [29]. From our previous discussion it is apparent that if $d_s \leq 2$, diffusion is recurrent. This means that diffusion follows an anomalous pattern described by (2.10); the mean squared displacement grows as $\left\langle z^2(t)\right\rangle \propto t^{\gamma}$ with the exponent $\gamma \neq 1$. To deal with this, a consistent generalization of the diffusion equation (2.18) could have a fractional-order temporal derivative such as

$$\frac{\partial^{\gamma} p(z,t)}{\partial t^{\gamma}} = \mathcal{D}_{\gamma} \frac{\partial^2 p(z,t)}{\partial z^2},$$

where $\mathcal{D}_{\gamma}$ is the fractional diffusion coefficient and the fractional order γ depends on d_w, the fractal dimension of the walk. The previous fractional diffusion equation generalizes Fick's second law, and therefore it allows scientists to describe complex systems with anomalous behavior in much the same way as simpler systems [29].

Also, in order to appreciate the extent of spatial heterogeneity, Berding [30] introduced a heterogeneity function for reaction–diffusion systems evolving to spatially inhomogeneous steady-state conditions. The same author discusses particular applications and compares specific reaction–diffusion mechanisms with regard to their potential for heterogeneity.

3

Nonlinear Dynamics

A wonderful harmony arises from joining together the seemingly unconnected.

Heraclitus of Ephesus (544-483 BC)

Series of measurements from many physiological processes appear random. On the other hand, we are used to thinking that the determinants of variability cannot be known because of the multiplicity and interconnectivity of the factors affecting the phenomena. This idea relies on the classical view of randomness, which requires that a complex process have a large (perhaps infinite) number of degrees of freedom that are not directly observed but whose presence is manifested through fluctuations. However, over the last two decades, scientists from various fields of research have shown that randomness generated by deterministic dynamic processes exhibits spectra practically indistinguishable from spectra of pure random processes. This is referred to as *chaotic behavior*, a specific subtype of nonlinear dynamics, which is the science dealing with the analysis of dynamic systems [31, 32].

The paradox with the term "chaos" is the contradiction between its meaning in colloquial use and its mathematical sense. Routinely, we use the word chaos in everyday life as a synonym for randomness having catastrophic implications. In mathematics, however, "chaos" refers to irregular behavior of a process that appears to be random, but is not. Accordingly, this apparent random-looking behavior poses a fundamental dilemma regarding the origin of randomness in a set of irregular observations from a dynamic process: Is the system chaotic or not? In other words, does the irregular behavior of the observations arise from noise or chaos?

Figure 3.1 illustrates the difference between random and chaotic systems:

- Subplot (A) shows a series of uniformly distributed random numbers between 0 and 1.
- In (B), the plot was generated by the logistic map, a deterministic model of the form $y_{i+1} = 4y_i(1 - y_i)$.

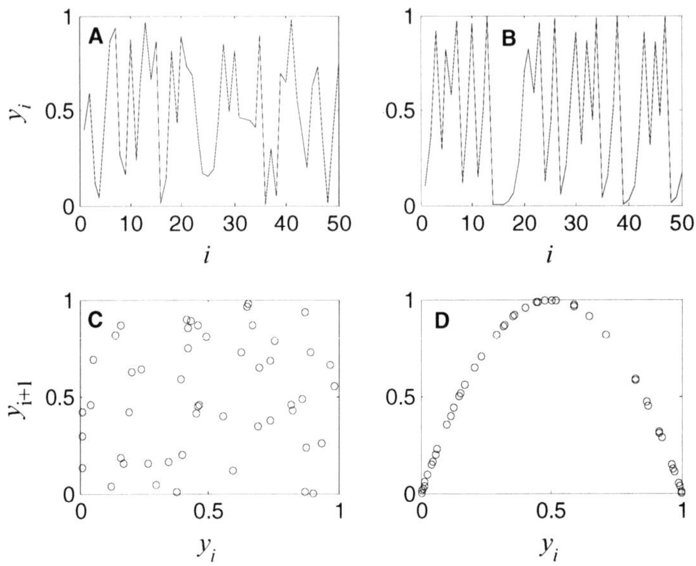

Figure 3.1: The difference between random (**A**, **C**) and chaotic (**B**, **D**) processes pictured as a series of numbers (**A**, **B**) and as pseudophase plots (**C**, **D**).

It is impossible to distinguish the two models visually. The subplots C and D are the socalled *pseudophase* plots of the two sequences of plots A and B, respectively: each y_i is plotted against its consequent y_{i+1}. The random sequence (A) produces scattered points (C) showing that there is no correlation between successive points. In contrast, the points of the deterministic sequence (B) lie in a well-formed line (D).

The key property in this complex, unpredictable, random-like behavior is nonlinearity. When a system (process, or model, or both) consists only of linear components, the response is proportional to its stimulus and the cumulative effect of two stimuli is equal to the summation of the individual effects of each stimulus. This is the superposition principle, which states that every linear system can be studied by breaking it down into its components (thus reducing complexity). In contrast, for nonlinear systems, the superposition principle does not hold; the overall behavior of the system is not at all the same as the summation of the individual behaviors of its components, making complex, unpredictable behavior a possibility. Nevertheless, not every nonlinear system is chaotic, which means that nonlinearity is a necessary but not a sufficient condition for chaos.

The basic ideas of chaos were introduced more than a hundred years ago; however, its significance and implications were realized relatively recently because chaos was studied in detail after the wide dissemination of computers in the 1970s. Although its study started from the fields of mathematics, astronomy, and physics, scientists from almost every field became interested in these ideas. The life sciences are good candidates for chaos due to the complexity of biological processes, although many consider the advanced mathematics and modeling techniques used a drawback. However, during the last 20 years the science of chaos has evolved into a truly interdisciplinary field of research that has changed the way scientists look at phenomena.

3.1 Dynamic Systems

A dynamic system is a deterministic system whose state is defined at any time by the values of several variables $\underline{y}(t)$, the so-called *states* of the system, and its evolution in time is determined by a set of rules. These rules, given a set of initial conditions $\underline{y}(0)$, determine the time evolution of the system in a unique way. This set of rules can be either

- differential equations of the form

  $$\dot{\underline{y}}(t) = \underline{g}\left(\underline{y}, t, \theta\right),$$

  and the system is called a *flow*, or

- discrete equations in which every consequent generation of the variables $\underline{y}$ is given by an equation of the form

  $$\underline{y}_{i+1} = \underline{g}\left(\underline{y}_i, \theta\right),$$

  where $\underline{y}_i$ stands for the ith generation of the variable $\underline{y}$, and then the system is called a *map*.

In the above definitions, θ represents a set of parameters of the system, having constant values. These parameters are also called *control parameters*. The set of the system's variables forms a representation space called the *phase space* [32]. A point in the phase space represents a unique state of the dynamic system. Thus, the evolution of the system in time is represented by a curve in the phase space called *trajectory* or *orbit* for the flow or the map, respectively. The number of variables needed to describe the system's state, which is the number of initial conditions needed to determine a unique trajectory, is the dimension of the system. There are also dynamic systems that have infinite dimension. In these cases, the processes are usually described by differential equations with partial derivatives or time-delay differential equations, which can be considered as a set of infinite in number ordinary differential equations. The fundamental property of the phase space is that trajectories can never intersect themselves or each other. The phase space is a valuable tool in dynamic systems analysis since it is easier to analyze the properties of a dynamic system by determining

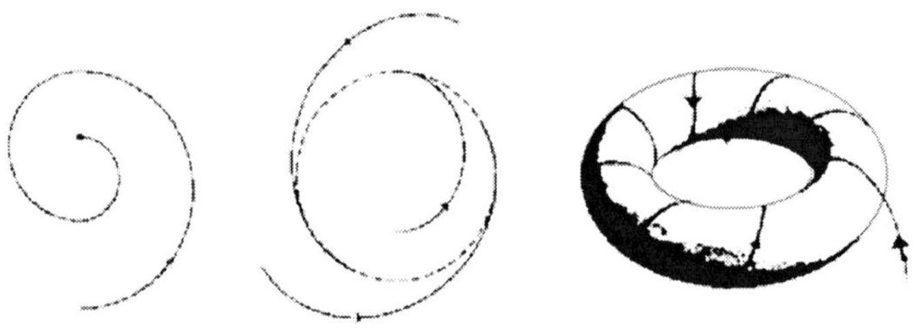

Figure 3.2: A schematic representation of various types of attractors. Reprinted from [33] with permission from Springer.

topological properties of the phase space than by analyzing the time series of the values of the variables directly.

Stable limit sets in the phase space are of supreme importance in experimental and numerical settings because they are the only kind of limit set that can be observed naturally, that is, by simply letting the system run (cf. Appendix A).

3.2 Attractors

Dynamic systems are classified in two main categories: conservative and nonconservative systems. Conservative systems have the property of conserving the volume that is formed by an initial set of points in phase space as times goes by, although the shape of the volume may change. In other words, a volume in phase space resembles an incompressible liquid. On the other hand, nonconservative systems do not possess this property and an initial volume in phase space, apart from changing its shape, may also grow or shrink. In the latter case (when the volume shrinks) the system is called *dissipative*. Most processes in nature, including biological processes, are dissipative.

The trajectories of dissipative dynamic systems, in the long run, are confined in a subset of the phase space, which is called an *attractor* [32], i.e., the set of points in phase space where the trajectories converge. An attractor is usually an object of lower dimension than the entire phase space (a point, a circle, a torus, etc.). For example, a multidimensional phase space may have a point attractor (dimension 0), which means that the asymptotic behavior of the system is an equilibrium point, or a limit cycle (dimension 1), which corresponds to periodic behavior, i.e., an oscillation. Schematic representations for the point, the limit cycle, and the torus attractors, are depicted in Figure 3.2. The point attractor is pictured on the left: regardless of the initial conditions, the system ends up in the same equilibrium point. In the middle, a limit cycle is shown: the system always ends up doing a specific oscillation. The torus attractor on the right is the 2-dimensional equivalent of a circle. In fact, a circle can be called a 1-torus, the 2-dimensional torus can be called a 2-torus, and there is also the 3-torus and generally the m-torus. The trajectory on a 2-torus is a 2-dimensional oscillation with the ratio of the frequencies of the two oscillations being irrational. Because the trajectory never passes through the same point twice, in infinite time it fills the entire surface of the torus. This type of trajectory is called *quasi-periodic*. Being an attractor, the torus attracts all trajectories to fall on its surface.

Even the states of systems with infinite dimension, like systems described by partial differential equations, may lie on attractors of low dimension. The phase space of a system may also have more than one attractor. In this case the asymptotic behavior, i.e., the attractor at which a trajectory ends up, depends on the initial conditions. Thus, each attractor is surrounded by an attraction basin, which is the part of the phase space in which the trajectories from all initial conditions end up.

3.3 Bifurcation

A dynamic system may exhibit qualitatively different behavior for different values of its control parameters θ. Thus, a system that has a point attractor for some value of a parameter may oscillate (limit cycle) for some other value. The critical value where the behavior changes is called a *bifurcation point*, and the event a *bifurcation* [32]. More specifically, this kind of bifurcation, i.e., the transition from a point attractor to a limit cycle, is referred to as *Hopf bifurcation*.

Consider the 1-dimensional map

$$y_{i+1} = g\left(y_i, \theta\right) = \theta y_i \left(1 - y_i\right). \tag{3.1}$$

This difference equation is called a *logistic map*, and represents a simple deterministic system, where given a y_i one can calculate the consequent point y_{i+1} and so on. We are interested in solutions $y_i \geq 0$ with $\theta > 0$. This model describes the dynamics of a single species population [32]. For this map, the fixed points y^* on the first iteration are solutions of

$$y_1^* = \theta y_1^* \left(1 - y_1^*\right),$$

namely

$$y_{1A}^* = 0 \quad y_{1B}^* = \left(\theta - 1\right)/\theta,$$

with the corresponding characteristic multipliers (cf. Appendix A)

$$\xi_{1A} = \theta \quad \xi_{1B} = 2 - \theta.$$

As θ increases from zero but with $0 < \theta < 1$, the only realistic fixed point that is nonnegative is y_{1A}^*, which is stable since $0 < \xi_{1A} < 1$. The first bifurcation comes on y_{1A}^* for $\theta = 1$. When $1 < \theta < 3$, on the one hand, the fixed point y_{1A}^* becomes unstable since $\xi_{1A} > 1$, and on the other hand, the positive fixed point y_{1B}^* is stable since $-1 < \xi_{1B} < 1$. Although there are two steady states, for any initial condition different from $y = 0$, the system will end up after a few steps in

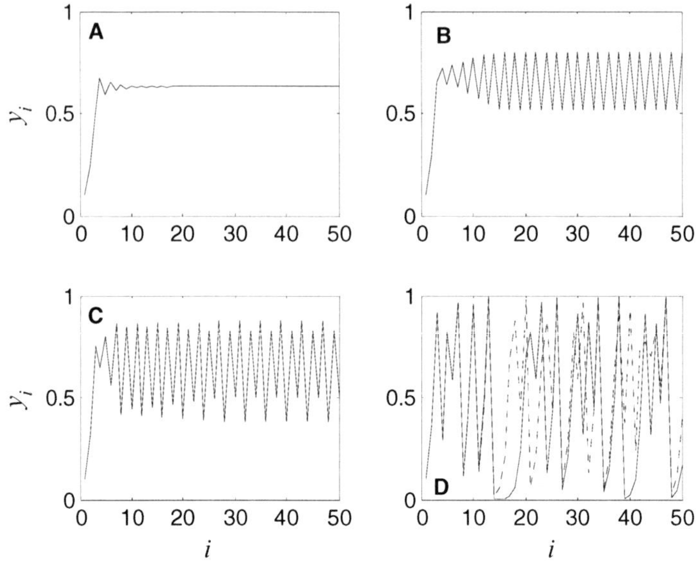

Figure 3.3: The logistic map, for various values of the parameter θ. (**A**) $\theta = 2.7$, (**B**) $\theta = 3.2$, (**C**) $\theta = 3.5$, (**D**) two chaotic trajectories for $\theta = 4$ are coplotted. The initial condition for all solid line plots (**A** to **D**) is $y_0 = 0.1$.

y_{1B}^* (Figure 3.3 A, fixed point of period 1 for $\theta = 2.7$). The second bifurcation comes at y_{1B}^* at $\theta = 3$ where $\xi_{1B} = -1$, and so locally, near y_{1B}^*, we have a periodic solution.

To see what is happening when θ passes through the bifurcation value $\theta = 3$, we examine the stability at the second iteration. The second iteration can be thought of as a first iteration in a model where the iterative time step is 2. The fixed points are solutions of

$$y_2^* = \theta^2 y_2^* (1 - y_2^*) [1 - \theta y_2^* (1 - y_2^*)] .$$

This equation leads to the following solutions:

$$y_{2A}^* = 0, \quad y_{2B}^* = \tfrac{\theta-1}{\theta}, \quad y_{2C}^* = \tfrac{\theta+1-\sqrt{\theta^2-2\theta-3}}{2\theta}, \quad y_{2D}^* = \tfrac{\theta+1+\sqrt{\theta^2-2\theta-3}}{2\theta},$$

when $3 < \theta < 1 + \sqrt{6}$. The corresponding characteristic multipliers are

$$\xi_{2A} = \theta^2, \quad \xi_{2B} = (2-\theta)^2, \quad \xi_{2C} = \xi_{2D} = -\theta^2 + 2\theta + 4.$$

Hence, $\xi_{2A} > 1$, $\xi_{2B} > 1$, $-1 < \xi_{2C} < 1$, and $-1 < \xi_{2D} < 1$. Thus, the y_{2C}^* and y_{2D}^* of the second iteration are stable. What this means is that there is a stable equilibrium of the second iteration, i.e., if we start at y^*_{2C} or y^*_{2D}, for example, we come back to it after 2 iterations. What happens now is that for any initial condition, except $y = 0$ and $y = (\theta - 1)/\theta$, the system after a few steps will end up forming a never-ending succession of the two values of y^*_{2C} and y^*_{2D} (Figure 3.3 B, fixed points of period 2 for $\theta = 3.2$).

As θ continues to increase ($1+\sqrt{6} < \theta$), the characteristic multipliers ξ_{2C} and ξ_{2D} pass through $\xi = -1$, and so these 2-period solutions become unstable. At this stage, we look at the fourth iterate and we find, as might now be expected, that a 4-cycle periodic solution appears (Figure 3.3 C, fixed point of period 4 for $\theta = 3.5$). The period doubles repeatedly and goes to infinity as one approaches a critical point θ_c at which instability sets in for all periodic solutions, e.g., for the model (3.1), $\theta_c \approx 3.5699456$. Above θ_c all fixed points are unstable and the system is chaotic.

The bifurcation situation is illustrated in Figure 3.4, where the stable fixed points y^* are plotted as a function of the parameter θ. These bifurcations are called *pitchfork bifurcations*, for obvious reasons from the picture they generate in Figure 3.4. For example, if $3 < \theta < 1 + \sqrt{6}$, then the periodic solution is between the two y^* that are the intersections of the vertical line through the θ value and the curve of equilibrium points. From Figure 3.4, we note that the difference between the values of θ at which two successive bifurcations take place decreases. It was actually found that the ratio of two successive intervals of θ between successive bifurcations is universally constant, namely $\delta = 4.66920161$, not only for this specific system, but for all systems of this kind, and it is referred to as the *Feigenbaum constant* [32]. Although we have concentrated here on the logistic map, this kind of behavior is typical of maps with dynamics like (3.1); that is, they all exhibit bifurcations to higher periodic solutions eventually leading to chaos.

So, apart from the regular behavior, which is either steady-state, periodic, or quasi-periodic behavior (trajectory on a torus, Figure 3.2), some dynamic systems exhibit chaotic behavior, i.e., trajectories follow complicated aperiodic patterns that resemble randomness. Necessary but not sufficient conditions in order for chaotic behavior to take place in a system described by differential equations are that it must have dimension at least 3, and it must contain nonlinear terms. However, a system of three nonlinear differential equations need not exhibit chaotic behavior. This kind of behavior may not take place at all, and when it does, it usually occurs only for a specific range of the system's control parameters θ.

3.4 Sensitivity to Initial Conditions

As pointed out, for $\theta > \theta_c$ there exist infinitely many unstable steady states of period 1, 2, 4, 8, ... and no stable steady states. This means that almost any initial condition leads to an aperiodic trajectory that looks random as in Figure 3.3 D, but actually the behavior is chaotic. In this figure, two chaotic orbits for $\theta = 4$ are coplotted. Only the initial conditions of the two trajectories differ

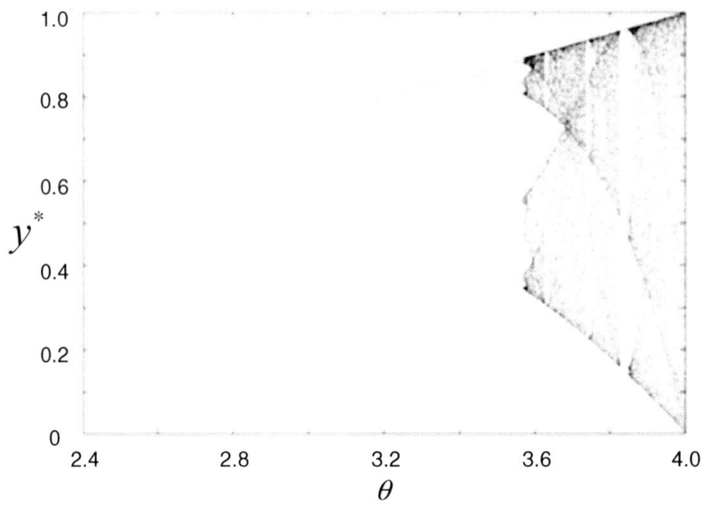

Figure 3.4: The bifurcation diagram of the logistic map.

slightly. For the solid line the initial condition is $y = 0.1$, whereas for the dashed line it is $y = 0.10001$. Although the difference is extremely small, the effect is not at all negligible. The orbits follow an indistinguishable route only for the first 10 steps. Thereafter, they deviate dramatically. Thus, sensitivity to the initial conditions, together with its main consequence of long-term unpredictability, is exhibited.

Hence, the main characteristic of chaotic behavior is the sensitivity to initial conditions. This means that nearby trajectories, whose initial conditions are only slightly different, follow completely different evolutions in time. This property has the implication of unpredictability of the time evolution of the system in the long run due to our inability to know the initial conditions with infinite accuracy. The deviation of two initially neighboring trajectories increases exponentially with time, i.e., proportional to $\exp(\lambda t)$, where the exponent λ is called the *Lyapunov exponent* [32, 34]. Lyapunov exponents are a generalization of the eigenvalues at an equilibrium point and of characteristic multipliers. They depend on the initial conditions and they can be used to determine the stability of quasi-periodic and chaotic behavior as well as of equilibrium points and periodic solutions. For a flow, the Lyapunov exponents are equal to the real parts of the eigenvalues at the equilibrium point, and for a map, they are equal to the magnitudes of the characteristic multipliers at the fixed point. A dynamic system has the same number of Lyapunov exponents as its dimension. The Lyapunov exponents express the deviation of initially nearby trajectories in each "direction." So, a Lyapunov exponent may be negative for a stable "direction," which expresses the exponential approach of two nearby trajectories, and positive for exponential deviation, which expresses the divergence of two nearby trajectories. A system of high dimension may have Lyapunov exponents of all signs and is considered chaotic if at least one of them is positive, which states that at least in one "direction" there exists sensitivity to the initial conditions.

Because chaotic systems may have both negative and positive Lyapunov exponents, their asymptotic behavior can be limited in an attractor as well, where the negative exponents express the convergence to the attractor and the positive the exponential divergence (chaotic behavior) within the attractor. These chaotic attractors are not elementary topological entities with integer dimensions like a point, a circle, or a torus. Instead they have a fractal dimension, which defines an extremely complicated object of infinite detail, though confined in a finite space. This kind of attractor is called a *strange attractor* [32], and the integer dimension of the entire phase space in which the attractor lives is called the *embedding dimension* of the attractor. The two concepts, exponential divergence of initially neighboring trajectories and confinement in a compact space, appear contradictory. However, the fractal structure of the strange attractor makes their coexistence feasible.

3.5 Reconstruction of the Phase Space

The concepts of nonlinear dynamics do not apply only to abstract mathematical models that are described by maps or flows. Useful results can be obtained from observations gathered from real processes as well. Real-life observations, like biological signals, are usually time series of measured quantities. Instead of studying a time series statistically, the idea is to consider it as if it came out of a dynamic system. Then, one tries to reconstruct its phase space (pseudophase space in the case of observed data, when the state variables are unknown) and see whether any structure is detectable, either visually or using certain mathematical and numerical tools [35–37]. The absence of any structure in phase space (e.g., a scatter of points) means that the system is random (Figure 3.1 C). However, the presence of structure is evidence of the dynamic origin of the time series and the existence of an attractor (Figure 3.1 D). The dimension of the attractor can give us information about the dynamic behavior of the whole system. If, for example, the dimension of the attractor is not an integer, it corresponds to a strange attractor and the system exhibits chaotic behavior. The embedding dimension of the attractor, which is actually the dimension of the reconstructed phase space and in the case of a strange attractor should be the next greater integer of the fractal dimension, gives the least number of independent variables, or states, needed to describe the system.

The phase space reconstruction of a time series is accomplished by the method of delays. An embedding dimension d_e is chosen, plus a time delay t°, and then the phase space is constructed using as variables $y(t)$, $y(t+t^\circ)$, $\ldots$, $y(t+(d_e-1)\,t^\circ)$, for all t. It is evident that the choice of d_e and t° is crucial for the reconstruction. There are certain theorems and tests that help in the proper choice of these parameters, but experience and trial are also valuable

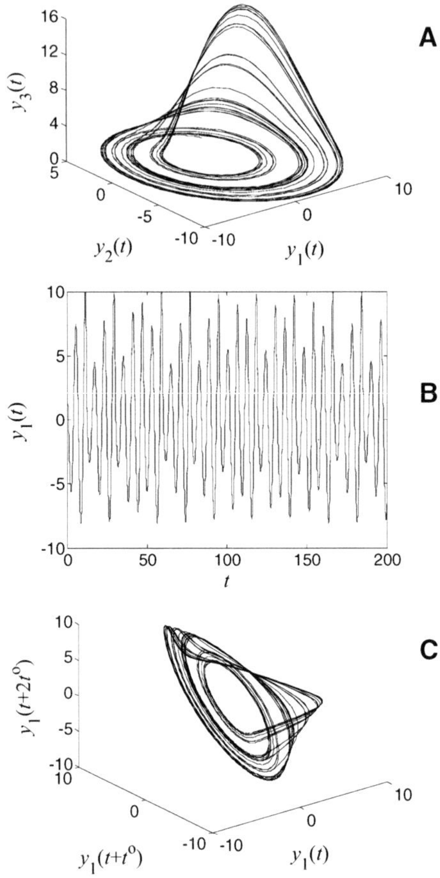

Figure 3.5: The Rössler strange attractor. (**A**) The phase space. (**B**) The state variable $y_1(t)$. (**C**) Reconstruction in the pseudophase space.

tools. It must be mentioned though that due to the automated character of the algorithms, the danger of misleading results always exists. During the past years an overuse of these techniques was noticed and many of the results obtained by this rationale were either wrong or led to erroneous conclusions due to poor application of the techniques and algorithms.

Example 1 *The Rössler Strange Attractor*

Figure 3.5 illustrates the model of the *Rössler strange attractor* [32]. The set of nonlinear differential equations is

$$\begin{aligned} \dot{y}_1 &= -y_2 - y_3, & y_1(0) &= 3, \\ \dot{y}_2 &= y_1 + 0.2y_2, & y_2(0) &= 3, \\ \dot{y}_3 &= 0.4 + y_1 y_3 - 5.7y_3, & y_3(0) &= 0. \end{aligned}$$

The single trajectory plotted in the 3-dimensional phase space never passes through the same point a second time, yet it never leaves a compact volume, thus forming a fractal object of infinite detail (fractal dimension ≈ 2.07), Figure 3.5 A. The state variable y_1 plotted in Figure 3.5 B as a function of time exhibits obvious aperiodicity. In Figure 3.5 C, the Rössler attractor is reconstructed in pseudophase space with the method of delays, making use only of the data from the y_1 variable, as if y_1 were an observable quantity and nothing more of the underlying dynamics were known. Of course, here the dimension of the system is also known and one does not have to try other values for the dimension. Every value of $y_1(t)$ is plotted against $y_1(t+t^\circ)$ and $y_1(t+2t^\circ)$ with lag time $t^\circ = 1$. The reconstructed phase space is not identical to the original one; however, the main topology and features are depicted adequately. ∎

3.6 Estimation and Control in Chaotic Systems

A key factor in modeling is parameter estimation. One usually needs to fit the established model to experimental data in order to estimate the parameters of the model both for simulation and control. However, a task so common in a classical system is quite difficult in a chaotic one. The sensitivity of the system's behavior to the initial conditions and the control parameters makes it very hard to assess the parameters using tools such as least squares fitting. However, efforts have been made to deal with this problem [38]. For nonlinear data analysis, a combination of statistical and mathematical tests on the data to discern inner relationships among the data points (determinism vs. randomness), periodicity, quasiperiodicity, and chaos are used. These tests are in fact nonparametric indices. They do not reveal functional relationships, but rather directly calculate process features from time-series records. For example, the calculation of the dimensionality of a time series, which results from the phase space reconstruction procedure, as well as the Lyapunov exponent are such nonparametric indices. Some others are also commonly used:

- *Correlation dimension.* The correlation dimension is calculated by measuring the Hausdorff dimension according to the method of Grassberger [36, 39]. The dimension of the system relates to the fewest number of independent variables necessary to specify a point in the state space [40]. With random data, the dimension increases with increase of the embedding space. In deterministic data sets, the dimension levels off, even though the presence of noise may yield a slow rise.

- *Singular value decomposition* and eigenvalues of the singular value matrix phase plots. By applying singular value decomposition to the embedded matrix one can improve the appearance of the trajectories in phase space by separating out the noise and the different frequencies from each other, which is important when one is working with experimental data [37, 41]. The eigenvalues give a strong indication of the dimension of the system. A random system shows no demarcation of values, whereas a deterministic system does, as the embedding dimension increases. Each column of data is equivalent to an independent variable; by plotting one column vector vs. another, one can construct the phase space and observe the flows with arrows indicating the direction [42].

The above indices contrast with those destined for linear data analysis:

- The *autocorrelation* (or correlation) function is obtained by multiplying each $y(t)$ by $y(t - t^{\circ})$, where t° is a time delay, and summing the products over all points [43]. Examination of the sum plotted as a function of t° reveals the level of dependency of data points on their neighbors. The correlation time is the value of t° for which the value of the correlation function falls to $\exp(-1)$. When the correlation function falls abruptly to zero, that indicates that the data are without a deterministic component; a slow fall to zero is a sign of stochastic or deterministic behavior; when the data slowly drop to zero and show periodic behavior, then the data are highly correlated and are either periodic or chaotic in nature [37, 43].
- Following a fast Fourier transform of the data, the *power spectrum* shows the power (the Fourier transform squared) as a function of frequency. Random and chaotic data sets fail to demonstrate a dominant frequency. Periodic or quasi-periodic data sets will show one or more dominant frequencies [37].

Chaotic systems are characterized by extreme sensitivity to tiny perturbations. This phenomenon is also known as the *butterfly effect*. This famous term was coined by Lorenz [44], who noticed that long-term prediction of the weather using his system of differential equations was impossible. Lorenz observed that tiny differences in the initial conditions start to grow at a greater and greater speed, until the predictions become nonsense. In an analogous manner, the flapping of a single butterfly's wing today will produce a tiny change in the state of the atmosphere, which in the long run will diverge from that which would otherwise exist in the unperturbed state.

The butterfly effect is often regarded as a troublesome property, and for many years it was generally believed that chaotic motions are neither predictable nor controllable. Von Neumann around 1950 first reported a differing view that small, carefully chosen, preplanned atmospheric disturbances could lead after some time to desired large-scale changes in the weather. Using this chaotic sensitivity, recent work demonstrates that the butterfly effect permits the use of tiny feedback perturbations to control trajectories in chaotic systems, a capability without counterpart in nonchaotic systems [45]. Indeed, it is possible to accomplish this only because the chaotic systems are characterized by exponential growth of small disturbances. This exponential growth implies that we can reach any accessible target extremely quickly, using only a small perturbation.

The relevant research fits broadly into two categories [46]. First, one may ask to select a desired behavior among an infinite variety of behaviors naturally present in chaotic systems, and then stabilize this behavior by applying only tiny changes to an accessible system parameter. Second, one can use the sensitivity of chaotic systems to direct trajectories rapidly to a desired state and steer the system to a general target in state space (not necessarily a periodic orbit). This means that chaotic systems can achieve great flexibility in their ultimate performance.

The presence of chaos may be a great advantage for control in a variety of situations. Typically, in a nonchaotic system, small controls can only change the system dynamics slightly. Short of applying large controls or greatly modifying the system, we are stuck with whatever system performance already exists. In a chaotic system, on the other hand, we are free to choose among a rich variety of dynamic behaviors. Thus, we anticipate that it may be advantageous to design chaos into systems, allowing such variety without requiring large controls or the design of separate systems for each desired behavior.

3.7 Physiological Systems

The application of nonlinear dynamics in physiological systems proposes a new basis in the way certain pathological phenomena emerge. The main characteristic is that a pathological symptom is considered as a sudden qualitative change in the temporal pattern of an illness, such as when a bifurcation takes place. This change can be caused either by endogenous factors or by an exterior stimulus that changes one or more critical control parameters. According to this rationale, therapeutic strategies should aim to invert the progress of the disease and restore normal physiological conditions by interfering with the control parameters. This is in contrast to the classical approach, in which the effort is focused on eliminating the symptoms with a linear rationale that relates the therapeutic stimulus to the effect through a proportionality. This is a general concept also referred to as *dynamical disease*, a term introduced by Mackey and Glass [31, 47–49] (cf. also Section 11.1.2). It is widely appreciated that chaotic behavior dominates physiological systems. Moreover, periodic or other nonchaotic states are considered pathological, whereas the chaotic behavior is considered to be the normal, healthy state. The reason for this has to be associated with a fundamental advantage of nonlinear over classical systems. Indeed, small variations of the control parameters may offer finer, more rapid, and more energy-efficient controllability of the system compared to linear systems [50]. This may be the reason why nature prefers chaos to regularity, and of course the latter is a good enough reason for applied biological sciences such as biopharmaceutics, pharmacokinetics, and pharmacodynamics to adopt this rationale to a greater extent.

Part II

MODELING IN BIOPHARMACEUTICS

Most drugs on the market today are taken orally. Provided a drug is well absorbed, this route of administration results in effective therapy with a minimum of inconvenience to the patient. Therefore, much effort in pharmaceutical research has been devoted to developing oral dosage forms that can deliver the drug to the systemic circulation in a timely and efficient manner. *Biopharmaceutics* refers to the study of the effect of formulation characteristics, the physicochemical properties of the drug, and the physiology of drug absorption. The main object of this discipline is to close the gap between the in vitro data collected in the laboratory and the in vivo performance of the pharmaceutical formulation. Since drug molecules can penetrate the epithelial barrier of the gastrointestinal tract only if they are in solution, drug dissolution and release are the most important biopharmaceutical processes for oral drug delivery. In parallel, the physiological conditions prevailing in the gastrointestinal tract play a significant role and should be taken into account when drug absorption is studied.

Most research on gastrointestinal absorption is based on the concept of homogeneity, that is, the description of average behavior. Some of the most often used paradigms are those borrowed from chemical engineering literature to model hydrodynamics, permeability, and absorption. For example, in the field of dissolution testing, a well-stirred (homogeneous) dissolution medium is used to mimic the in vivo conditions [51]. Calculations associated with the effective intestinal permeability or the unstirred water layer thickness in permeability studies assume that the hydrodynamics of the solution in the intestinal segment obey the well-stirred model [52]. The tank and tube models, often used for the analysis of drug dissolution and uptake in the gastrointestinal tract [53–55], are accompanied with the assumptions of perfect mixing and homogeneous flow, respectively.

One can argue, however, that the assumptions of homogeneity and well-stirred media are not only obvious, but that they are in fact contrary to the evidence, given the anatomical and physiological complexity of the gastrointestinal tract. In vivo drug dissolution, release, and uptake are heterogeneous processes since they take place at interfaces of different phases, e.g., liquid–solid and liquid–membrane boundaries, while diffusion, which is the principal mechanism of all processes, operates under topological constraints. In addition, all processes occur in heterogeneous environments, e.g., variable stirring conditions in the lumen.

Therefore, a proper analysis of drug dissolution, release, and uptake should take into account the heterogeneous character of these processes. In this second part of the book, we consider these important biopharmaceutical processes, placing emphasis on their heterogeneous features.

4

Drug Release

An equation relating the rate of release of solid drugs suspended in ointment bases into perfect sinks is derived. ... The amount of drug released ... is proportional to the square root of time.

Takeru Higuchi
School of Pharmacy, University of Wisconsin, Madison
Journal of Pharmaceutical Sciences 50:874-875 (1961)

The term "release" encompasses several processes that contribute to the transfer of drug from the dosage form to the bathing solution (e.g., gastrointestinal fluids, dissolution medium). The objective of this chapter is to present the spectrum of mathematical models that have been developed to describe drug release from controlled-release dosage forms. These devices are designed to deliver the drug at a rate that is governed more by the dosage form and less by drug properties and conditions prevailing in the surrounding environment. The release mechanism is an important factor in determining whether both of these objectives can be achieved. Depending on the release mechanism, the controlled release systems can be classified into

1. diffusion-controlled,
2. chemically controlled and
3. swelling-controlled.

By far, diffusion is the principal release mechanism, since apart from the diffusion-controlled systems, diffusion takes place at varying degrees in both chemically and swelling-controlled systems. The mathematical modeling of release from diffusion-controlled systems relies on the fundamental Fick's law (2.11), (2.16) with either concentration-independent or concentration-dependent diffusion coefficients. Depending on the formulation characteristics of the device, various types of diffusion can be conceived, i.e., diffusion through an inert matrix, a hydrogel, or a membrane. For chemically controlled systems, the rate of drug release is controlled by

- the degradation and in some cases the dissolution of the polymer in erodible systems or
- the rate of the hydrolytic or enzymatic cleavage of the drug–polymer chemical bond in pendant chain systems.

For swelling-controlled systems the swelling of the polymer matrix after the inward flux of the liquid bathing the system induces the diffusion of drug molecules towards the bathing solution.

In the following sections of this chapter we present the mathematical models used to describe drug release from hydroxypropyl methylcellulose (HPMC) controlled-release dosage forms. HPMC is the most widely used hydrophilic polymer for oral drug delivery systems. Since HPMC exhibits high swellability, drug release from HPMC-based systems is the result of different simultaneously operating phenomena. In addition, different types of HPMC are commercially available and therefore a universal pattern of drug release from HPMC-based systems cannot be pointed out. Accordingly, a wide spectrum of models has been used to describe drug release kinetics from HPMC-based matrix tablets. The sequential presentation below of the mathematical models presented attempts to provide hints to their interrelationships, along with their time evolution, and avoids a strict classification, e.g., empirical vs. mechanistically based models. The last part of the chapter is devoted to the rapidly emerging applications of Monte Carlo simulation in drug release studies. Finally, a brief mention of applications of nonlinear dynamics to drug release phenomena is made at the end of the chapter.

4.1 The Higuchi Model

In 1961 Higuchi [56] analyzed the kinetics of drug release from an ointment assuming that the drug is homogeneously dispersed in the planar matrix and the medium into which drug is released acts as a perfect sink, Figure 4.1. Under these pseudo-steady-state conditions, Higuchi derived (4.1) for the cumulative amount $q(t)$ of drug released at time t:

$$q(t) = \mathcal{A}\sqrt{\mathcal{D}(2c_0 - c_s)\, c_s t}, \qquad c_0 > c_s, \tag{4.1}$$

where $\mathcal{A}$ is the surface area of the ointment exposed to the absorbing surface, $\mathcal{D}$ is the diffusion coefficient of drug in the matrix medium, and c_0 and c_s are the initial drug concentration and the solubility of the drug in the matrix, respectively. Although a planar matrix system was postulated in the original analysis [56], modified forms of (4.1) were published [57–59] for different geometries and matrix characteristics, e.g., granular matrices.

Equation (4.1) is frequently written in simplified form:

$$\frac{q(t)}{q_\infty} = k\sqrt{t}, \tag{4.2}$$

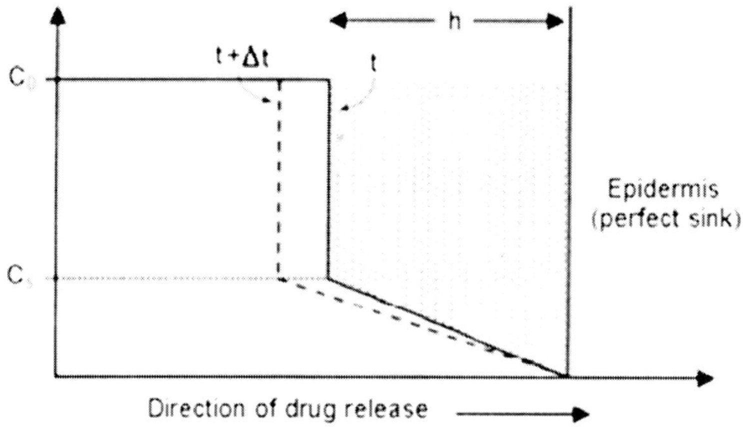

Figure 4.1: The spatial concentration profile of drug (*solid line*) existing in the ointment containing the suspended drug in contact with a perfect sink according to Higuchi's assumptions. The broken line indicates the temporal evolution of the profile, i.e., a snapshot after a time interval Δt. For the distance h above the exposed area, the concentration gradient $(c_0 - c_s)$ is considered constant assuming that c_0 is much higher than c_s.

where q_∞ is the cumulative amount of drug released at infinite time and k is a composite constant with dimension time$^{-1/2}$ related to the drug diffusional properties in the matrix as well as the design characteristics of the system. For a detailed discussion of the assumptions of the Higuchi derivation in relation to a valid application of (4.2) to real data, the reader can refer to the review of Siepmann and Peppas [60].

Equation (4.2) reveals that the fraction of drug released is linearly related to the square root of time. However, (4.2) cannot be applied throughout the release process since the assumptions used for its derivation are not obviously valid for the entire release course. Additional theoretical evidence for the time limitations in the applicability of (4.2) has been obtained [10] from an exact solution of Fick's second law of diffusion for thin films of thickness δ under perfect sink conditions, uniform initial drug concentration with $c_0 > c_s$, and assuming constant diffusion coefficient of drug $\mathcal{D}$ in the polymeric film. In fact, the short-time approximation of the exact solution is

$$\frac{q(t)}{q_\infty} = 4\sqrt{\frac{\mathcal{D}t}{\pi\delta^2}} = k'\sqrt{t}, \tag{4.3}$$

where $k' = 4\sqrt{\mathcal{D}/\pi\delta^2}$. Again, the proportionality between the fraction of drug released and the square root of time is justified, (4.3). These observations have led to a rule of thumb, which states that the use of (4.2) for the analysis of release data is recommended only for the first 60% of the release curve ($q(t)/q_\infty \leq 0.60$). This arbitrary recommendation does not rely on strict theoretical and experimental findings and is based only on the fact that completely different physical conditions have been postulated for the derivation of the equivalent (4.2) and (4.3), while the underlying mechanism in both situations is classical diffusion. In this context, a linear plot of the cumulative amount of drug released $q(t)$ or the fraction of drug released $q(t)/q_\infty$ (utilizing data up to 60% of the release curve) vs. the square root of time is routinely used in the literature as an indicator for diffusion-controlled drug release from a plethora of delivery systems.

4.2 Systems with Different Geometries

One of the first physicochemical studies [61] dealing with diffusion in glassy polymers published in 1968 can be considered as the initiator of the realization that two apparently independent mechanisms of transport, a Fickian diffusion and a case-II transport, contribute in most cases to the overall drug release. Fick's law governs the former mechanism, while the latter reflects the influence of polymer relaxation on the molecules' movement in the matrix [62]. The first studies on this topic [63, 64] were focused on the analysis of Fickian and non-Fickian diffusion as well as the coupling of relaxation and diffusion in glassy polymers. The models used to describe drug release from different geometries are quoted below:

1. Fickian diffusional release form a thin polymer film. Equation (4.3) gives the short-time approximation of the fractional drug released from a thin film of thickness δ.

2. Case II release from a thin polymer film. The fractional drug release $q(t)/q_\infty$ follows zero-order kinetics [65, 66] according to

$$\frac{q(t)}{q_\infty} = \frac{2k_0}{c_0\delta}t, \tag{4.4}$$

   where k_0 is the Case-II relaxation constant and c_0 is the drug concentration, which is considered uniform.

3. Case II radial release from a cylinder. The following equation describes the fractional drug released, $q(t)/q_\infty$, when case II drug transport with radial release from a cylinder of radius ρ is considered [66]:

$$\frac{q(t)}{q_\infty} = \frac{2k_0}{c_0\rho}t - \left(\frac{k_0}{c_0\rho}t\right)^2. \tag{4.5}$$

4. Case II 1-dimensional radial release from a sphere. For a sphere of radius ρ with Case II 1-dimensional radial release, the fractional drug released,

$q(t)/q_\infty$, is given [66] by

$$\frac{q(t)}{q_\infty} = \frac{3k_0}{c_0\rho}t - 3\left(\frac{k_0}{c_0\rho}t\right)^2 + \left(\frac{k_0}{c_0\rho}t\right)^3. \tag{4.6}$$

5. Case II radial and axial release from a cylinder. We quote below a detailed analysis of Case II radial and axial release from a cylinder [67] since (4.4) and (4.5) are special cases of the general equation derived in this section.

The analysis of Case II drug transport with axial and radial release from the cylinder depicted in Figure 4.2 is based on two assumptions:

- a boundary is formed between the glassy and rubbery phases of the polymer, and
- the movement of this boundary takes place under constant velocity.

First, the release surface is determined. A cylinder of height $2L$ that is allowed to release from all sides can be treated as a cylinder of height L that can release from the round side and the top only, Figure 4.2. This second case is easier to analyze and is also implied in [66] for the release of drug from a thin film of thickness $L'/2$. If the big cylinder of Figure 4.2 is cut in half across the horizontal line, two equal cylinders, each of height L, are formed. If drug release from the two newly formed areas (top and bottom) of the two small cylinders is not considered, the two cylinders of height L' exhibit the same release behavior as the big cylinder, i.e., $q(t)_{2L} = 2q(t)_L$ and $q_{\infty,2L} = 2q_{\infty,L}$; consequently,

$$\frac{q(t)_{2L}}{q_{\infty,2L}} = \frac{q(t)_L}{q_{\infty,L}}.$$

This proportionality demonstrates that the analysis of the release results can describe both of the following cases: either a cylinder of height L that releases from the round and top surfaces, or a cylinder of height $2L$ that releases from all sides, Figure 4.2.

At zero time, the height and radius of the cylinder are L and ρ, respectively, Figure 4.2. After time t the height of the cylinder decreases to L' and its radius to ρ' assuming Case II drug transport for both axial and radial release. The decrease rate of radius ρ' and height L' of the cylinder it can be written

$$\dot{\rho}' = \dot{L}' = -\frac{k_0}{c_0}, \tag{4.7}$$

where k_0 is the Case II relaxation constant and c_0 is the drug concentration (considered uniform). The assumed value of the penetration layer speed is implied from the analysis of the cases studied in [65,66], which are simpler than the present case. Initial conditions for the above equations are simply $\rho'(0) = \rho$ and $L'(0) = L$.

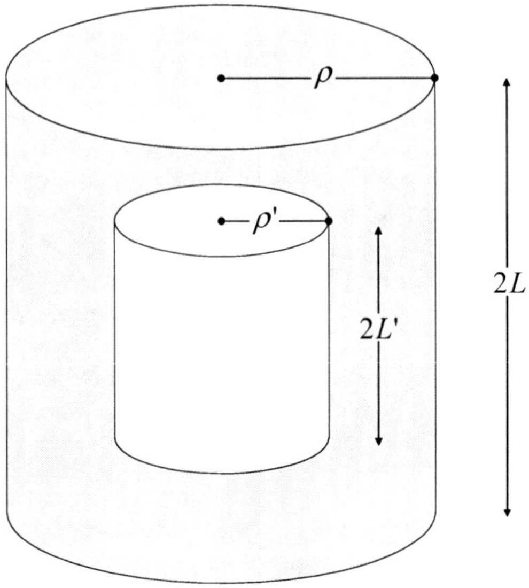

Figure 4.2: Case II drug transport with axial and radial release from a cylinder of height $2L$ and radius ρ at $t = 0$. Drug release takes place from all sides of the big cylinder. The drug mass is contained in the grey region. After time t the height of the cylinder is reduced to $2L'$ and its radius to ρ' (*small cylinder*).

After integration of (4.7), we obtain the following equations as well as the time for which each one is operating:

$$\begin{array}{ll} \rho' = \rho - (k_0/c_0)\, t, & t \leq (c_0/k_0)\, \rho, \\ L' = L - (k_0/c_0)\, t, & t \leq (c_0/k_0)\, L. \end{array} \tag{4.8}$$

This means that the smaller dimension of the cylinder (ρ or L) determines the duration of the phenomenon.

The amount of drug released at any time t is given by the following mass-balance equation:

$$q(t) = c_0 \pi \left(\rho^2 L - \rho'^2 L'\right). \tag{4.9}$$

Substituting (4.8) into (4.9), the following expression for mass $q(t)$ as a function of time t is obtained:

$$q(t) = c_0 \pi \left[\rho^2 L - \left(\rho - \frac{k_0}{c_0} t\right)^2 \left(L - \frac{k_0}{c_0} t\right)\right].$$

And for the mass released at infinite time, we can write

$$q_\infty = c_0 \pi \rho^2 L.$$

From the previous equations, the fraction released $q(t)/q_\infty$ as a function of time t is obtained:

$$\frac{q(t)}{q_\infty} = \left(\frac{2k_0}{c_0\rho} + \frac{k_0}{c_0 L}\right) t - \left(\frac{k_0^2}{c_0^2\rho^2} + \frac{2k_0^2}{c_0^2\rho L}\right) t^2 + \frac{k_0^3}{c_0^3\rho^2 L} t^3. \tag{4.10}$$

This equation describes the entire fractional release curve for Case II drug transport with axial and radial release from a cylinder. Again, (4.10) indicates that the smaller dimension of the cylinder (ρ or L) determines the total duration of the phenomenon. When $\rho \gg L$, (4.10) can be approximated by

$$\frac{q(t)}{q_\infty} = \frac{k_0}{c_0 L} t,$$

which is identical to (4.4) with the difference of a factor of 2 due to the fact that the height of the cylinder is $2L$. When $\rho \ll L$, (4.10) can be approximated by

$$\frac{q(t)}{q_\infty} = \frac{2k_0}{c_0\rho} t \quad \left(\frac{k_0}{c_0\rho} t\right)^2,$$

which is also identical to (4.5). These results demonstrate that the previously obtained (4.4) and (4.5) are special cases of the general solution (4.10).

4.3 The Power-Law Model

Peppas and coworkers [64,68] introduced a semiempirical equation (the so-called power law) to describe drug release from polymeric devices in a generalized way:

$$\frac{q(t)}{q_\infty} = k t^\lambda, \tag{4.11}$$

where k is a constant reflecting the structural and geometric characteristics of the delivery system expressed in dimensions of time$^{-\lambda}$, and λ is a release exponent the value of which is related to the underlying mechanism(s) of drug release. Equation (4.11) enjoys a wide applicability in the analysis of drug release studies and the elucidation of the underlying release mechanisms. Apart from its simplicity, the extensive use of (4.11) is mainly due to the following characteristics:

- Both Higuchi equations (4.1) and (4.3), which describe Fickian diffusional release from a thin polymer film, are special cases of (4.11) for $\lambda = 0.5$; also, (4.4) is a special case of (4.11) for $\lambda = 1$.

- It can describe adequately the first 60% of the release curve when (4.5) and (4.6) govern the release kinetics [66,67].

- The value of the exponent λ obtained from the fitting of (4.11) to the first 60% of the experimental release data, from polymeric-controlled delivery systems of different geometries, is indicative of the release mechanism, Table 4.1.

Table 4.1: Values of the exponent λ in (4.11) and the corresponding release mechanisms from polymeric-controlled delivery systems of various geometries [65].

Exponent λ			**Release mechanism**
Thin film	Cylinder	Sphere	
0.5	0.45	0.43	Fickian diffusion
$0.5 < \lambda < 1.0$	$0.45 < \lambda < 0.89$	$0.43 < \lambda < 0.85$	Anomalous transport
1.0	0.89	0.85	Case II transport

From the values of λ listed in Table 4.1, only the two extreme values 0.5 and 1.0 for thin films (or slabs) have a physical meaning. When $\lambda = 0.5$, pure Fickian diffusion operates and results in diffusion-controlled drug release. It should be recalled here that the derivation of the relevant (4.3) relies on short-time approximations and therefore the Fickian release is not maintained throughout the release process. When $\lambda = 1.0$, zero-order kinetics (Case II transport) are justified in accord with (4.4). Finally, the intermediate values of λ (cf. the inequalities in Table 4.1) indicate a combination of Fickian diffusion and Case II transport, which is usually called *anomalous transport*.

It is interesting to note that even the more realistic model adhering to the Case II radial and axial drug release from a cylinder, (4.10), can be described by the power-law equation. In this case, pure Case II drug transport and release is approximated (Table 4.1) by the following equation:

$$\frac{q(t)}{q_{\infty}} \approx kt^{0.89}. \tag{4.12}$$

A typical example of comparison between (4.10) and (4.12) when $\rho < L$ is shown in Figure 4.3. One should note the resemblance, along the first 60% of the curves, to the kinetic profiles derived from these equations.

4.3.1 Higuchi Model vs. Power-Law Model

Drug release data are frequently plotted as percent (or fractional) drug released vs. $t^{1/2}$. This type of plot is usually accompanied by linear regression analysis using $q(t)/q_{\infty}$ as dependent and $t^{1/2}$ as independent variable. This routinely applied procedure can lead to misinterpretations regarding the diffusional mechanism, as is shown below using simulation studies [69].

Simulated data were generated from (4.11) using values for λ and k ranging from 0.4 to 0.65 and from 0.05 to 0.5, respectively. The range of λ values is the neighborhood of the Higuchi exponent 0.5, which is the theoretical value for a diffusion-controlled release process. Moreover, values of λ in the range 0.4–0.65 are frequently quoted in the literature for the discernment of drug release mechanisms (pure diffusion, anomalous transport, and combination) from HPMC matrix devices of different geometries [65,66]. The values assigned to k are similar to the estimates obtained when (4.3) is fitted to drug release data, whereas

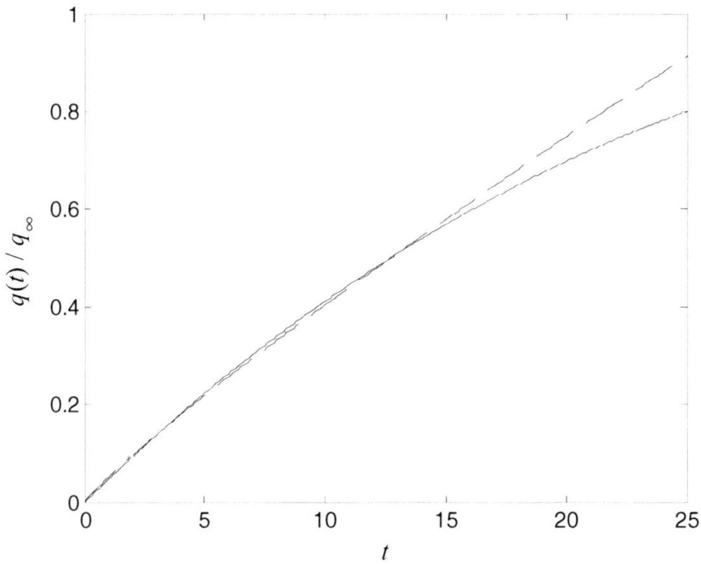

Figure 4.3: Fractional drug release $q(t)/q_\infty$ vs. time (arbitrary units) for Case II transport with axial and radial release from a cylinder. Comparison of the solutions presented by (4.10) with $k_0 = 0.01$, $c_0 = 0.5$, $\rho = 1$, $L = 2.5$ (*dashed line*) and (4.12) with $k = 0.052$ (*solid line*).

k has dimension of time$^{-1/2}$. The constraint $q(t)/q_\infty \leq 1$ was used for each set generated from (4.11). The duration of the simulated release experiment was arbitrarily set equal to 8 ($t \leq 8$). Therefore, the number of the simulated data generated from (4.11) varied according to the specific value assigned to k using in all cases a constant time step, 0.01. The pairs of data $(q(t)/q_\infty, t)$ generated from (4.11) were further analyzed using linear regression analysis in accord with (4.3).

Table 4.2 shows the results of linear regression analysis ($q(t)/q_\infty$ vs. $t^{1/2}$) for the data generated from (4.11). As expected, the theoretically correct sets of data ($\lambda = 0.5$) exhibited ideal behavior (intercept$=0$, $R^2 = 1$). Judging from the determination coefficient R^2 values in conjunction with the number of data points utilized in regression, all other sets of data with $\lambda \neq 0.5$ are also described nicely if one does not apply a more rigorous analysis, e.g., plot of residuals. It is also worthy of mention that the positive intercepts were very close to zero and only in two cases ($k = 0.4$, $\lambda = 0.4$; $k = 0.5$, $\lambda = 0.4$) were they found to be in the range 0.10–0.11. In parallel, any negative intercepts were very close to the origin of the axes.

These observations indicate that almost the entire set of data listed in Table

Table 4.2: Results of linear regression $q(t)/q_\infty$ vs. $t^{1/2}$ for data generated from (4.11). (a) Estimates not statistically significant different from zero were obtained. (b) Number of data points utilized in regression.

k	λ	intercept	slope	R^2	N^b
	0.40	0.01287	0.03668	0.9970	800
	0.45	0.006719	0.04305	0.9993	800
0.05	0.50	0^a	0.05	1	800
	0.55	−0.00576	0.05760	0.9994	800
	0.60	−0.01545	0.06571	0.9976	800
	0.65	−0.02436	0.07501	0.9950	800
	0.40	0.0772	0.02201	0.9970	800
	0.45	0.04031	0.2583	0.9993	800
0.30	0.50	0^a	0.3	1	800
	0.55	−0.04418	0.3456	0.9994	800
	0.60	−0.08866	0.3925	0.9976	743
	0.65	−0.1258	0.4349	0.9949	637
	0.40	0.1030	0.2935	0.9970	800
	0.45	0.05270	0.3451	0.9993	766
0.40	0.50	0^a	0.4	1	625
	0.55	−0.04676	0.4513	0.9994	529
	0.60	−0.08829	0.4987	0.9976	460
	0.65	−0.1253	0.5422	0.9948	4409
	0.40	0.1117	0.3800	0.9969	565
	0.45	0.05243	0.4424	0.9993	466
0.50	0.50	0^a	0.5	1	400
	0.55	−0.04649	0.5525	0.9993	352
	0.60	−0.0878	0.6002	0.9975	317
	0.65	−0.1245	0.6432	0.9947	290

4.2 and generated from (4.11) can be misinterpreted as obeying (4.3). Under real experimental conditions the discernment of kinetics is even more difficult when linear regression of $q(t)/q_\infty$ vs. $t^{1/2}$ is applied. This is so if one takes into account

- the usually small number of experimental data points available,
- the constraint for the percentage of drug released, $q(t)/q_\infty \leq 0.60$,
- the experimental error of data points,
- the high variability or lack of data points at the early stages of the experiment, and
- the possible presence of a delay in time.

Therefore, it is advisable to fit (4.11) directly to experimental data using nonlinear regression. Conclusions concerning the release mechanisms can be based on the estimates for λ and the regression line statistics [69].

4.4 Recent Mechanistic Models

Although the empirical and semiempirical models described above provide adequate information for the drug release mechanism(s), better insight into the release process can be gained from mechanistic models. These models have the advantage of being more accurate and predictive. However, mechanistic models are more physically realistic and therefore mathematically more complex since they describe all concurrent physicochemical processes, e.g., diffusion, dissolution, swelling. Additionally, they require the use of time- and/or position-, direction-dependent diffusivities. This mathematical complexity is the main disadvantage of the mechanistic models since explicit analytical solutions of the partial differential equations cannot be derived. In this case, one has to rely on numerical solutions and less frequently on implicit analytical solutions.

Although the emphasis of this section will be on the most recent mechanistic approaches, the work of Fu et al. [70] published in 1976 should be mentioned since it deals with the fundamental release problem of a drug homogeneously distributed in a cylinder. In reality, Fu et al. [70] solved Fick's second law equation assuming constant cylindrical geometry and no interaction between drug molecules. These characteristics imply a constant diffusion coefficient in all three dimensions throughout the release process. Their basic result in the form of an analytical solution is

$$\frac{q(t)}{q_\infty} = 1 - \frac{8}{h^2\rho^2}\left[\sum_{i=1}^{\infty} \alpha_i^{-2} \exp\left(-\mathcal{D}\alpha_i^2 t\right)\right]\left[\sum_{j=1}^{\infty} \beta_j^{-2} \exp\left(-\mathcal{D}\beta_j^2 t\right)\right],$$

where $\beta_j = (2j+1)\,\pi/\,(2h)$, α_i are the roots of the equation $J_0(\rho\alpha) = 0$, and J_0 is the zero-order Bessel function. Here, h denotes the half-length, ρ the radius of the cylinder, and, i and j are integers. Note that for small t the series is very slowly converging. Even keeping 100 terms of the above series is still not a good enough approximation of $q(t)/q_\infty$, for $t \approx 0$. For long times all terms with high values of α and β decay rapidly and only the term with the lowest value survives. The series reduces to a simple exponential after some time.

Gao et al. [71, 72] developed a mathematical model to describe the effect of formulation composition on the drug release rate for hydroxypropyl methylcellulose-based tablets. An effective drug diffusion coefficient $\mathcal{D}'$, was found to control the rate of release as derived from a steady-state approximation of Fick's law in one dimension:

$$\frac{q(t)}{q_\infty} = \frac{\mathcal{A}}{V}\sqrt{\frac{\mathcal{D}'t}{\pi}},$$

where $\mathcal{A}$ is the surface area and V the volume available for release, while $\mathcal{D}'$ corresponds to the quotient $\mathcal{D}/\tau$, where $\mathcal{D}$ is the classical drug diffusion coefficient in the release medium and τ is the tortuosity of the diffusing matrix.

In a series of papers Narasimhan and Peppas [73–75] developed models that take into account the dissolution of the polymer carrier. According to the theory, the polymer chain, at the surface of the system, disentangles (above a critical water concentration) and diffuses into the release medium. The polymer's dissolution rate constant and the decreasing with time surface area of the device control the kinetics of the polymer mass loss. Symmetry planes in axial and radial direction, placed at the center of the matrix, for the water and drug concentration profiles allow the development of an elegant mathematical analysis. Fick's second law of diffusion for cylindrical geometry is used to model both water and drug diffusion. Since both the composition and the dimensions of the device change with time while the diffusion coefficients for both species are considered to be dependent on the water content, the complex partial differential equations obtained are solved numerically. The model has been used successfully to describe the effect of the initial theophylline loading of HPMC-based tablets on the resulting drug release rate.

Recently, a very sophisticated mechanistic model called the *sequential layer* model was presented [76–81]; the model considers inhomogeneous polymer swelling, drug dissolution, polymer dissolution, and water and drug diffusion with nonconstant diffusivities and moving boundary conditions. The raptation theory was used for the description of polymer dissolution, while water and drug diffusion were described using Fick's second law of diffusion. An exponential dependence of the diffusion coefficients on the water content was taken into account. Moving boundaries were considered since the polymer swells, the drug and the polymer dissolve, thereby making the interface matrix/release medium not stationary. The model was applied successfully in the elucidation of the swelling and drug release behavior from HPMC matrices using chlorpheniramine maleate, propranolol HCl, acetaminophen, theophylline, and diclophenac as model drugs.

4.5 Monte Carlo Simulations

In a Monte Carlo simulation we attempt to follow the time evolution of a model that does not proceed in some rigorously predefined fashion, e.g., Newton's equations of motion. Monte Carlo simulations are appropriate for models whose underlying mechanism(s) are of a stochastic nature and their time evolution can be mimicked with a sequence of random numbers, which is generated during the simulation. The repetitive Monte Carlo simulations of the model with different sequences of random numbers yield results that agree within statistical error but are not identical. The goal is to understand the stochastic component of the physical process making use of the perfect control of "experimental" conditions in the computer-simulation experiment, examining every aspect of the system's configuration in detail. Since the mass transport phenomena, e.g., drug

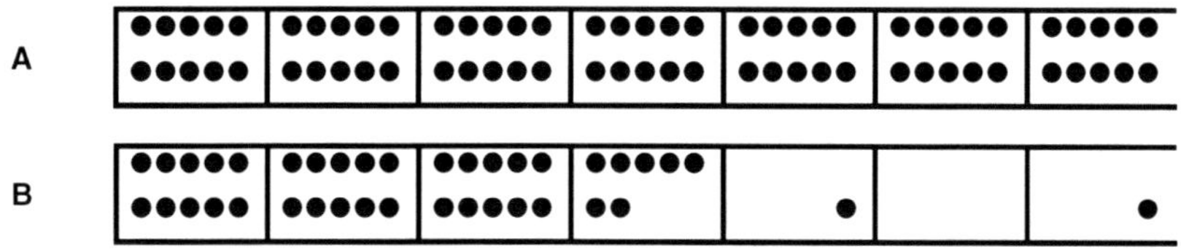

Figure 4.4: Schematic of a system used to study diffusion under the Higuchi assumptions. (**A**) Initial configuration of the system, (**B**) evolution after time t. Particles are allowed to leak only from the right side of the system. Reprinted from [82] with permission from Springer.

diffusion and the chemical processes, e.g., polymer degradation encountered in drug release studies, are random processes, Monte Carlo simulations are used to elucidate the release mechanisms. In the next section we demonstrate the validity of the Higuchi law using Monte Carlo simulations and in the following two sections we focus on the use of Monte Carlo simulations for the description of drug release mechanisms based on Fickian diffusion from Euclidean or fractal spaces. Finally, the last portion of this section deals with Monte Carlo simulations of drug release from bioerodible microparticles.

4.5.1 Verification of the Higuchi Law

The presuppositions for the application of the Higuchi law (4.2) have been discussed in Section 4.1. However, it is routinely quoted in the literature without a rigorous proof that only the first 60% of the release curve data should be utilized for a valid application of (4.2). Recently, this constraint has been verified for the Higuchi model using Monte Carlo computer simulations [82] (cf. Appendix B).

To mimic the conditions of the Higuchi model, a 1-dimensional matrix of 200 sites has been constructed, Figure 4.4. Each site is labeled with the number of particles it currently hosts. Initially all sites have 10 particles, i.e., the total number n_0 of particles monitored is 2000. Drug molecules move inside the matrix by the mechanism of Fickian diffusion and cannot move to a site unless this site is empty. Thus, the system is expected to behave as if its "concentration" were much higher than its "solubility," which is the basic assumption made in the theoretical derivation of the Higuchi equation. The matrix can leak only from the site at its edge in full analogy with Figure 4.1. The diffusive escape process is simulated by selecting a particle at random and moving it to a randomly selected nearest-neighbor site. If the new site is an empty site then the move is allowed and the particle is moved to this new site. If the new site is already occupied, the move is rejected. A particle is removed from the lattice as soon as it migrates to the leak site. After each particle move, time is incremented by arbitrary time units, the Monte Carlo microSteps (MCS), during

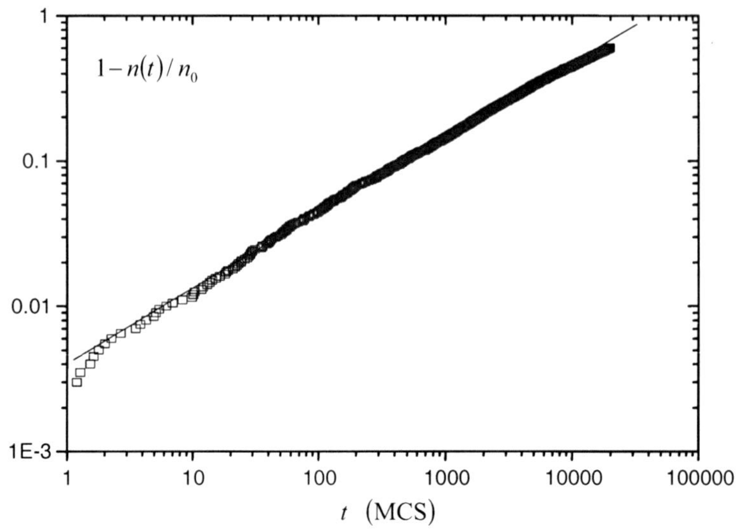

Figure 4.5: Log-log plot of $1-n(t)/n_0$ vs. time. Simulation results are indicated as points using the first 60% of the release data. The slope of the fitted line is 0.51 and corresponds to the exponent of the Higuchi equation. The theoretical prediction is 0.50.

which the movement takes place. One MCS is the smallest time unit in which an event can take place. The increment is chosen to be $1/n(t)$, where $n(t)$ is the number of particles remaining in the system. This is a typical approach in Monte Carlo simulations. The number of particles that are present inside the cylinder as a function of time is monitored until the cylinder is completely empty of particles. Figure 4.5 shows the simulation results for the first 60% of the release data; the slope of the line is 0.51 very close to the value 0.50 expected by the Higuchi equation.

The simulation results presented in Figure 4.5 provide an indirect proof of the valid use of the first 60% of the release data in line with (4.2). Needless to say, the Monte Carlo simulations in Figure 4.5 do not apply to the diffusion problem associated with the derivation of (4.3).

4.5.2 Drug Release from Homogeneous Cylinders

The general problem that we will focus on in this section is the escape of drug molecules[1] from a cylindrical vessel. Initially, theoretical aspects are presented demonstrating that the Weibull function can describe drug release kinetics from cylinders, assuming that the drug molecules move inside the matrix by a Fickian

[1] The terms "drug molecule" and "particle" will be used in this section interchangeably.

diffusion mechanism. Subsequently, Monte Carlo simulations will be used to substantiate the theoretical result and provide a link between the Weibull model and the physical kinetics of the release process [82].

Theoretical Aspects

A simple approximate solution is sought for the release problem, which can be used to describe release even when interacting particles are present. The particles are assumed to move inside the vessel in a random way. The particle escape rate is expected to be proportional to the number $n(t)$ of particles that exist in the vessel at time t. The rate will also depend on another factor, which will show how "freely" the particles are moving inside the vessel, how easily they can find the exits, how many of these exits there are, etc. This factor is denoted by g. Hence, a differential equation for the escape rate can be written

$$\dot{n}(t) = -agn(t),$$

where a is a proportionality constant and the negative sign means that $n(t)$ decreases with time. If the factor g is kept constant, it may be included in a and in this case the solution of the previous equation is

$$n(t) = n_0 \exp(-at)$$

using the initial condition $n(0) = n_0$. The last equation is similar to the asymptotic result derived by Fu et al. [70] for pure Fickian diffusion inside a cylinder for long times (cf. Section 4.4).

It stands to reason to assume that the factor g should be a function of time since as time elapses a large number of drug molecules leave the vessel and the rest can move more freely. Thus, in general one can write that $g = g(t)$ and the previous differential equation becomes

$$\dot{n}(t) = -ag(t)\,n(t). \tag{4.13}$$

A plausible assumption is to consider that $g(t)$ has the form $g(t) \propto t^{-\mu}$.

We are interested in supplying a short-time approximation for the solution of the previous equation. There are two ways to calculate this solution. The direct way is to make a Taylor expansion of the solution. The second, more physical way, is to realize that for short initial time intervals the release rate $\dot{n}(t)$ will be independent of $n(t)$. Thus, the differential equation (4.13) can be approximated by $\dot{n}(t) = -ag(t)$. Both ways lead to the same result.

- For $\mu = 1/2$, (4.13) leads to $n(t) \propto \sqrt{t}$ (as a short-time approximation) exactly as predicted by the Higuchi law.
- For $\mu = 0$ we obtain, again as a short-time approximation, the result $n(t) = n_0 - at$ corresponding to *ballistic* exit (zero-order kinetics).

The above imply that choosing $g(t) = t^{-\mu}$ is quite reasonable. In this case (4.13) will be

$$\dot{n}(t) = -at^{-\mu} n(t).$$

Solving this equation we obtain

$$n(t) = n_0 \exp\left(-at^b\right), \tag{4.14}$$

where $b = 1 - \mu$.

The above reasoning shows that the stretched exponential function (4.14), or Weibull function as it is known, may be considered as an approximate solution of the diffusion equation with a variable diffusion coefficient due to the presence of particle interactions. Of course, it can be used to model release results even when no interaction is present (since this is just a limiting case of particles that are weakly interacting).

It is clear that it cannot be proven that the Weibull function is the best choice of approximating the release results. There are infinitely many choices of the form $g(t)$ and some of them may be better than the Weibull equation. This reasoning merely indicates that the Weibull form will probably be a good choice. The simulation results below show that it is indeed a good choice. The above reasoning is quite important since it provides a physical model that justifies the use of the Weibull function in order to fit experimental release data.

Simulations

A brief outline of the Monte Carlo techniques used for the problem of drug release from cylinders is described in Appendix B. The results obtained for cylinders of different dimensions are shown in Figure 4.6. In all cases it is possible to achieve a quite accurate fitting of the simulation results for $n(t)$ using the Weibull function [82]. It turns out that the exponent b takes values in the range 0.69 to 0.75. Figure 4.7 shows that the fitting is very accurate especially at the beginning, and it remains quite good until all of the drug molecules are released. The number of particles that have escaped from the matrix is equal to

$$\widetilde{n}(t) = n_0 - n(t) = n_0 \left[1 - \exp\left(-at^b\right)\right], \tag{4.15}$$

where a and b are parameters that have to be experimentally determined.

Ritger and Peppas [65, 66] have shown that the power law (4.11) describes accurately the first 60% of the release data. It is easy to show that the two models (4.11) and (4.14) coincide for small values of t. Note that $\widetilde{n}(t)/n_0$ is directly linked to $q(t)/q_\infty$. From the Taylor expansion of $\exp(-\chi)$, we can say that for small values of χ we have $\exp(-\chi) \approx 1 - \chi$. From (4.15), setting $\chi = at^b$, one gets

$$\widetilde{n}(t)/n_0 = at^b$$

for small values of at^b, which has the same form as the power-law model. For this approximation to hold, the quantity at^b has to be small. This does not

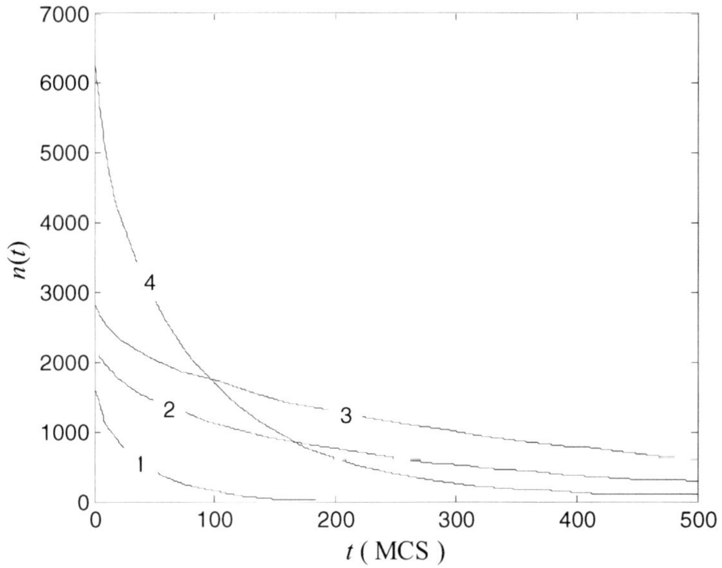

Figure 4.6: Number of particles inside a cylinder as a function of time. (**1**) Cylinder with height of 31 sites and diameter 16 sites. Number of drug molecules $n_0 = 1750$. (**2**) Cylinder with height 7 sites and diameter 31 sites. Number of drug molecules $n_0 = 2146$. (**3**) Cylinder with height 5 sites and diameter 41 sites. Number of drug molecules $n_0 = 2843$. (**4**) Cylinder with height 51 sites and diameter 21 sites. Number of drug molecules $n_0 = 6452$.

mean that t itself must be small. As long as a is small, t may take larger values and the approximation will still be valid.

A comparison of the simulation results and fittings with the Weibull and the power-law model is presented in Figure 4.7. Obviously, the Weibull model describes quite well all release data, while the power law diverges after some time. Of course both models can describe equally well experimental data for the first 60% of the release curve.

The Physical Connection Between a, b and the System Geometry

The parameters a and b are somehow connected to the geometry and size of the matrix that contains the particles. This connection was investigated by performing release simulations for several cylinder sizes and for several initial drug concentrations [82]. The Weibull function was fitted to the simulated data to obtain estimates for a and b. If one denotes by N_{leak} the number of leak sites and by N_{tot} the total number of sites, in the continuum limit the ratio $N_{\text{leak}}/N_{\text{tot}}$

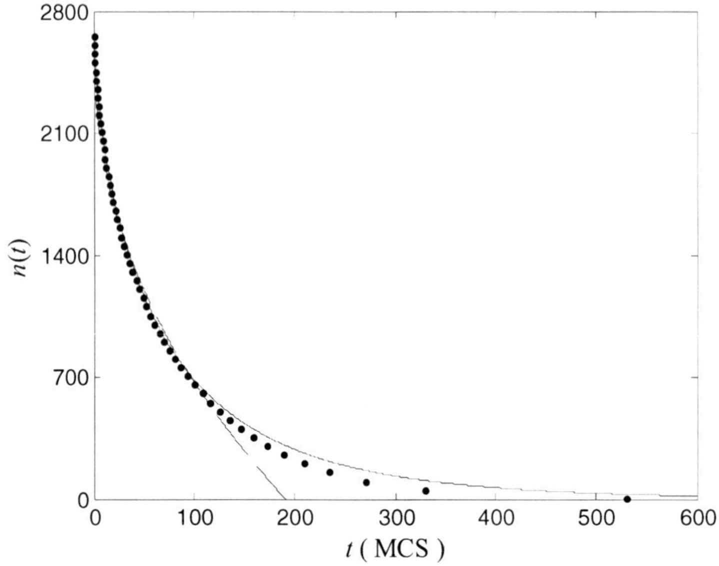

Figure 4.7: Number of particles inside a cylinder as a function of time with initial number of drug molecules $n_0 = 2657$. Simulation for cylinder with height 21 sites and diameter 21 sites (*dotted line*). Plot of curve $n(t) = 2657 \exp\left(-0.049t^{0.72}\right)$, Weibull model fitting (*solid line*). Plot of curve $n(t) = 2657\left(1 - 0.094t^{0.45}\right)$, power-law fitting (*dashed line*).

is proportional to the leak surface of the system. Plots of a vs. $N_{\text{leak}}/N_{\text{tot}}$ (not shown) were found to be linear and independent of the initial drug concentration; this implies that a is proportional to the specific leak surface, i.e., the surface to volume ratio. The slopes of the straight lines were found to be in the range 0.26–0.30 [82]. The value of the slope can be related to the mathematical model presented in the theoretical section since the number of particles escaping at time dt was assumed to be proportional to $an(t)$; thus, the simulation results can be summarized as $an(t) = 0.28\,(N_{\text{leak}}/N_{\text{tot}})\,n(t)$. Assuming a uniform distribution of particles, $N_{\text{leak}}/N_{\text{tot}}$ is the probability that a particle is at a site that is just one step from the exit. Accordingly, $(N_{\text{leak}}/N_{\text{tot}})\,n(t)$ is the mean number of particles that are able to escape at a given instant of time. Since there are 6 neighboring sites in the 3-dimensional space, the probability for a particle to make the escaping step is $1/6$ (≈ 0.17). It is quite close to the 0.28 value of the simulation. The difference is due to the fact that after some time, the distribution of particles is no longer uniform. There are more empty cells near the exits than inside, so the mean number of particles that are able to escape at a given instant is rather less than $(N_{\text{leak}}/N_{\text{tot}})\,n(t)$. This explains the higher value of the slope.

The plot of b values obtained from release simulations for several cylinder sizes and initial drug concentrations vs. $N_{\text{leak}}/N_{\text{tot}}$ (not shown) was also linear [82] with a slope practically independent of the initial concentration, $b = 0.65 + 0.4\,(N_{\text{leak}}/N_{\text{tot}})$. There are two terms contributing to b; one depends on $N_{\text{leak}}/N_{\text{tot}}$ and the other does not. Actually b is expected to be proportional to the specific surface, since a high specific surface means that there are a lot of exits, so finding an exit is easier. The constant term depends on the ability of the particles to move inside the matrix, the interaction between the particles, etc. The linear relationship yields the value of $b = 0.69$ when the exits cover the entire surface of the cylinder ($N_{\text{leak}} = N_{\text{tot}}$).

4.5.3 Release from Fractal Matrices

Apart from the classical mechanisms of release, e.g., Fickian diffusion from a homogeneous release device (cf. Sections 4.5.1 and 4.5.2) or Case II release there are also other possibilities. For example, the gastrointestinal fluids can penetrate the release device as it is immersed in the gastrointestinal tract fluids, creating areas of high diffusivity. Thus, the drug molecules can escape from the release device through diffusion from these high diffusivity "channels." Now, the dominant release mechanism is diffusion, but in a complex disordered medium. The same is true when the polymer inside the release device is assuming a configuration resembling a disordered medium. This is a model proposed for HPMC matrices [83]. Several diffusion properties have to be modified when we move from Euclidean space to fractal and disordered media.

The Pioneering Work of Bunde et al.

The problem of the release rate from devices with fractal geometry was first studied by Bunde et al. [84]. As such a structure a percolation cluster at the critical point, assuming cyclic boundary conditions, embedded on a 2-dimensional square lattice, was used. The concentration of open sites is known to be approximately $p = 0.593$ (cf. Section 1.7). The fractal dimension of the percolation fractal is known to be 91/48. The simulation starts with a known initial drug concentration $c_0 = 0.5$ and with randomly distributed drug molecules inside the fractal matrix. The drug molecules move inside the fractal matrix by the mechanism of diffusion. Excluded volume interactions between the particles, meaning that two molecules cannot occupy the same site at the same time, were also assumed. The matrix can leak from the intersection of the percolation fractal with the boundaries of the square box where it is embedded. Bunde et al. [84] specifically reported that the release rate of drug in a fractal medium follows a power law and justified their finding as follows: "the nature of drug release drastically depends on the dimension of the matrix and is different depending on whether the matrix is a normal Euclidean space or a fractal material such as a polymer, corresponding to the fact that the basic laws of physics are quite different in a fractal environment."

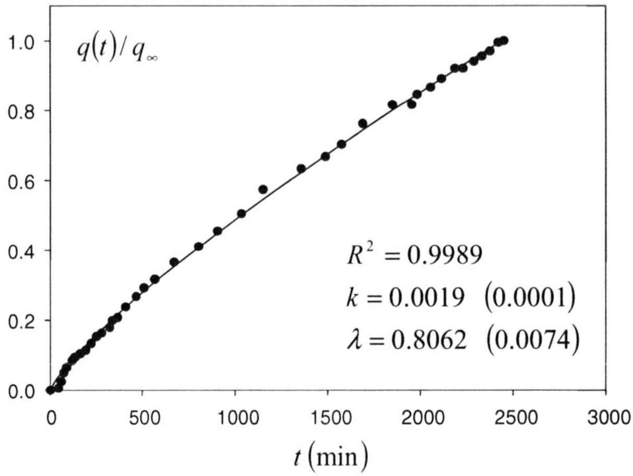

Figure 4.8: Fitting of (4.11) to the entire set of fluoresceine release data from HPMC matrices [85].

Can the Power Law Describe the "Entire" Release Curve?

Based on the findings of Bunde et al. [84], one can also conceive that the entire, classical % release vs. time curves from devices of fractal geometry should also follow a power law with (a different) characteristic exponent. Although the power law has been extensively used for the description of the initial 60% of the release data, it has also been shown that the power law can describe the entire drug release profile of several experimental data [69]. Typical examples of fittings of (4.11) to experimental data of drug release from HPMC matrices along with the estimates obtained for k and λ are shown in Figures 4.8, 4.9, and 4.10 [69]. In all cases, the entire release profile was analyzed and the fitting results were very good. All these experimental results were explained [69] on the basis of the Bunde at al. [84] findings. However, it will be shown below that the conclusion that the release rate follows a power law is accurate only for infinite problems. For problems in which the finite size is inherent, as happens to be the case in drug release studies, a power law is valid only in the initial stages of the release process.

The Weibull Function Describes Drug Release from Fractal Matrices

Kosmidis et al. [87] reexamined the random release of particles from fractal polymer matrices using the percolation cluster at the critical point, Figure 4.11, following the same procedure as proposed by Bunde at al. [84]. The intent of the study was to derive the details of the release problem, which can be used

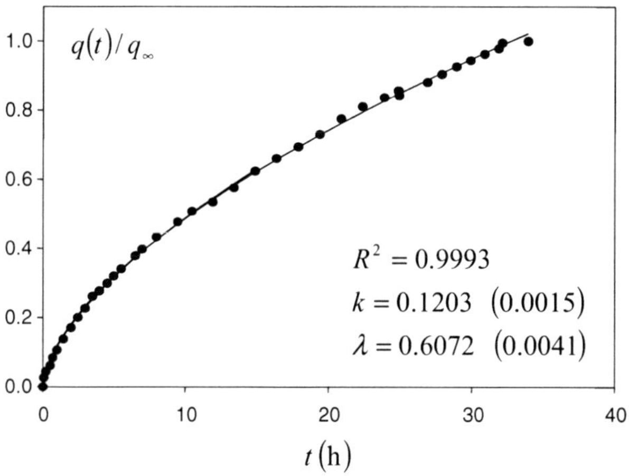

Figure 4.9: Fitting of (4.11) to the entire set of buflomedil pyridoxal release data from HPMC matrices [86].

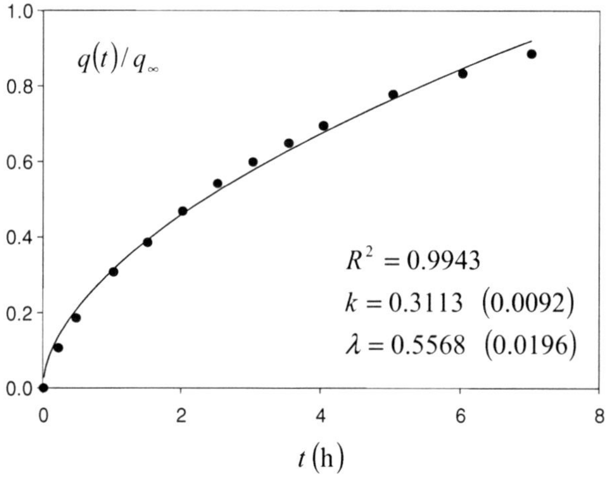

Figure 4.10: A typical example of fitting (4.11) to chlorpheniramine maleate release data from HPMC K15M matrix tablets (tablet height 4 mm; tablet radius ratio 1 : 1) [81].

to describe release when particles escape not from the entire boundary but just from a portion of the boundary of the release device under different interactions between the particles that are present.

The release problem can be seen as a study of the kinetic reaction $A+B \rightarrow B$ where the A particles are mobile, the B particles are static, and the scheme describes the well-known trapping problem [88]. For the case of a Euclidean matrix the entire boundary (i.e., the periphery) is made of the trap sites, while for the present case of a fractal matrix only the portions of the boundary that are part of the fractal cluster constitute the trap sites, Figure 4.11. The difference between the release problem and the general trapping problem is that in release, the traps are not randomly distributed inside the medium but are located only at the medium boundaries. This difference has an important impact in real problems for two reasons:

- Segregation is known to play an important role in diffusion in disordered media (cf. Section 2.5.1). In the release problem the traps are *segregated* from the beginning, so one expects to observe important effects related to this segregation.
- The problem is inherently a finite-size problem. Results that otherwise would be considered as finite-size effects and should be neglected are in this case essential. At the limit of infinite volume there will be no release at all. Bunde et al. [84] found a power law also for the case of trapping in a model with a trap in the middle of the system, i.e., a classical trapping problem. In such a case, which is different from the model examined here, it is meaningful to talk about finite-size effects. In contrast, release from the surface of an infinite medium is impossible.

The fractal kinetics treatment of the release problem goes as follows [87]. The number of particles present in the system (vessel) at time t is $n(t)$. Thus, the particle escape rate will be proportional to the fraction g of particles that are able to reach an exit in a time interval $\mathrm{d}t$, i.e., the number of particles that are sufficiently close to an exit. Initially, all molecules are homogeneously distributed over the percolation cluster. Later, due to the fractal geometry of the release system segregation effects will become important [16]. Accordingly, g will be a function of time, so that $g(t)$ will be used to describe the effects of segregation (generation of depletion zones), which is known to play an important role when the medium is disordered instead of homogeneous [16].

We thus expect a differential equation of the form of (4.13) to hold, where a is a proportionality constant, $g(t)\,n(t)$ denotes the number of particles that are able to reach an exit in a time interval $\mathrm{d}t$, and the negative sign denotes that $n(t)$ decreases with time. This is a kinetic equation for an $A+B \rightarrow B$ reaction. The constant trap concentration $[B]$ has been absorbed in the proportionality constant a. The basic assumption of fractal kinetics [16] is that $g(t)$ has the form $g(t) \propto t^{-\mu}$. In this case, the solution is supplied by (4.14).

The form of this equation is a stretched exponential. In cases in which a system can be considered as infinite (for example, release from percolation

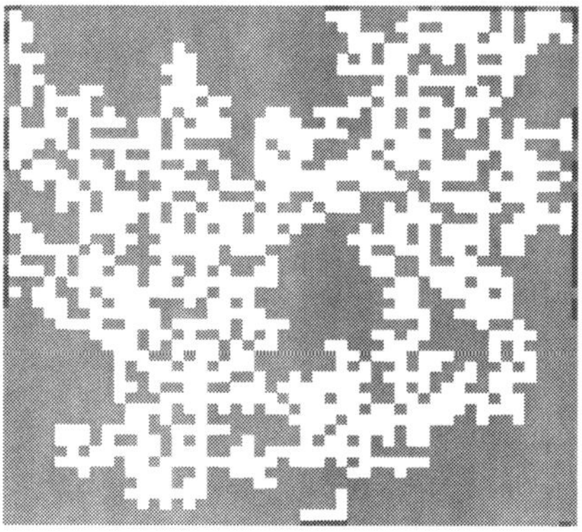

Figure 4.11: A percolation fractal embedded on a 2-dimensional square lattice of size 50×50. Cyclic boundary conditions were used. We observe, especially on the boundaries, that there are some small isolated clusters, but these are not isolated since they are actually part of the largest cluster because of the cyclic boundary conditions. Exits (release sites) are marked in dark gray, while all lighter grey areas are blocked areas. Reprinted from [87] with permission from American Institute of Physics.

fractals from an arbitrary site located at the middle of the volume) then the number of particles $n(t)$ inside the system is practically unchanged. Treating $n(t)$ as constant and letting $g(t) \propto t^{-\mu}$ in the right-hand side of (4.13), will lead to a power law for the quantity $\dot{n}(t)$. Since most physical problems belong to this class it is widely believed that the release rate from fractal matrices follows a power law. In the case of release from the periphery and if we want to study the system until all particles have escaped, as is often the case for practical applications, then (4.14) is of practical importance.

The above reasoning shows that the stretched exponential function (4.14), or Weibull function as it is known, may be considered as an approximate solution of the release problem. The advantage of this choice is that it is general enough for the description of drug release from vessels of various shapes, in the presence or absence of different interactions, by adjusting the values of the parameters a and b. Monte Carlo simulation methods were used to calculate the values of the parameters a and (mainly) the exponent b [87].

Simulations

The drug molecules move inside the fractal matrix by the mechanism of diffusion, assuming excluded volume interactions between the particles. The matrix can leak at the intersection of the percolation fractal with the boundaries of the square box where it is embedded, Figure 4.11.

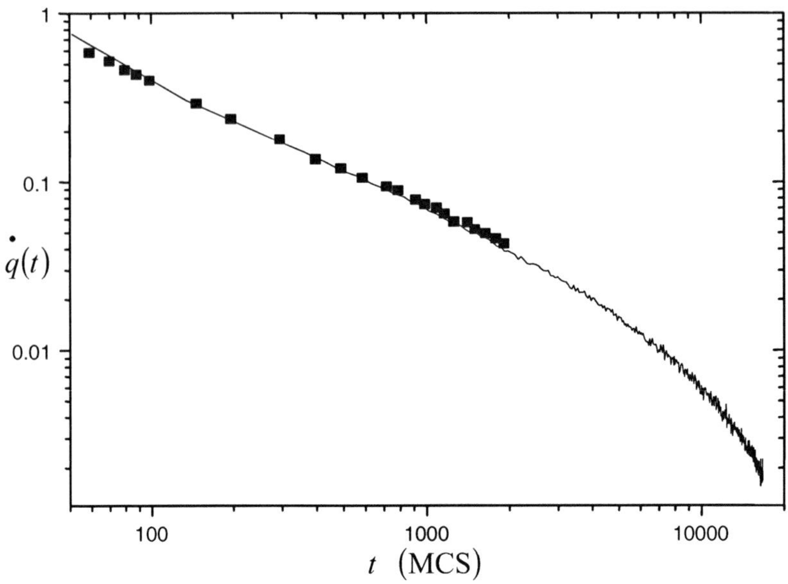

Figure 4.12: Plot of the release rate $\dot{q}(t)$ vs. time. The lattice size is 50×50 and the initial concentration of particles is $c_0 = 0.50$. Points are the results given in [84], while the line is the result of the simulation in [87].

The diffusion process is simulated by selecting a particle at random and moving it to a randomly selected nearest neighbor site. If the new site is an empty site then the move is allowed and the particle is moved to this new site. If the new site is already occupied, the move is rejected since excluded volume interactions are assumed. A particle is removed from the lattice as soon as it migrates to a site lying within the leak area. After each particle move, time is incremented. As previously, the increment is chosen to be $1/n(t)$, where $n(t)$ is the number of particles remaining in the system. This is a typical approach in Monte Carlo simulations, and it is necessary because the number of particles continuously decreases, and thus, the time unit is MCS characterizing the system is the mean time required for all $n(t)$ particles present to move one step. The number of particles that are present inside the matrix as a function of time until a fixed number of particles (50 particles) remains in the matrix is monitored. The results are averaged using different initial random configurations over 100 realizations. The release rate $\dot{q}(t)$ is calculated by counting the number of particles that diffuse into the leak area in the time interval between t and $t+1$.

Figure 4.12 shows simulation results (line) for the release of particles from a fractal matrix with initial concentration $c_0 = 0.50$, on a lattice of size 50×50. The simulation stops when more than 90% of the particles have been released from the matrix. This takes about 20,000 MCS. In the same figure the data by

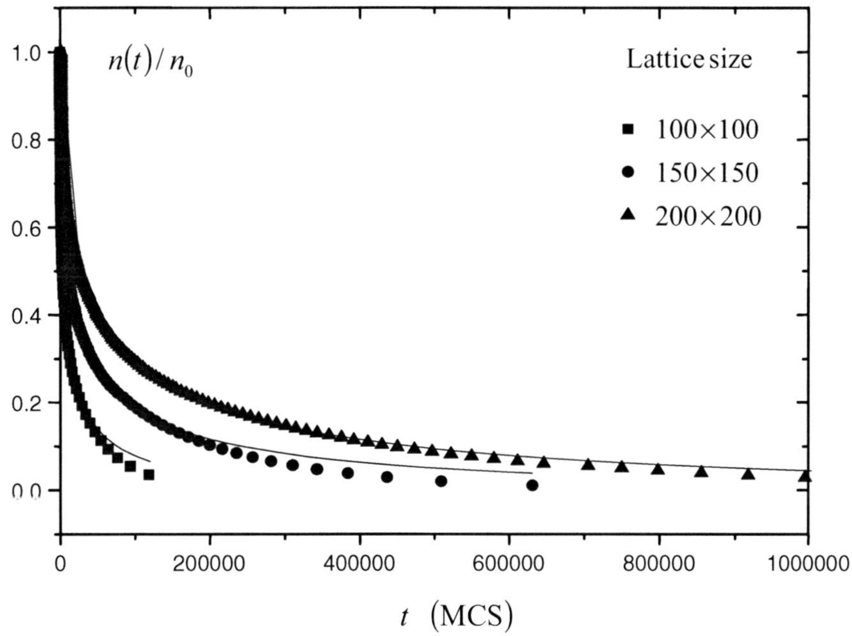

Figure 4.13: Plot of the number of particles (normalized) remaining in the percolation fractal as a function of time t for lattice sizes 100×100, 150×150, and 200×200. $n(t)$ is the number of particles that remain in the lattice at time t and n_0 is the initial number of particles. Simulation results are represented by points. The solid lines represent the results of nonlinear fitting with a Weibull function.

Bunde at al. [84] (symbols), which cover the range 50–2,000 MCS, are included. Because of the limited range examined in that study, the conclusion was reached that the release rate $\dot{q}(t)$ is described by a power law, with an exponent value between 0.65 and 0.75 [84]. With the extended range examined, Figure 4.12, this conclusion is not true, since in longer times $\dot{q}(t)$ deviates strongly from linearity, as a result of the finiteness of the problem.

In Figure 4.13, $n(t)/n_0$ is plotted as a function of time for different lattice sizes. The data were fitted with a Weibull function (4.14), where the parameter a ranges from 0.05 to 0.01 and the exponent b from 0.35 to 0.39. It has been shown [82] that (4.14) also holds in the case of release from Euclidean matrices. In that case the value of the exponent b was found to be $b \approx 0.70$.

These results reveal that the same law describes release from both fractal and Euclidean matrices. The release rate is given by the time derivative of (4.14). For early stages of the release, calculating the derivative of (4.14) and performing a Taylor series expansion of the exponential will result in a power law for the release rate, just as Bunde at al. [84] have observed. If we oversimplify the release problem by treating it as a classical kinetics problem, we would expect a pure exponential function[2] instead of a stretched exponential (Weibull) function. The stretched exponential arises due to the segregation of the particles because of the fractal geometry of the environment. Concerning the release from Euclidean matrices [82], it has been demonstrated that the stretched exponential functional form arises due to the creation of a concentration gradient near the releasing boundaries. Note that although the functional form describing the release is the same in Euclidean and fractal matrices, the value of the exponent b is different, reflecting the slowing down of the diffusion process in a disordered medium. However, these results apparently point to a universal release law given by the Weibull function. The above considerations substantiate the use of the Weibull function as a more general form for drug release studies.

4.6 Discernment of Drug Release Kinetics

In the two previous sections the Weibull function was shown to be successful in describing the entire release profile assuming Fickian diffusion of drug from fractal as well as from Euclidean matrices. Since specific values were found for the exponent b for each particular case, a methodology based on the fitting results of the Weibull function (4.14) to the entire set of experimental %-release-time data can be formulated for the differentiation of the release kinetics [89]. Basically, successful fittings with estimates for b higher than one (sigmoid curves) rule out the Fickian diffusion of drug from fractal or Euclidean spaces and indicate a complex release mechanism. In contrast, successful fittings with estimates for b lower than one can be interpreted in line with the results of the Monte Carlo simulations of Sections 4.5.2 and 4.5.3. The exponent b of the Weibull function using the entire set of data was associated with the mechanisms of diffusional release as follows:

- $b < 0.35$: Not found in simulation studies [82, 87]. May occur in highly disordered spaces much different from the percolation cluster.
- $b \approx 0.35$–0.39: Diffusion in fractal substrate morphologically similar to the percolation cluster [87].
- $0.39 < b < 0.69$: Diffusion in fractal or disordered substrate different from the percolation cluster. These values were not observed in Monte Carlo simulation results [82,87]. It is, however, plausible to assume this possibility since there has to be a crossover from fractal to Euclidian dimension.
- $b \approx 0.69$–0.75: Diffusion in normal Euclidean space [82].
- $0.75 < b < 1$: Diffusion in normal Euclidean substrate with contribution of another release mechanism. In this case, the power law can describe the entire set of data of a combined release mechanism (cf. below).

[2] The classical kinetics solution is obtained by solving (4.13) in case of $g(t) = 1$.

- $b = 1$: First-order release obeying Fick's first law of diffusion; the rate constant a controls the release kinetics, and the dimensionless solubility–dose ratio determines the final fraction of dose dissolved [90].
- $b > 1$: Sigmoid curve indicative of complex release mechanism. The rate of release increases up to the inflection point and thereafter declines.

When Fickian diffusion in normal Euclidean space is justified, further verification can be obtained from the analysis of 60% of the release data using the power law in accord with the values of the exponent quoted in Table 4.1. Special attention is given below for the values of b in the range 0.75–1.0, which indicate a combined release mechanism. Simulated pseudodata were used to substantiate this argument assuming that the release obeys exclusively Fickian diffusion up to time $t = 90$ (arbitrary units), while for $t > 90$ a Case II transport starts to operate too; this scenario can be modeled using

$$\frac{q(t)}{q_\infty} = 1 - \exp\left(-0.05t^{0.70}\right) + \begin{cases} 0 & \text{for} \quad t \leq 90, \\ 0.004\,(t-90)^{0.89} & \text{for} \quad t > 90. \end{cases} \tag{4.16}$$

Also, the following equation was used to simulate concurrent release mechanisms of Fickian diffusion and Case II transport throughout the release process:

$$\frac{q(t)}{q_\infty} = 1 - \exp\left(-0.05t^{0.70}\right) + 0.004t^{0.89}. \tag{4.17}$$

Pseudodata generated from (4.16) and (4.17) are plotted in Figure 4.14 along with the fitted functions

$$y(t) = 0.0652t^{0.5351} \quad \text{and} \quad y(t) = 0.0787t^{0.5440}.$$

The nice fittings of the previous functions to the release data generated from (4.16) and (4.17), respectively, verify the argument that the power law can describe the entire set of release data following combined release mechanisms. In this context, the experimental data reported in Figures 4.8 to 4.10 and the nice fittings of the power-law equation to the entire set of these data can be reinterpreted as a combined release mechanism, i.e., Fickian diffusion and a Case II transport.

4.7 Release from Bioerodible Microparticles

In bioerodible drug delivery systems various physicochemical processes take place upon contact of the device with the release medium. Apart from the classical physical mass transport phenomena (water imbibition into the system, drug dissolution, diffusion of the drug, creation of water-filled pores) chemical reactions (polymer degradation, breakdown of the polymeric structure once the system becomes unstable upon erosion) occur during drug release.

The mathematical model developed by Siepmann et al. [91] utilizes Monte Carlo techniques to simulate both the degradation of the ester bonds of the

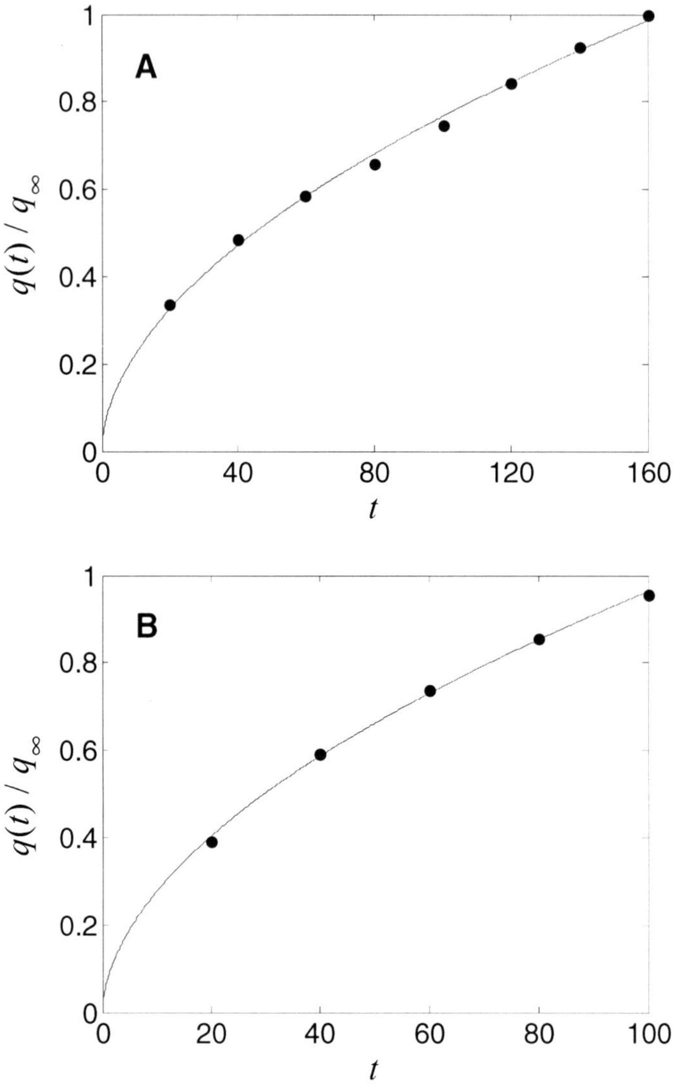

Figure 4.14: (**A**) Points are simulation data produced using (4.16). The solid line is the fitting of the power law (4.11) to data. Best-fitting parameters are $k = 0.0652$ for the proportionality constant and $\lambda = 0.5351$ for the exponent. (**B**) Points are simulation data produced using (4.17). The solid line is the fitting of the power law (4.11) to data. Best-fitting parameters are $k = 0.0787$ for the proportionality constant and $\lambda = 0.5440$ for the exponent. Time is expressed in arbitrary units.

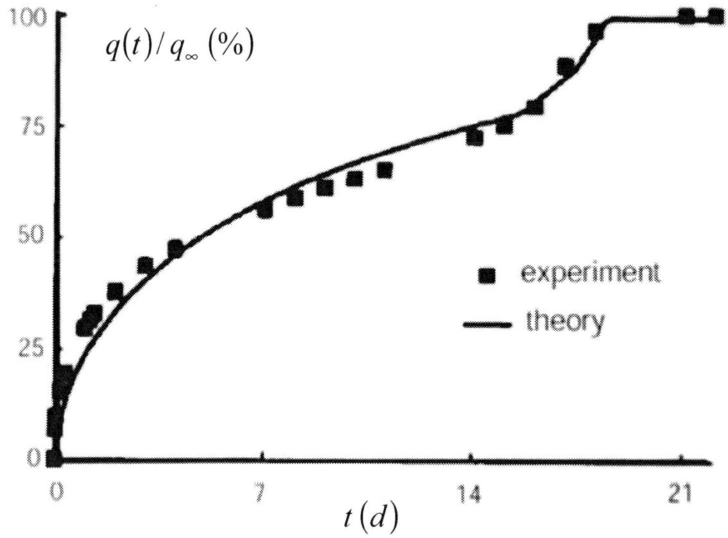

Figure 4.15: Triphasic drug release kinetics from PLGA-based microparticles in phosphate buffer pH 7.4: experimental data (*symbols*) and fitted theory (*curve*). Reprinted from [91] with permission from Springer.

polymer poly-lactic-co-glycolic acid (PLGA) and the polymer's erosion (cleavage of the polymer chains throughout the PLGA matrix). Both phenomena are considered random, and the lifetime of the pixel representing the polymer's degradation is calculated as a function of a random variable obeying a Poisson distribution. The modeling of the physical processes (dissolution and diffusion) takes into account the increase of porosity of the matrix with time because of the polymer's erosion. This information is derived from the Monte Carlo simulations of the polymer's degradation-erosion and allows the calculation of the time- and position-dependent axial and radial diffusivities of the drug. Further, the diffusional mass transport processes are described using Fick's second law with spatially and temporally dependent diffusion coefficients. The numerical solution of the partial differential equation describing the kinetics of the three successive phases of drug release (initial burst, zero-order- and second rapid release) was found to be in agreement with the experimental release data of 5-fluorouracil loaded PLGA microparticles, Figure 4.15 [91]. This model has been further used to investigate the effect of the size of the biodegradable microparticles on the release rate of 5-fluorouracil [92].

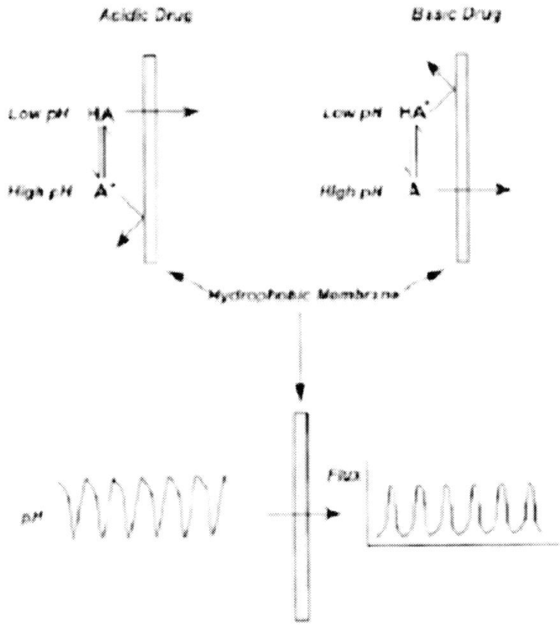

Figure 4.16: Illustration of conversion of pH oscillations to oscillations in drug flux across a lipophilic membrane. Reprinted from [96] with permission from Wiley–Liss Inc., a subsidiary of John Wiley and Sons, Inc.

4.8 Dynamic Aspects in Drug Release

Although the development of controlled drug delivery systems is usually based on the simple notion "a constant delivery is optimal," there are well-known exceptions. For example, drug administration in a periodic, pulsed manner is desirable for endogenous compounds, e.g., hormones [93]. The most classical example is the administration of insulin to diabetic patients in order to maintain blood glucose levels at an approximately constant level [94]. In reality, the pancreas behaves as a feedback controller, which changes its output with time in response to food intake or changes in metabolic activity. Hence, the delivery system should not simply maintain insulin levels within an acceptable physiological range to counterbalance the failure of the patient's pancreas to secrete sufficient insulin, but it should also mimic the normal pancreas's feedback controlling function. In other words, the delivery system should secrete insulin according to the (bio)sensed glucose levels in an automatic, periodic manner. These two steps, sensing and delivery, are the basic features of all self-regulated delivery systems regardless the variable, e.g., glucose, temperature, pressure, that is monitored to control the delivery of a pharmacological agent [95].

Since all these systems behave like autonomous oscillators fueled either di-

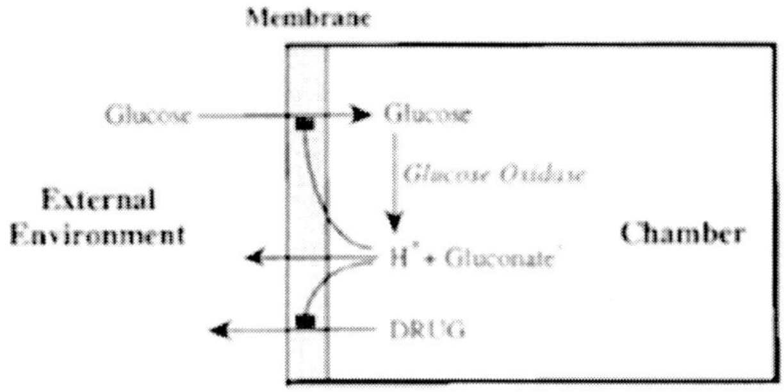

Figure 4.17: Schematic of pulsating drug delivery device based on feedback inhibition of glucose transport to glucose oxidase through a hydrogel membrane. Changes in permeability to glucose are accompanied by modulation of drug permeability. Reprinted from [97] with permission from American Institute of Physics.

rectly or indirectly by the variable monitored, the factors involved in the production of pulsatile oscillations have been studied thoroughly. One of the most studied means for driving the periodic delivery of drugs is the utilization of chemical pH oscillators [96, 98, 99]. It was demonstrated that periodic drug delivery could be achieved as a result of the effect of pH on the permeability of acidic or basic drugs through lipophilic membranes. The model system of Giannos et al. [98] comprises a thin ethylene vinyl-acetate copolymer membrane separating a sink from an iodate-thiosulfate-sulfite pH oscillator compartment into which drugs like nicotine or benzoic acid are introduced. In the work of Misra and Siegel [96,99] a model system consisting of the bromate-sulfite-marble pH oscillator in a continuously stirred tank reactor is used, along with acidic drugs of varying concentration. Figure 4.16 provides a schematic for the periodic flux of a drug through the membrane according to the pH oscillations. In one of the studies, Misra and Siegel [96] provided evidence that low concentrations of acidic drugs can attenuate and ultimately quench chemical pH oscillators by a simple buffering mechanism. In the second study, Misra and Siegel [99] demonstrated that multiple, periodic pulses of drug flux across the membrane can be achieved when the concentration of the drug is sufficiently low.

Another approach for periodically modulated drug release is based on an enzyme/hydrogel system, which, due to negative chemomechanical feedback instability, swells and de-swells regularly in the presence of a constant glucose level [100]. The enzyme glucose oxidase catalyses the conversion of glucose to gluconate and hydrogen ions; the latter affect the permeability of the poly(N-isopropylacrylamide-co-methacrylic acid) hydrogel membrane to glucose since the hydrogel swells with increasing pH and de-swells with decreasing pH, Figure 4.17. This system has been studied extensively from a dynamic point of view [97, 101]. It was found that the model allows, depending on system parameters and external substrate concentration, two separate single steady states, double steady state, and permanently alternating oscillatory behavior.

5

Drug Dissolution

The rate at which a solid substance dissolves in its own solution is proportional to the difference between the concentration of that solution and the concentration of the saturated solution.

Arthur A. Noyes and Willis R. Whitney
Massachusetts Institute of Technology, Boston
Journal of the American Chemical Society 19:930-934 (1897)

The basic step in drug dissolution is the reaction of the solid drug with the fluid and/or the components of the dissolution medium. This reaction takes place at the solid–liquid interface and therefore dissolution kinetics are dependent on three factors, namely the flow rate of the dissolution medium toward the solid–liquid interface, the reaction rate at the interface, and the molecular diffusion of the dissolved drug molecules from the interface toward the bulk solution, Figure 5.1. As we stated in Section 2.4.2, a process (dissolution in our case) can be either diffusion or reaction limited depending on which is the slower step. The relative importance of interfacial reaction and molecular diffusion (steps 2 and 3 in Figure 5.1, respectively) can vary depending on the hydrodynamic conditions prevailing in the microenvironment of the solid. This is so since both elementary steps 2 and 3 in Figure 5.1 are heavily dependent on the agitation conditions. For example, diffusion phenomena become negligible when externally applied intense agitation in in vitro dissolution systems gives rise to forced convection. Besides, the reactions at the interface (step 2) and drug diffusion (step 3) in Figure 5.1 are dependent on the composition of the dissolution medium. Again, the relative importance can vary according to the drug properties and the specific composition of the medium. It is conceivable that our limited knowledge of the hydrodynamics under in vivo conditions and the complex and position- and time-dependent composition of the gastrointestinal fluids complicates the study of dissolution phenomena in particular when one attempts to develop in vitro–in vivo correlations.

Early studies in this field of research formulated two main models for the interpretation of the dissolution mechanism: the diffusion layer model and the

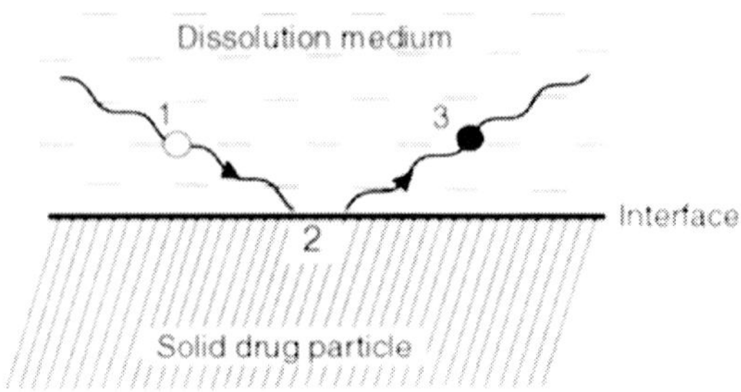

Figure 5.1: The basic steps in the drug dissolution mechanism. (**1**) The molecules (○) of solvent and/or the components of the dissolution medium are moving toward the interface; (**2**) adsorption–reaction takes place at the liquid–solid interface; (**3**) the dissolved drug molecules (•) move toward the bulk solution.

interfacial barrier model. Both models assume that there is a stagnant liquid layer in contact with the solid, Figure 5.2. According to the diffusion layer model (Figure 5.2 A), the step that limits the rate at which the dissolution process occurs is the rate of diffusion of the dissolved drug molecules through the stagnant liquid layer rather than the reaction at the solid–liquid interface. For the interfacial barrier model (Figure 5.2 B), the rate-limiting step of the dissolution process is the initial transfer of drug from the solid phase to the solution, i.e., the reaction at the solid–liquid interface.

Although the diffusion layer model is the most commonly used, various alterations have been proposed. The current views of the diffusion layer model are based on the so-called *effective diffusion boundary layer*, the structure of which is heavily dependent on the hydrodynamic conditions. In this context, Levich [102] developed the convection–diffusion theory and showed that the transfer of the solid to the solution is controlled by a combination of liquid flow and diffusion. In other words, both diffusion and convection contribute to the transfer of drug from the solid surface into the bulk solution. It should be emphasized that this observation applies even under moderate conditions of stirring.

5.1 The Diffusion Layer Model

Noyes and Whitney published [103] in 1897 the first quantitative study of a dissolution process. Using water as a dissolution medium, they rotated cylinders of benzoic acid and lead chloride and analyzed the resulting solutions at various time points. They found that the rate $\dot{c}(t)$ of change of concentration $c(t)$ of dissolved species was proportional to the difference between the saturation solubility c_s of the species and the concentration existing at any time t. Using

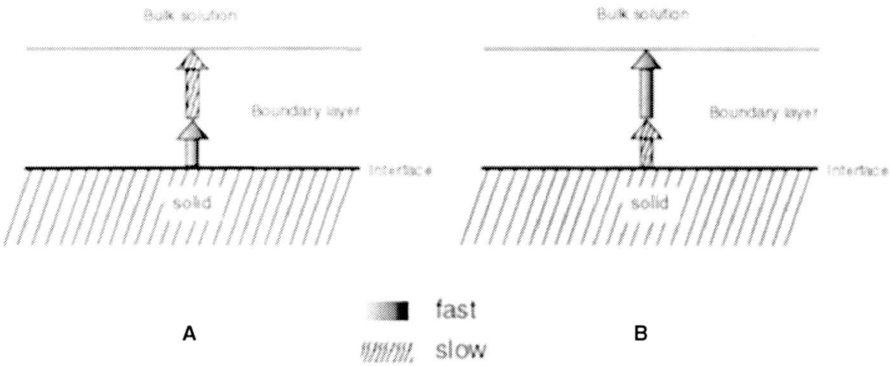

Figure 5.2: Schematic representation of the dissolution mechanisms according to: (**A**) the diffusion layer model, and (**B**) the interfacial barrier model.

k as a proportionality constant, this can be expressed as

$$\dot{c}(t) = k[c_s - c(t)] \quad c(0) = 0. \tag{5.1}$$

Although it was not stated in the original article of Noyes and Whitney, it should be pointed out that the validity of the previous equation relies on the assumption that the amount used, q_0, is greater than or equal to the amount required to saturate the dissolution medium, q_s. Later on, (5.1) was modified [102, 104] and expressed in terms of the dissolved amount of drug $q(t)$ at time t while the effective surface area $\mathcal{A}$ of the solid was taken into account:

$$\dot{q}(t) = \frac{\mathcal{DA}}{\delta}\left[c_s - \frac{q(t)}{V}\right] \quad q(0) = 0, \tag{5.2}$$

where $\mathcal{D}$ is the diffusion coefficient of the substance, δ is the effective diffusion boundary layer thickness adjacent to the dissolving surface, and V is the volume of the dissolution medium. In this case, the first-order rate constant k (dimension of time^{-1}) appearing in (5.1) and governing the dissolution process is

$$k = \frac{\mathcal{DA}}{\delta V}. \tag{5.3}$$

The integrated form of (5.2) gives the cumulative mass dissolved at time t:

$$q(t) = c_s V[1 - \exp(-kt)]. \tag{5.4}$$

The limit $t \to \infty$ defines the total drug amount, $q_s = c_s V$, that could be eventually dissolved in the volume V assuming that the amount used q_0 is greater than q_s. Thus, we can define the accumulated fraction of the drug in solution at time t as the ratio $q(t)/q_s$. Equation (5.4) expressed in terms of concentration ($c(t) = q(t)/V$) leads to the most useful form for practical purposes:

$$c(t) = c_s [1 - \exp(-kt)]. \tag{5.5}$$

Equation (5.5) is the classical equation quoted in textbooks indicating the exponential increase of concentration $c(t)$ approaching asymptotically the saturation solubility c_s.

Also, (5.1) indicates that initially ($t \to 0$) when $c(t)$ is small ($c(t) \leq 0.15 c_s$) in comparison to c_s:

$$\dot{c}(t)\Big|_{t\to 0} = kc_s.$$

If this applies then we consider that sink conditions exist. Under sink conditions the concentration $c(t)$ increases linearly with time,

$$c(t) = kc_s t \quad t \to 0, \tag{5.6}$$

and the dissolution rate is proportional to saturation solubility:

$$\dot{q}(t)\Big|_{t\to 0} = Vkc_s.$$

5.1.1 Alternative Classical Dissolution Relationships

The aforementioned analysis demonstrates that these classical concepts are in full agreement with Fick's first law of diffusion and the equivalent expressions in Sections 2.3 and 2.4. However, there are obvious deficiencies of the classical description of dissolution since the validity of (5.3) presupposes that all terms in this equation remain constant throughout the dissolution process. For example, the drug surface area $\mathcal{A}$ of powders and immediate release formulations is decreasing as dissolution proceeds. In fact, a dramatic reduction of the surface area takes place whenever the dose is not used in large excess, i.e., the drug mass divided by product of the volume of the dissolution medium and the drug's solubility is less than 10. This problem has been realized over the years and equations that take into account the diminution of the surface area have been published. For example, Hixson and Crowell [105] developed the following equation, which is usually called the *cube-root law*, assuming that dissolution occurs from spherical particles with a mono-disperse size distribution under sink conditions:

$$q_0^{1/3} - [q(t)]^{1/3} = k_{1/3} t, \tag{5.7}$$

where q_0 and $q(t)$ are the initial drug amount and the drug amount at time t after the beginning of the process, respectively, and $k_{1/3}$ is a composite cube-root rate constant. Alternatively, when sink conditions do not apply, the following equation (usually called the law of 2/3) can be used:

$$[q(t)]^{-2/3} - q_0^{-2/3} = k_{2/3} t, \tag{5.8}$$

where $k_{2/3}$ is a composite rate constant for the law of 2/3.

Although these approaches demonstrate the important role of the drug material's surface and its morphology on dictating the dissolution profile, they still suffer from limitations regarding the shape and size distribution of particles as well as the assumptions on the constancy of the diffusion layer thickness δ and the drug's diffusivity $\mathcal{D}$ throughout the process implied in (5.5), (5.6), (5.7), and (5.8). In reality, the parameters δ and $\mathcal{D}$ cannot be considered constant during the entire course of the dissolution process when poly-disperse powders are used and/or an initial phase of poor deaggregation of granules or poor wetting of formulation is encountered. In addition, the diffusion layer thickness appears to depend on particle size. For all aforementioned reasons, (5.5), (5.6), (5.7), and (5.8) have been proven adequate in modeling dissolution data only when the presuppositions of constancy of terms in (5.3) are fulfilled.

5.1.2 Fractal Considerations in Drug Dissolution

Drug particles are classically represented as ideal smooth spheres when dissolution phenomena are considered. The surface area of a spherical smooth object is a multiple of the scale, e.g., cm^2, and has a topological dimension $d_t = 2$. If one knows the radius ρ, the surface area of the sphere is $4\pi\rho^2$. However, many studies indicate that the surfaces of most materials are fractal [106]. The measured surface areas of irregular and rough surfaces increase with decreasing scale according to the specific surface structure. These surfaces have fractal dimensions d_f lying between the topological and the embedding dimensions: $2 < d_f < 3$.

Since the surface area of solids in dissolution studies is of primary importance, the roughness of the drug particles has been the subject of many studies. For example, Li and Park [107] used atomic force microscopy to determine the fractal properties of pharmaceutical particles. Moreover, analysis of the surface ruggedness of drugs, granular solids, and excipients using fractal geometry principles has been applied extensively [108–111]. Most of these studies underline the importance of surface ruggedness on dissolution. It is also interesting to note that considerations of the surface roughness are not restricted to the macroscopic level. The same concepts can also be applied to microscopic levels. A typical example is the importance of the surface roughness of proteins in binding phenomena [112].

Farin and Avnir [113] were the first to use fractal geometry to determine effects of surface morphology on drug dissolution. This was accomplished by the use of the concept of fractal reaction dimension d_r [114], which is basically the effective fractal dimension of the solid particle toward a reaction (dissolution in this case). Thus, (5.7) and (5.8) were modified [113] to include surface roughness effects on the dissolution rate of drugs for the entire time course of dissolution (5.9) and under sink conditions (5.10):

$$[q(t)]^{-\alpha} - q_0^{-\alpha} = \alpha k_{1/3}^{*} t, \tag{5.9}$$

$$q_0^{1-\alpha} - [q(t)]^{1-\alpha} = q_s (1-\alpha) k_{1/3}^{*} t, \tag{5.10}$$

where $\alpha = d_r/3$ and q_s is the drug amount that could be dissolved in the volume of the dissolution medium and $k_{1/3}^*$ is the dissolution rate constant of the modified cube root. Although the previous equations describe quantitatively the dissolution of solids with fractal surfaces, their application presupposes that the value of d_r is known.

According to the classical scaling laws, an estimate of d_r can be obtained from the slope of a log-log plot of the initial rate of dissolution $\dot{q}(t)\Big|_{t\to 0}$ vs. the radius ρ of the various particle sizes. This kind of calculation relies on the fundamental proportionality

$$\dot{q}(t)\Big|_{t\to 0} \propto \mathcal{A} \propto \rho^{d_r - 3},$$

where $\mathcal{A}$ is the effective surface area; the slope of $\log \dot{q}(t)\Big|_{t\to 0}$ vs. $\log \rho$ corresponds to $d_r - 3$, in agreement with the relationship for measurements regarding areas in Section 1.4.2. However, this approach for the calculation of d_r requires the execution of a number of experiments with a variety of particles of well-defined size and shape characteristics, which can also exhibit different d_r values.

For the aforementioned reasons, a simpler method requiring only a dissolution run with particles of a given size has been proposed for the estimation of d_r [115]. As can be seen from (5.9) and (5.10), on plotting the values of the left-hand side against time t, one can obtain the value of $k_{1/3}^*$ from the slope of the straight line. In practice, this involves choosing a starting value for d_r, e.g., 2, and, using an iterative method, searching for the linearity demanded by the previous equations for the experimental data pairs $(q(t), t)$. When this has been found, one knows values both for $k_{1/3}^*$ and d_r.

5.1.3 On the Use of the Weibull Function in Dissolution

In 1951, Weibull [116] described a more general function that can be applied to all common types of dissolution curves. This function was introduced in the pharmaceutical field by Langenbucher in 1972 [117] to describe the accumulated fraction of the drug in solution at time t, and it has the following form:[1]

$$\frac{q(t)}{q_\infty} = 1 - \exp\left[-(\lambda t)^{\mu}\right], \tag{5.11}$$

where q_∞ is the total mass that can be eventually dissolved and λ, μ are constants. The *scale parameter* λ defines the time scale of the process, while the *shape parameter* μ characterizes the shape of the curve, which can be exponential ($\mu = 1$), S-shaped ($\mu > 1$), or exponential with a steeper initial slope ($\mu < 1$), Figure 5.3.

[1] In the phamaceutical literature the exponential in the Weibull function is written as $\exp(-\lambda t^{\mu})$ and therefore λ has dimension time$^{-\mu}$. In the version used herein (equation 5.11), the dimension of λ is time^{-1}.

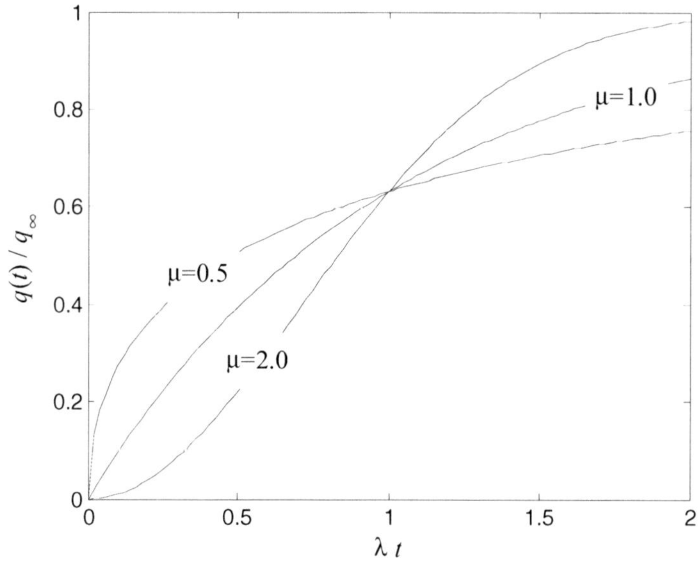

Figure 5.3: Accumulated fraction of drug dissolved, $q(t)/q_{\infty}$ as a function of λt according to the Weibull distribution function (5.11).

It is also worthy of mention that a gamma distribution function proposed by Djordjevic [118] for modeling in vitro dissolution profiles implies a relevant type of time dependency for the amount of drug dissolved.

The successful use of the Weibull function in modeling the dissolution profiles raises a plausible query: What is the rationale of its success? The answer will be sought in the relevance of the Weibull distribution to the kinetics prevailing during the dissolution process.

The basic theory of chemical kinetics originates in the work of Smoluchowski [119] at the turn of the twentieth century. He showed that for homogeneous reactions in 3-dimensional systems the rate constant is proportional to the diffusion coefficient. In dissolution studies this proportionality is expressed with $k \propto \mathcal{D}$, where k is the intrinsic dissolution rate constant. In addition, both $\mathcal{D}$ and k are time-independent in well-stirred, homogeneous systems. However, that is not true for lower dimensions and disordered systems in chemical kinetics. Similarly, homogeneous conditions may not prevail during the entire course of the dissolution process in the effective diffusion boundary layer adjacent to the dissolving surface. It is very difficult to conceive that the geometric and hydrodynamic characteristics of this layer are maintained constant during the entire course of drug dissolution. Accordingly, the drug's diffusional properties change with time and the validity of use of a classical rate constant k in (5.1) is questionable.

It stands to reason that an instantaneous yet time dependent rate coefficient $k(t)$ governing dissolution under inhomogeneous conditions can be written as

$$k(t) = k_{\circ} \left(\frac{t}{t_{\circ}} \right)^{-\gamma} \qquad \text{with} \qquad t \neq 0, \tag{5.12}$$

where $k_{\circ}$ is a rate constant not dependent on time, $t_{\circ}$ is a time scale parameter, and γ is a pure number. In a simpler form ($t_{\circ} = 1$), the previous relation is used in chemical kinetics to describe phenomena that take place under dimensional constraints or understirred conditions [16]. It is used here to describe the time dependency of the dissolution rate "constant" that originates from the change of the parameters involved in (5.3) during the dissolution process, i.e., the reduction of the effective surface area $\mathcal{A}$ and/or the inhomogeneous hydrodynamic conditions affecting δ and subsequently $\mathcal{D}$.

Using (5.12) to replace k in (5.1), also changing the concentration variables to amounts $Vc(t) = q(t)$, $V\mathrm{d}c(t) = \mathrm{d}q(t)$, and using, instead of $c_s V = q_s$, for generality purposes $c_{\infty} V = q_{\infty}$ (which applies to both $q_{\infty} = q_s$, or $q_{\infty} = q_0$), we obtain

$$\dot{q}(t) = k_{\circ} \left(\frac{t}{t_{\circ}} \right)^{-\gamma} [q_{\infty} - q(t)], \qquad q(t_0) = 0,$$

and after integration,

$$\frac{q(t)}{q_{\infty}} = 1 - \exp\left\{ -\frac{k_{\circ} t_{\circ}}{1-\gamma} \left[\left(\frac{t}{t_{\circ}} \right)^{1-\gamma} - \left(\frac{t_0}{t_{\circ}} \right)^{1-\gamma} \right] \right\}.$$

Taking the limit as t_0 approaches zero, for $\gamma < 1$ we get the following equation:

$$\frac{q(t)}{q_{\infty}} = 1 - \exp\left[-\frac{k_{\circ} t_{\circ}}{1-\gamma} \left(\frac{t}{t_{\circ}} \right)^{1-\gamma} \right]. \tag{5.13}$$

This equation is identical to the Weibull equation (5.11) for

$$\lambda = \frac{1}{t_{\circ}} \left(\frac{k_{\circ} t_{\circ}}{1-\gamma} \right)^{1/(1-\gamma)} \qquad \text{and} \qquad \mu = 1 - \gamma.$$

Furthermore, (5.13) collapses to the "homogeneous" (5.4) when $\gamma = 0$. These observations reveal that the parameter μ of (5.11) can be interpreted in terms of the heterogeneity of the process. For example, an S-shaped dissolution curve with $\mu > 1$ in (5.11) for an immediate release formulation can now be interpreted as a heterogeneous dissolution process (with $\gamma < 0$ in equation 5.13), whose rate increases with time during the upwards, concave initial limb of the curve and decreases after the point of inflection. This kind of behavior can be associated with an initial poor deaggregation or poor wetting.

Most importantly, it was shown that the structure of the Weibull function captures the time-dependent character of the rate coefficient governing the dissolution process. These considerations agree with Elkoski's [120] analysis of the

Weibull function and provide an indirect, physically based interpretation [121] for its superiority over other approaches for the analysis of dissolution data. In other words, drug dissolution is a typical example of a heterogeneous process since, as dissolution proceeds, homogeneous conditions cannot be maintained in the critical region of the microenvironment of drug particles. Thus, drug dissolution exhibits fractal-like kinetics like other heterogeneous processes (e.g., adsorption, catalysis) since it takes place at the boundary of different phases (solid–liquid) under topological constraints.

5.1.4 Stochastic Considerations

The dissolution process can be interpreted stochastically since the profile of the accumulated fraction of amount dissolved from a solid dosage form gives the probability of the residence times of drug molecules in the dissolution medium. In fact, the accumulated fraction of the drug in solution, $q(t)/q_\infty$, has a statistical sense since it represents the cumulative distribution function of the random variable *dissolution time* T, which is the time up to dissolution for an individual drug fraction from the dosage form. Hence, $q(t)/q_\infty$ can be defined statistically as the probability that a molecule will leave the formulation prior to t, i.e., that the particular dissolution time T is smaller than t:

$$q(t)/q_\infty = \Pr[\text{leave the formulation prior to } t] = \Pr[T < t].$$

Conversely,

$$1 - q(t)/q_\infty = \Pr[\text{survive in the formulation to } t] = \Pr[T \geq t].$$

Since $q(t)/q_\infty$ is a distribution function, it can be characterized by its statistical moments. The first moment is defined as the *mean dissolution time* (MDT) and corresponds to the expectation of the time up to dissolution for an individual drug fraction from the dosage form:

$$MDT = E[T] = \int_0^\infty t\frac{dq(t)}{q_\infty} = \frac{ABC}{q_\infty}, \tag{5.14}$$

where q_∞ is the asymptote of the dissolved amount of drug and ABC is the area between the cumulative dissolution curve and the horizontal line that corresponds to q_∞, Figure 5.4.

Since the fundamental rate equation of the diffusion layer model has the typical form of a first-order rate process (5.1), using (5.4) and (5.14), the MDT is found equal to the reciprocal of the rate constant k:

$$MDT = \frac{1}{k}. \tag{5.15}$$

As a matter of fact, all dissolution studies, which invariably rely on (5.1) and do not make dose considerations, utilize (5.15) for the calculation of the MDT. However, the previous equation applies only when the entire available amount

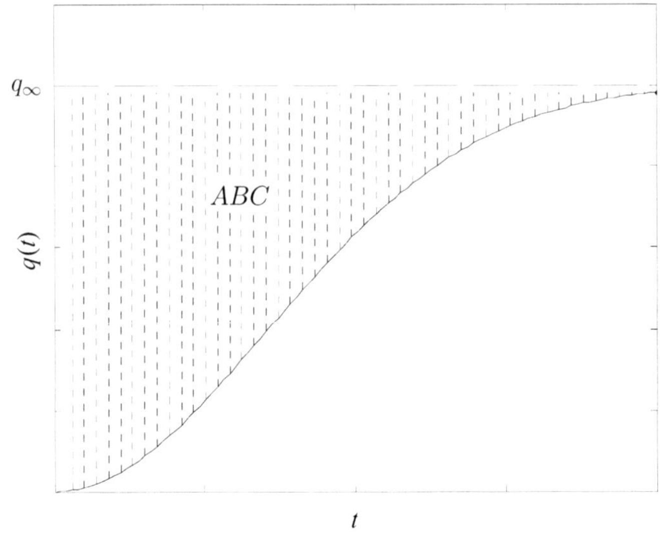

Figure 5.4: The cumulative dissolution profile $q(t)$ as a function of time. The symbols are defined in the text.

of drug (dose) q_0 is dissolved. Otherwise, the mean dissolution time of the dose is not defined, i.e., MDT is infinite.

In fact, it will be shown below that MDT is dependent on the *dose–solubility ratio* if one takes into account the dose q_0 actually utilized [90]. Also, it will be shown that the widely used (5.15) applies only to a special limiting case. Multiplying both parts of (5.1) by V/q_0 (volume of the dissolution medium/actual dose), one gets the same equation in terms of the fraction of the actual dose of drug dissolved, $\varphi(t) \triangleq q(t)/q_0$:

$$\dot{\varphi}(t) = k\left[\tfrac{1}{\theta} - \varphi(t)\right], \quad \varphi(0) = 0, \tag{5.16}$$

where θ is the dose–solubility ratio

$$\theta \triangleq \frac{q_0}{c_s V} = \frac{q_0}{q_s} \tag{5.17}$$

expressed as a dimensionless quantity. Equation (5.16) has two solutions:

- When $\theta \leq 1$ ($q_0 \leq q_s$), which means that the entire dose is eventually dissolved:

$$\varphi(t) = \begin{cases} \frac{1}{\theta}\left[1 - \exp(-kt)\right] & \text{for} \quad t < t_\circ, \\ 1 & \text{for} \quad t \geq t_\circ, \end{cases}$$

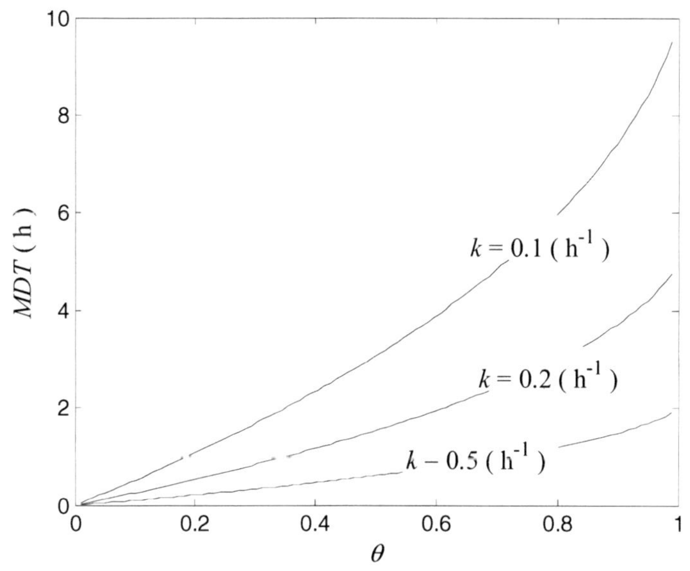

Figure 5.5: Plot of MDT vs. θ using (5.18) for different values of k.

where $t_\circ = -\frac{\ln(1-\theta)}{k}$ is the time at which dissolution terminates ($\varphi(t_\circ) = 1$). Similarly to (5.14), the MDT is

$$MDT = \int_0^{t_\circ} t \mathrm{d}\varphi(t) = \frac{\theta + (1-\theta)\ln(1-\theta)}{k\theta}. \tag{5.18}$$

This equation reveals that the MDT depends on both k and θ. Figure 5.5 shows a plot of MDT as a function of θ for three different values of the rate constant k. Note that (5.15) is obtained from (5.18) for $\theta = 1$ (the actual dose is equal to the amount needed to saturate the volume of the dissolution medium). In other words, the classically used (5.15) is a special case of the general equation (5.18).

- When $\theta > 1$ ($q_0 > q_s$), which means that only a portion of the dose is dissolved and the drug reaches the saturation level $1/\theta$:

$$\varphi(t) = \frac{1}{\theta}\left[1 - \exp(-kt)\right].$$

The MDT is infinite because the entire dose is not dissolved. Therefore, the term mean saturation time, MDT_s, [122] has been suggested as more appropriate when we refer only to the actually dissolved portion of dose, in order to get a meaningful time scale for the portion of the dissolved drug dose:

$$MDT_s = \int_0^\infty t \frac{d\varphi(t)}{1/\theta} = \frac{1}{k}, \tag{5.19}$$

which is independent of θ.

This analysis demonstrates that when $\theta \leq 1$, dose–solubility considerations should be taken into account in accord with (5.18) for the calculation of MDT; the MDT is infinite when $\theta > 1$. Equation (5.15) can be used to obtain an estimate for MDT only in the special case $\theta = 1$. Finally, (5.19) describes the MDT_s of the fraction of dose dissolved when $\theta > 1$.

5.2 The Interfacial Barrier Model

In the interfacial barrier model of dissolution it is assumed that the reaction at the solid–liquid interface is not rapid due to the high free energy of activation requirement and therefore the reaction becomes the rate-limiting step for the dissolution process (Figure 5.1), thus, drug dissolution is considered as a reaction-limited process for the interfacial barrier model. Although the diffusion layer model enjoys widespread acceptance since it provides a rather simplistic interpretation of dissolution with a well-defined mathematical description, the interfacial barrier model is not widely used because of the lack of a physically-based mathematical description.

In recent years two novel models [122,123] have appeared that were proposed to describe the heterogeneous features of drug dissolution. They are considered here as continuous (in well-stirred media) or discrete (in understirred media) reaction-limited dissolution models. Their derivation and relevance is discussed below.

5.2.1 A Continuous Reaction-Limited Dissolution Model

Lansky and Weiss [122] proposed a novel model by considering the reaction of the undissolved solute with the free solvent yielding the dissolved drug complexed with solvent:

$$[\text{undissolved drug}] + [\text{free solvent}] \rightarrow [\text{dissolved drug complexed with solvent}].$$

Further, global concentrations as a function of time for the reactant species of the above reaction were considered, assuming that the solvent is not in excess and applying classical chemical kinetics. The following equation was found to describe the rate of drug dissolution in terms of the fraction of drug dissolved:

$$\dot{\varphi}(t) = k^* [1 - \varphi(t)] [1 - \theta\varphi(t)], \quad \varphi(0) = 0, \tag{5.20}$$

where $\varphi(t)$ denotes the fraction of drug dissolved up to time t, and θ is the dimensionless dose–solubility ratio (5.17); k^* is a fractional (or relative) dissolution rate constant with dimensions time^{-1}. The fractional dissolution rate is a decreasing function of the fraction of dissolved amount $\varphi(t)$, as has also been observed for the diffusion layer model (5.16). However, (5.20) reveals a form of second-order dependency of the reaction rate on the dissolved amount $\varphi(t)$. In reality, a classical second-order dependency is observed for $\theta = 1$. These are unique features, which are not encountered in models dealing with diffusion-limited dissolution. All the above characteristics indicate that (5.20) describes the continuous-homogeneous character of the reaction of the solid with the solvent or the component(s) of the dissolution medium, i.e., a reaction-limited dissolution process in accord with the interfacial barrier model.

The solution of (5.20) for $\theta \neq 1$ yields the monotonic function

$$\varphi(t) = \frac{\exp\left[k^* (1-\theta) t\right] - 1}{\exp\left[k^* (1-\theta) t\right] - \theta}, \tag{5.21}$$

and for $\theta = 1$,

$$\varphi(t) = \frac{k^* t}{k^* t + 1},$$

with the same asymptotes as found above for the diffusion-layer model, i.e., $\varphi(\infty) = 1$ for $\theta \leq 1$ and $\varphi(\infty) = 1/\theta$ for $\theta > 1$. It is interesting to note that both MDT and MDT_s for the model of the previous equation depend on the dose–solubility ratio θ when $\theta \neq 1$. Thus, the MDT for $\theta < 1$ is

$$MDT = -\frac{1}{k^* \theta} \ln(1-\theta), \tag{5.22}$$

while the MDT_s for $\theta > 1$ is

$$MDT_s = \frac{1}{k^*} \ln\left(\frac{\theta}{\theta - 1}\right). \tag{5.23}$$

For $\theta = 1$ the MDT is infinite. It should be noted that the MDT for the diffusion layer model depends also on θ for $\theta < 1$ while the MDT_s is equal to $1/k$ when $\theta \geq 1$, (5.18) and (5.19). However, this dependency is different in the two models, cf. (5.18), (5.19), and (5.22), (5.23).

5.2.2 A Discrete Reaction-Limited Dissolution Model

Dokoumetzidis and Macheras [123] developed a population growth model for describing drug dissolution under heterogeneous conditions. In inhomogeneous media, Fick's laws of diffusion are not valid, while global concentrations cannot be used in the dissolution rate equation. In order to face the problem of complexity and circumvent describing the system completely, the reaction of the solid with the solvent or the component(s) of the dissolution medium was described as the "birth" of the population of the dissolved drug molecules from the corresponding population of solid drug particles, Figure 5.6. In this context, only instants of the system's behavior are considered and what happens in the meanwhile is ignored. The jump from one instant to the next is done by a logical rule, which is not a physical law, but an expression that gives realistic results

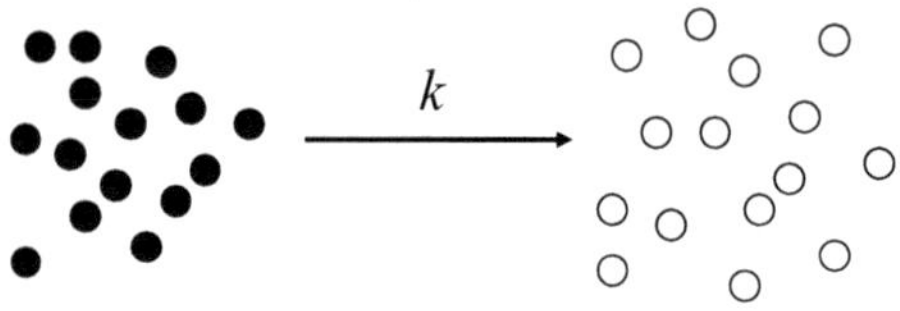

Figure 5.6: A discrete, reaction-limited dissolution process interpreted with the population growth model of dissolution.

based on logical assumptions. The variable of interest (mass dissolved) is not considered as a continuous function of time, but is a function of a discrete time index specifying successive "generations."

Defining s_i and y_i as the populations of the drug molecules in the solid state and in solution in the ith generation ($i = 0, 1, 2, \ldots$), respectively, the following finite difference equation describes the change of y_i between generations i and $i + 1$:

$$y_{i+1} = y_i + ks_i = y_i + k(q_0 - y_i), \quad y_0 = 0,$$

where k is a proportionality constant that controls the reaction of the solid particles with the solvent or the components of the dissolution medium, and q_0 is the population of the drug molecules in the solid state corresponding to dose (Figure 5.6). The growth of y_i is not unlimited since the solubility of drug in the medium restricts the growth of y_i. Thus, the rate of dissolution decreases as the population of the undissolved drug molecules decreases as reaction proceeds. For each one of the drug particles of the undissolved population, the solubility q_s (expressed in terms of the amount needed to saturate the medium in the neighborhood of the particle) is used as an upper "local" limit for the population growth of the dissolved drug molecules. Accordingly, the growth rate is a function of the population level and can be assumed to decrease with increasing population in a linear manner:

$$k \to k(y_i) = k\left(1 - \frac{y_i}{q_s}\right),$$

where q_s is the saturation level of the population, i.e., the number of drug molecules corresponding to saturation solubility. Thus, the previous recursion relation is replaced with the nonlinear discrete equation

$$y_{i+1} = y_i + k(q_0 - y_i)\left(1 - \frac{y_i}{q_s}\right), \quad y_0 = 0.$$

This equation can be normalized in terms of dose by dividing both sides by q_0 and written more conveniently using $y_i/q_0 = \varphi_i$, $y_{i+1}/q_0 = \varphi_{i+1}$, and

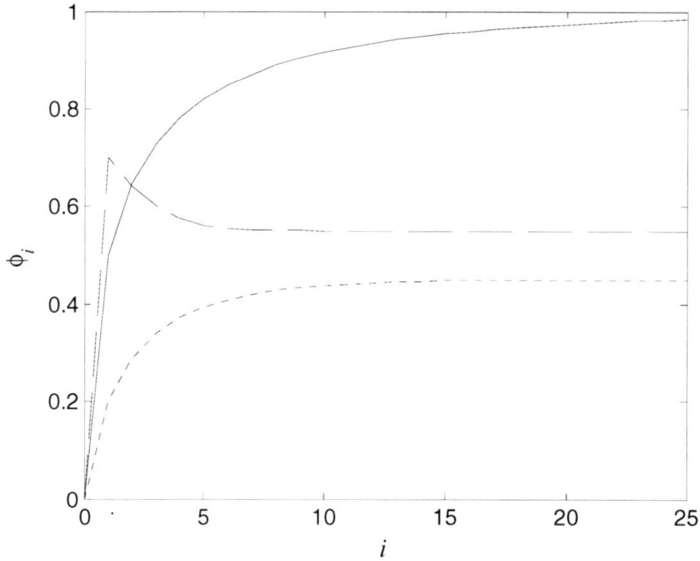

Figure 5.7: Plot of the dissolved fraction φ_i as a function of generations i using (5.24) with $k = 0.5$, $\theta = 0.83$ (*solid line*); $k = 0.7$, $\theta = 1.82$ (*dashed line*); $k = 0.2$, $\theta = 2.22$ (*dotted line*).

$\theta = q_0/q_s$:

$$\varphi_{i+1} = \varphi_i + k\left(1 - \varphi_i\right)\left(1 - \theta\varphi_i\right), \quad \varphi_0 = 0, \tag{5.24}$$

where φ_i and φ_{i+1} are the dissolved fractions of drug dose at generations i and $i+1$, respectively. The previous discrete equation, if written as

$$\varphi_{i+1} - \varphi_i = k\left(1 - \varphi_i\right)\left(1 - \theta\varphi_i\right), \quad \varphi_0 = 0, \tag{5.25}$$

becomes equivalent to its continuous analogue (5.20). As expected, (5.25) has the two classical fixed point, $\varphi_A^* = 1$ when $\theta \leq 1$ and $\varphi_B^* = 1/\theta$ when $\theta > 1$, Figure 5.7. All discrete features of (5.25) are in full analogy with the fractional dissolution rate differential equation (5.20), and it is for this reason that the two approaches are considered counterparts [122].

Since difference equations exhibit dynamic behavior [124, 125], the stability of the fixed points of (5.24) is explored according to the methodology presented in Appendix A. The absolute value of the derivative of the right-hand side of (5.24) is compared with unity for each fixed point. There are the following cases:

- If $\theta < 1$, the derivative is equal to $1 - k\left(1 - \theta\right)$ and the condition for stability of the fixed point $\varphi_A^* = 1$ is

$$0 < k < \frac{2}{1-\theta}.$$

- If $\theta > 1$, the derivative is equal to $1 - k(\theta - 1)$ and the condition for stability of the fixed point $\varphi_B^* = 1/\theta$ is

$$0 < k < \frac{2}{\theta - 1}.$$

- If $\theta = 1$, the derivative is equal to unity and therefore the fixed point $\varphi_A^* = 1$ is neither stable or unstable.

Because of the discrete nature of (5.25), the first step always gives $\varphi_1 = k$; hence, k is always lower than 1, i.e., the theoretical top boundary of φ_i. Comparing the second step $\varphi_2 = k + k(1-k)(1-\theta k)$ with the first one $\varphi_1 = k$, one can obtain the conditions $k > 1/\theta$ and $\theta > 1$, which ensure that the first step is higher than the following steps (Figure 5.7 B). The usual behavior encountered in dissolution studies, i.e., a monotonic exponential increase of φ_i reaching asymptotically 1, or the saturation level $1/\theta$, is observed when $\theta \leq 1$ (Figure 5.7 A) or when $k < 1/\theta$ for $\theta > 1$ (Figure 5.7 C), respectively.

As previously pointed out, when one uses (5.24) for $\theta > 1$ and values of k in the range $1/\theta < k < 2/(\theta - 1)$, the first step is higher than the plateau value followed by a progressive decline to the plateau (Figure 5.8 A, B). For $1/\theta$ and k values close enough, the descending part of the dissolution curve is smooth, concave either upward (Figure 5.8 B) or initially downward and then upward (Figure 5.8 A); this decline can also take the form of a fading oscillation when k is close to $2/(\theta - 1)$ (Figure 5.8 C, D). When k exceeds $2/|\theta - 1|$, the fixed points become unstable, bifurcating to a double-period stable fixed point. So we have both the unstable main point and the generated double-period stable point. This mechanism is called bifurcation and is common to dynamic systems (cf. Chapter 3).

Equation (5.25) can be used to estimate the proportionality constant k and θ from experimental data by plotting the fraction dissolved (φ_i) as a function of the generations i. Prior to plotting, the sampling times are transformed to generations defining arbitrarily a constant sampling interval as a "time unit." By doing so, an initial estimate for k can be obtained by reading the value of φ_i corresponding to the first datum point. When $\theta > 1$ an initial estimate for θ can be obtained from the highest value of the dissolved fraction at the end of the dissolution run. However, an estimate for θ cannot be obtained from visual inspection when $\theta \leq 1$ since $\varphi_A^* = 1$ in all cases. The initial estimates for k and θ can be further used as starting points in a computer fitting program to obtain the best parameter estimates.

The population growth model of dissolution utilizes the usual information available in dissolution studies, i.e., the amount dissolved at certain fixed intervals of time. The time points of all observations need to be transformed to

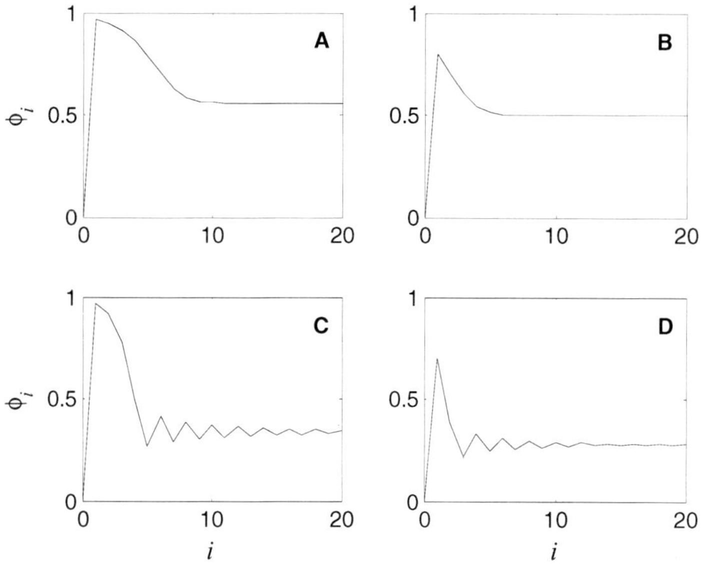

Figure 5.8: Plots of the dissolved fraction φ_i as a function of generations i using (5.24) with k and θ values satisfying the inequality $1/\theta < k < 2/(\theta - 1)$: (**A**) $k = 0.97$, $\theta = 1.79$; (**B**) $k = 0.8$, $\theta = 2.0$; (**C**) $k = 0.97$, $\theta = 2.94$; (**D**) $k = 0.7$, $\theta = 3.57$.

equally spaced values of time and furthermore to take the values $0, 1, 2, \ldots$. Since the model does not rely on diffusion principles it can be applied to both homogeneous and inhomogeneous conditions. This is of particular value for the correlation of in vitro dissolution data obtained under homogeneous conditions and in vivo observations adhering to the heterogeneous milieu of the gastrointestinal tract. The dimensionless character of k allows comparisons to be made for k estimates obtained for a drug studied under different in vitro and in vivo conditions, e.g., various dissolution media, fasted or fed state.

Example 2 *Danazol Data*

For the continuous model, a fitting example of (5.21) to actual experimental data of danazol [126] is shown in Figure 5.9. For the discrete model, a number of fitting examples are shown in Figure 5.10 for danazol dissolution data obtained by using 15 minutes as a "time unit." Table 5.1 lists the estimates for k and θ obtained from the computer analysis of danazol data utilizing an algorithm minimizing the sum of squared deviations between experimental and theoretical values obtained from (5.24). ∎

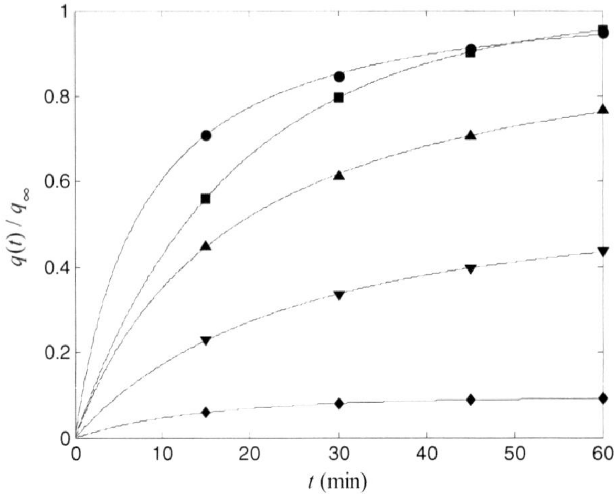

Figure 5.9: The fraction of dose dissolved as a function of time for the danazol data [126]. Symbols represent experimental points and the lines represent the fittings of (5.21) to data. Key (% sodium lauryl sulfate in water as dissolution medium): • 1.0; ■ 0.75; ▲ 0.50; ▼ 0.25; ♦ 0.10.

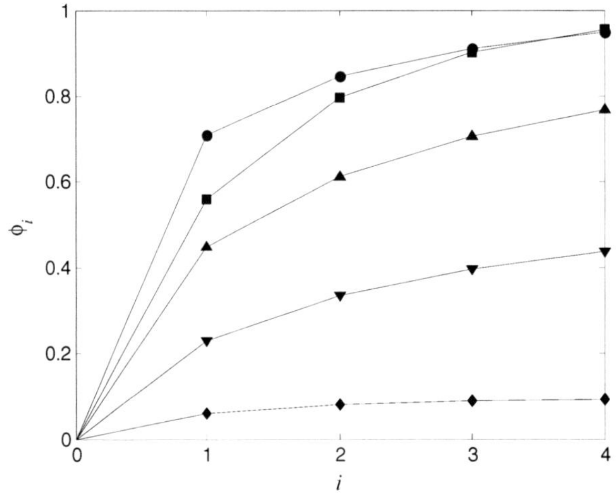

Figure 5.10: The fraction of dose dissolved φ_i as a function of generations i, where the solid line represents the fittings of (5.24) to danazol data [126]. Symbols represent experimental points transformed to the discrete time scale for graphing and fitting purposes assigning one generation equal to 15 minutes. Key (% sodium lauryl sulfate in water as dissolution medium): • 1.0; ■ 0.75; ▲ 0.50; ▼ 0.25; ♦ 0.10.

Table 5.1: Estimates for k and θ obtained from the fitting of (5.24) to danazol data, Figure 5.10. (a) Percentage of sodium lauryl sulfate in water, (b) Determination coefficient.

Dissolution medium[a]	k	θ	R^{2b}
0.10	0.06	10	0.993
0.25	0.23	1.82	0.9993
0.50	0.45	0.75	0.9999
0.75	0.56	0.08	0.9995
1.00	0.71	0.47	0.9996

5.2.3 Modeling Supersaturated Dissolution Data

The dissolution data are basically of monotonic nature (the drug concentration or the fraction of drug dissolved is increasing with time) and therefore the corresponding modeling approaches rely on monotonic functions. However, nonmonotonic dissolution profiles are frequently observed in studies dealing with coprecipitates of drugs with polymers and solid dispersion formulations [127, 128]. The dissolution profiles in these studies usually exhibit a supersaturation phenomenon, namely, an initial rapid increase of drug concentration to a supersaturated maximum followed by a progressive decline to a plateau value. This kind of behavior cannot be explained with the classical diffusion principles in accord with the diffusion layer model of dissolution. It seems likely that the initial sudden increase is associated with a rapid reaction of the solid particles with the dissolution medium. The dynamics of the difference equation for the population growth model of dissolution, (5.24), can capture this behavior and therefore can be used to model supersaturated dissolution data [129].

Example 3 *Nifedipine Data*

An example of fitting (5.24) to experimental data of a nifedipine solid dispersion formulation [128] is shown in Figure 5.11. Initially, the drug concentration values are transformed to the corresponding dissolved fractions of dose φ_i and plotted as a function of the generations i, obtained by using a "time unit" of 5 minutes. The transformation of sampling times to generations i is achieved by adopting the time needed to reach maximum concentration (equivalent to maximum fraction of dose dissolved) as the time unit of (5.24). Reading the maximum and lowest values of φ_i, one obtains initial estimates for parameters k and $1/\theta$, respectively. These values are further used as starting points in a computer program minimizing the sum of squared deviations between observed and predicted values to determine the best parameter estimates. The estimated parameter values for k and θ were found to be 0.323 and 4.06, respectively. The value of k denotes the maximum fraction of dose that is dissolved in a time interval equal to the time unit used. The value of θ corresponds to the reciprocal of the plateau value, which is the fraction of dose remaining in solution at steady state. ■

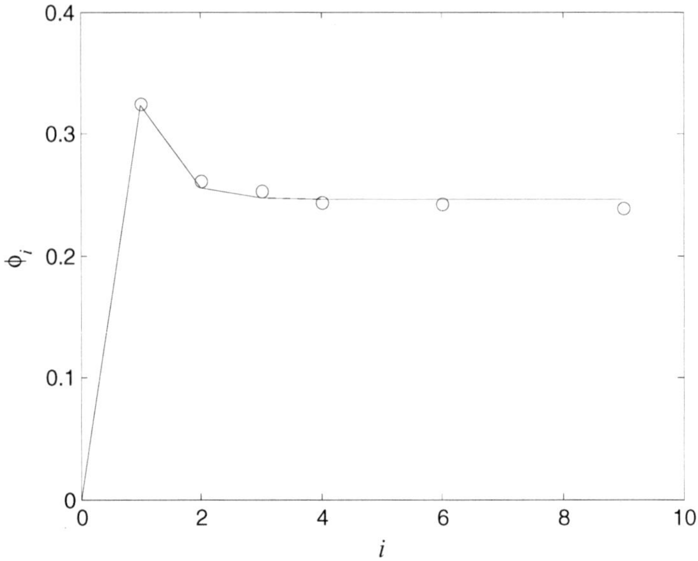

Figure 5.11: Plot of the dissolved fraction φ_i as a function of generations i (time step 5 min) using (5.24) for the dissolution of nifedipine solid dispersion with nicotinamide and polyvinylpyrolidone (1 : 3 : 1), in 900 ml of distilled water. Fitted line of (5.24) is drawn over the experimental data.

However, the use of (5.24) should not be considered as a panacea for modeling nonmonotonic dissolution curves. Obvious drawbacks of the model (5.24) are:

1. The data on the ascending limb of the dissolution curve, if any, should be ignored.

2. The time required to reach the maximum value of the dissolved fraction of drug should be adopted as the time interval between successive generations.

3. The time values of the data points that can be used for fitting purposes should be integer multiples of the time unit adopted.

Further, when k takes values much larger than $1/\theta$, (5.24) exhibits chaotic behavior following the period-doubling bifurcation (cf. Chapter 3). For example, (5.24) leads to chaos when $1/\theta = 0.25$ and k is greater than 0.855. Despite the aforementioned disadvantages, the model offers the sole approach that can be used to describe supersaturated dissolution data. In addition, the derivation of (5.24) relies on a model built from physical principles, i.e., a reaction-limited dissolution model. Other approaches based on empirical models, e.g., polynomial functions, could provide better fittings for supersaturated dissolution data but these approaches will certainly lack in physical meaning.

5.3 Modeling Random Effects

In all previous dissolution models described in Sections 5.1 and 5.2, the variability of the particles (or media) is not directly taken into account. In all cases, a unique constant (cf. Sections 5.1, 5.1.1, and 5.1.2) or a certain type of time dependency in the dissolution rate "constant" (cf. Sections 5.1.3, 5.2.1, and 5.2.2) is determined at the commencement of the process and fixed throughout the entire course of dissolution. Thus, in essence, all these models are deterministic. However, one can also assume that the above variation in time of the rate or the rate coefficient can take place randomly due to unspecified fluctuations in the heterogeneous properties of drug particles or the structure/function of the dissolution medium. Lansky and Weiss have proposed [130] such a model assuming that the rate of dissolution $k(t)$ is stochastic and is described by the following equation:

$$k(t) = \overline{k} + \sigma\xi(t),$$

where $\overline{k}$ is the deterministic part of the dissolution rate "constant," $\xi(t)$ is Gaussian white noise, and $\sigma > 0$ is its amplitude. According to the definition of this equation, the "constant" $\overline{k}$ represents the mean of $k(t)$.

The stochastic nature of $k(t)$ allows the description of the fraction of dose dissolved, $\varphi(t)$, in the form of a stochastic differential equation if coupled with the simplest dissolution model described by (5.16), assuming complete dissolution ($\theta = 1$):

$$\mathrm{d}\varphi(t) = \overline{k}[1 - \varphi(t)]\,\mathrm{d}t + \sigma\xi(t)[1 - \varphi(t)]\,\mathrm{d}B(t), \tag{5.26}$$

where the symbol $\varphi(t)$ is used here to denote the random nature of the process, while $\mathrm{d}B(t)$ comes from the Brownian motion since the noise $\xi(t)$ is the formal derivative of the Brownian motion, $B(t)$. The solution of (5.26) gives

$$\varphi(t) = 1 - \exp\left[-\left(\overline{k} + \frac{1}{2}\sigma^2\right)t - \sigma B(t)\right].$$

A discretized version of (5.26) can be used to perform Monte Carlo simulations using different values of σ and generate $\varphi(t)$-time profiles [130]. The random fluctuation of these profiles becomes larger as the value of σ increases.

Stochastic variation may be introduced in other models as well. In this context, Lansky and Weiss [130] have also considered random variation for the parameter k^* of the interfacial barrier model (5.20).

5.4 Homogeneity vs. Heterogeneity

Lansky and Weiss defined [131] the classical dissolution first-order model in terms of the fraction of dose dissolved, $\varphi(t)$ (equation 5.16 assuming $\theta = 1$),

$$\dot{\varphi}(t) = k[1 - \varphi(t)], \quad \varphi(0) = 0,$$

as the simplest homogeneous case, since the fractional dissolution rate function $k(t)$ derived from the above equation,

$$k(t) = \frac{\dot{\varphi}(t)}{1 - \varphi(t)},$$

is constant throughout the dissolution process. In physical terms, the homogeneous model dictates that each drug molecule has equal probability to enter solution during the entire course of the dissolution process. Plausibly, the various dissolution models have different time-dependent functional forms of $k(t)$. Accordingly, all these models were termed heterogeneous since the time dependence of the functions $k(t)$ denotes that the probability to enter solution is not identical for all drug molecules. To quantify the departure from the homogeneous case, Lansky and Weiss proposed [131] the calculation of the Kullback–Leibler information distance $Dist(f, \varphi)$ as a measure of heterogeneity of the function $f(t)$ from the homogeneous exponential model $\varphi(t)$ derived from the previous equation:

$$Dist(f, \varphi) = \int_0^\infty f(t) \ln \frac{f(t)}{\varphi(t)} \mathrm{d}t.$$

This measure of heterogeneity generalizes the notion of heterogeneity as a departure from the classical first-order model initially introduced [121] for the specific case of the Weibull function. In addition, the above equation can also be used for comparison between two experimentally obtained dissolution profiles [131].

The comparison of dissolution curves based on the calculation of $Dist(f, \varphi)$ is model-independent; however, other model-dependent comparative approaches have been proposed [132]. Caution should be exercised, though, when comparison of estimates of the parameters obtained from various models is attempted in the context of heterogeneity assessment. For example, the valid use of (5.15) for the homogeneous case presupposes that the amount needed to saturate the medium is exactly equal to the dose used in actual practice, i.e., $\theta = 1$ [132]. Recently, Lansky and Weiss presented [133] in a concise form the results of their recent studies [122, 130]. The empirical and semiempirical models for drug dissolution were reviewed and classified in five groups: first-order model with a time lag, models for limited solubility of drug, models of heterogeneous compound, Weibull and inverse Gaussian models, and models defined on a finite time window. In this contribution, the properties of models were investigated, the parameters were discussed, and the role of drug heterogeneity was studied.

5.5 Comparison of Dissolution Profiles

The comparison of dissolution profiles is of interest for both research and regulatory purposes. Several methods, which can be roughly classified as (1°) statistical approaches, (2°) model-dependent, and (3°) model-independent methods, have been reported in the literature for the comparison of dissolution profiles [134–136]. The statistical approaches are based on the analysis of variance, which is used to test the hypothesis that the two profiles are statistically similar. The model-dependent methods are mainly used for clarifying dissolution or release mechanisms under various experimental conditions and rely on the statistical comparison of the estimated parameters after fitting of a dissolution model (e.g., the Weibull model) to the raw data. The model-dependent methods can be applied to dissolution profiles with nonidentical dissolution sampling schemes, while the model-independent methods require identical sampling points since they are based on pairwise procedures for the calculation of indices (factors) from the individual raw data of two profiles. Two of these factors, namely, the *difference factor* f_1 and the *similarity factor* f_2, have been adopted by the regulatory agencies and have been included in the relevant dissolution Guidances for quality control testing [137–139]. Each one of these factors is calculated from the two mean dissolution profiles and is being used as a point estimate measure of the (dis)similarity of the dissolution profiles.

The difference factor f_1 [137] measures the relative error (as a percentage) between two dissolution curves over all time points:

$$f_1 = 100 \frac{\sum_{i=1}^{m} |R_i - T_i|}{\sum_{i=1}^{m} R_i}. \tag{5.27}$$

where m is the number of data points, R_i and T_i are the percentage of drug dissolved for the reference and test products at each time point i, respectively.

The similarity factor f_2 [137–139] is a logarithmic reciprocal transformation of the sum of squared errors and is a measurement of the similarity in the percentage dissolution between the two curves:

$$f_2 = 50 \log \left\{ 100 \left[1 + \frac{1}{m} \sum_{i=1}^{m} (R_i - T_i)^2 \right]^{-0.5} \right\}. \tag{5.28}$$

Both factors take values in the range 0–100 assuming that the percentage dissolved values for the two products are not higher than 100%. When no difference between the two curves exist, i.e., at all time points $R_i = T_i$, then $f_1 = 0$ and $f_2 = 100$. On the other hand, when the maximum difference between the two curves exists, i.e., at all time points $|R_i - T_i| = 100$, then $f_1 = 100$ and $f_2 = 0$.

The calculation of the factors from the mean profiles of the two drug products presupposes that the variability at each sample time point is low. Thus, for immediate release formulations, the FDA guidance [137] allows a coefficient of variation of no more than 20% for the early data points (e.g., 10 or 15 min), while a coefficient of variation less than 10% is required for the other time points.

According to the guidances [137, 139], when batches of the same formulation are compared, a difference up to 10% at all sample points is considered acceptable. On the basis of this boundary, the acceptable range of values derived from (5.27) and (5.28) for f_1 is 0–15 [137] and for f_2 is 50–100 [137, 139]. From a technical point of view, the following recommendations are quoted in the guidances [137, 139] for the calculation of f_1 and f_2 as point estimates:

1. a minimum of three time points (zero excluded),
2. 12 individual values for every time point for each formulation,
3. not more than one mean value of $> 85\%$ dissolved for each formulation.

Note that when more than 85% of the drug is dissolved from both products within 15 minutes, dissolution profiles may be accepted as similar without further mathematical evaluation. For the sake of completeness, one should add that some concerns have been raised regarding the assessment of similarity using the direct comparison of the f_1 and f_2 point estimates with the similarity limits [140–142]. Attempts have been made to bring the use of the similarity factor f_2 as a criterion for assessment of similarity between dissolution profiles in a statistical context using a bootstrap method [141] since its sampling distribution is unknown.

Although there are some differences between the European [139] and the US guidance [137, 138], e.g., the composition of the dissolution media, it should be pointed out that both recommend dissolution studies as quality assurance tests as well as for bioequivalence surrogate inference. The latter aspect is particularly well developed in the FDA guidance [138] in the framework of the biopharmaceutics classification system, which is treated in Section 6.6.1.

6

Oral Drug Absorption

The right drug for the right indication in the right dosage to the right patient.

Anonymous

The understanding and the prediction of oral drug absorption are of great interest for pharmaceutical drug development. Obviously, the establishment of a comprehensive framework in which the physicochemical properties of drug candidates are quantitatively related to the extent of oral drug absorption will accelerate the screening of candidates in the discovery/preclinical development phase. Besides, such a framework will certainly help regulatory agencies in developing scientifically based guidelines in accord with a drug's physicochemical properties for various aspects of oral drug absorption, e.g., dissolution, in vitro–in vivo correlations, biowaivers of bioequivalence studies.

However, the complex interrelationships among drug properties and processes in the gastrointestinal tract make the prediction of oral drug absorption a difficult task. In reality, drug absorption is a complex process dependent upon drug properties such as solubility and permeability, formulation factors, and physiological variables including regional permeability differences, pH, luminal and mucosal enzymes, and intestinal motility, among others. Despite this complexity, various qualitative and quantitative approaches have been proposed for the estimation of oral drug absorption. In all approaches discussed below the drug movement across the epithelial layer is considered to take place transcellularly since transcellular passive diffusion is the most common mechanism of drug transport.

The absorption models described in this chapter can be divided as follows:

- pseudoequilibrium models,
- mass-balance approaches,
- dynamic models,

- heterogeneous approaches, and
- models based on chemical structure.

The last section of this chapter is devoted to the regulatory aspects of oral drug absorption and in particular to the biopharmaceutics classification system and the relevant FDA guideline. At the very end of the chapter, we mention the difference between randomness and chaotic behavior as sources of the variability encountered in bioavailability and bioequivalence studies.

6.1 Pseudoequilibrium Models

These models assume that oral drug absorption takes place under equilibrium conditions. Spatial or temporal aspects of the drug dissolution, transit and uptake and the relevant physiological processes in the gastrointestinal tract are not taken into account. Only drug-related properties are considered as the key parameters controlling the absorption process.

6.1.1 The pH-Partition Hypothesis

Back in the 1940s, physiologists were the first to realize that in contrast to the capillary walls, with their large and unselective permeability, cell membranes present a formidable barrier to the diffusion of small molecules. A prominent scientist, M.H. Jacobs, in 1940 [143] was the first who studied the cell permeability of weak electrolytes and described quantitatively the nonionic membrane permeation of solutes. This observation initiated a number of specific studies [144–149] during the 1950s on the mechanisms of gastrointestinal absorption of drugs. The results of these studies formed the basis for the pH-partition hypothesis, which relates the dissociation constant, lipid solubility, and the pH at the absorption site with the absorption characteristics of various drugs throughout the gastrointestinal tract. Knowledge of the exact ionization of a drug is of primary importance since the un-ionized form of the drug, having much greater lipophilicity than the ionized form, is much more readily absorbed. Consequently, the rate and extent of absorption are principally related to the concentration of the un-ionized species. Since the pH in the gastrointestinal tract varies, the Henderson–Hasselbach equations for the ionization of acids,

$$\text{pH} = pK_a + \log\left[\frac{\text{ionized-concentration}}{\text{un-ionized-concentration}}\right],$$

and bases,

$$\text{pH} = pK_a + \log\left[\frac{\text{un-ionized-concentration}}{\text{ionized-concentration}}\right],$$

relate the fraction of the un-ionized species with the regional pH and the pK_a of the compound.

Most of the gastrointestinal absorption studies were found to be in accord with the principles of the pH-partition hypothesis. However, several deviations were noted and attributed to the unstirred water layer, the microclimate pH, and the mucus coat adjacent to the epithelial cell surface [150–152].

Although the pH-partition hypothesis relies on a quasi-equilibrium transport model of oral drug absorption and provides only qualitative aspects of absorption, the mathematics of passive transport assuming steady diffusion of the un-ionized species across the membrane allows quantitative permeability comparisons among solutes. As discussed in Chapter 2, (2.19) describes the rate of transport under sink conditions as a function of the permeability P, the surface area $\mathcal{A}$ of the membrane, and the drug concentration $c(t)$ bathing the membrane:

$$\dot{q}(t) = P\mathcal{A}c(t). \tag{6.1}$$

The proportionality between the rate of transport and permeability in (6.1) shows the importance of the latter parameter in the transcellular passive gastrointestinal absorption of drugs. Strictly speaking, one should utilize an estimate of the effective permeability (P_{eff}) [153] in (6.1) for predicting oral absorption potential of compounds. However, the methods for the estimation of P_{eff} are invasive, laborious, and time-consuming. Alternatively, various measures of lipophilicity such as the octanol/water partition coefficient ($\log P_c$) [154] and the distribution coefficient ($\log D$) [155] have been used as surrogates for predicting the oral absorption potential of compounds since permeability is mainly dependent on membrane partitioning.

6.1.2 Absorption Potential

In 1985 a major step in the theoretical analysis of oral drug absorption phenomena took place [156], when solubility and dose were also taken into account for the estimation of the absorption potential AP of a drug apart from the pH-partition hypothesis related parameters (lipophilicity, and degree of ionization). According to this concept, the AP is related proportionally to the octanol/water partition coefficient P_c, the fraction of the un-ionized species f_{un}, at pH= 6.5, and the physiological solubility c_s of the drug and inversely proportional to the dose q_0:

$$AP = \log\left[\frac{P_c f_{un} c_s V}{q_0}\right] = \log\left[\frac{P_c f_{un}}{\theta}\right]. \tag{6.2}$$

The logarithmic expression in the definition of AP has no physicochemical basis and is used for numerical reasons only; pH= 6.5 was selected as the representative pH of small intestines, where most of the absorption of drugs takes place. The incorporation of the terms P_c and f_{un} in the numerator of (6.2) means that the pH-partition hypothesis governs gastrointestinal absorption. Plausibly, AP was considered proportional to solubility and inversely proportional to the dose in accord with classical dissolution–absorption considerations. The volume term V corresponds to the small-intestinal volume content, which was set arbitrarily equal to 250 ml; moreover, the use of the term V makes the AP dimensionless. The ratio q_0/c_sV was defined as the dimensionless dose–solubility ratio in Section 5.1.4 and it was denoted by θ.

The validity of the approach based on (6.2) was proven when the fraction of dose absorbed, F_a, was found to increase with AP for several drugs with a wide variety of physicochemical properties and various degrees of extent of absorption [156]. Additional support for the AP concept was provided by a 3-dimensional plot of F_a as a function of the ionization-solubility/dose term (f_{un}/θ) and the octanol/water partition coefficient P_c [157]. In fact, because of the recent interest in the apparent permeability estimates P_{app} measured in the in vitro Caco-2 monolayer system, it was suggested that P_{app} can replace the octanol/water partition coefficient P_c in (6.2) [157].

Although the AP concept is a useful indicator of oral drug absorption, its qualitative nature does not allow the derivation of an estimate for F_a. A quantitative version of F_a as a function of AP was published in 1989. It was based on the equilibrium considerations used for the derivation of AP and the fundamental physicochemical properties in (6.2) with the implied competing intestinal absorption and nonabsorption processes [158]. This quantitative AP concept relies on (6.3), where a nonlogarithmic expression for AP is used:

$$F_a = \frac{(AP)^2}{(AP)^2 + f_{un}\,(1 - f_{un})}. \tag{6.3}$$

Based on physiological-physicochemical arguments, constraints were proposed for P_c, i.e., to be set equal to 1000 when $P_c > 1000$ and θ equal to 1 when $\theta < 1$. Equation (6.3) is, in reality, the first ever published explicit relationship between F_a and physicochemical drug properties. It was used to classify drugs according to their solubility, permeation, and ionization characteristics [158]. Moreover, (6.3) was monoparameterized :

$$F_a = \frac{Z^2}{1 + Z^2}, \tag{6.4}$$

where

$$Z = \frac{AP}{\sqrt{f_{un}\,(1 - f_{un})}}.$$

Equation (6.4) was used for fitting purposes using AP and F_a data reported in the literature and applying the constraints mentioned above for P_c and θ in the calculation of AP, Figure 6.1.

A number of modifications in the solubility and the partition coefficient terms of the AP have also been proposed in the literature [159–161]. According to these authors the modified absorption potential expressions can be considered better predictors of the passive absorption of drugs than the original AP. The most recent approach [161] relies on a single absorption parameter defined as the ratio of the octanol/water partition coefficient to the luminal oversaturation number. The latter is equal to the solubility-normalized dose for suspensions and equal to unity for solutions.

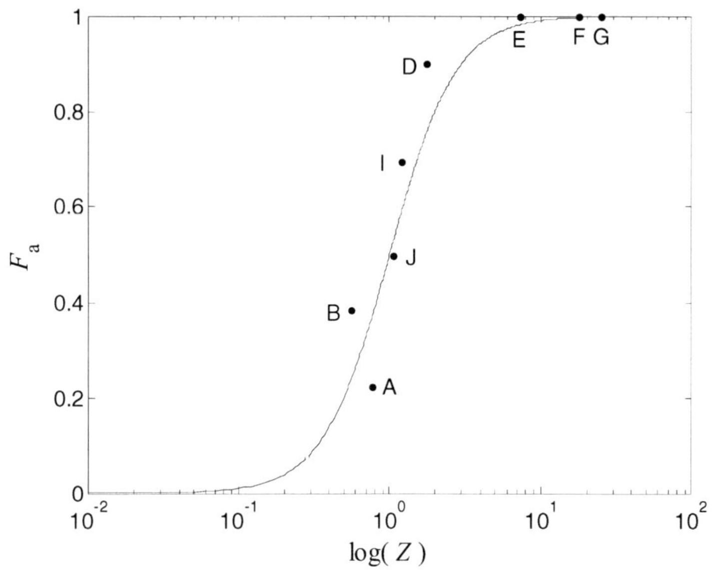

Figure 6.1: Plot of the fraction of dose absorbed for various drugs as a function of Z. Key: A acyclovir; B chlorothiazide solution; D hydrochlorothiazide; E phenytoin; F prednisolone; G digoxin (Lanoxicaps); I cimetidine; J mefenamic acid.

A relevant simple model was used to estimate the maximum absorbable dose (MAD) [162]. It takes into account the permeability, expressed in terms of a first-order rate constant k_a, the solubility c_s of the drug, and two physiological variables, the dissolution-intestinal volume V arbitrarily set to 250 ml, and the duration of gastrointestinal absorption t_a for 6 h:

$$MAD = V c_s k_a t_a.$$

This model assumes gastrointestinal absorption from a saturated solution of the drug (hypothetically maintained at a constant saturated value) for a time period equal to 6 h.

6.2 Mass Balance Approaches

These approaches place particular emphasis on the spatial aspects of the drug absorption from the gastrointestinal tract. The small intestine is assumed to be a cylindrical tube with fixed dimensions where the drug solution or suspension follows a homogeneous flow. Mass balance relationships under steady-state as-

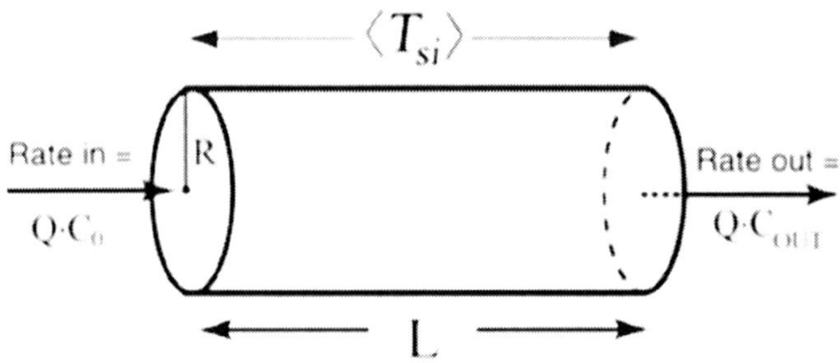

Figure 6.2: The small intestine is modeled as a homogeneous cylindrical tube of length L and radius R. c_0 and c_{out} are the inlet and outlet drug concentrations, respectively. The other symbols are defined in the text.

sumptions are used to estimate the fraction of dose absorbed as a function of the drug properties and of physiological parameters.

6.2.1 Macroscopic Approach

In the early 1990s the research group of G. Amidon in Ann Arbor applied mass balance approaches to the analysis of drug intestinal absorption [54, 55]. The small intestine is assumed to be a cylindrical tube with physiologically relevant dimensions (radius R and length L), while a constant volumetric flow rate Q describes the transit process of the intestinal contents, Figure 6.2. The *macroscopic approach* [54] refers mainly to highly soluble compounds. The incorporation of the dissolution step as an important part of the absorption process is treated in Section 6.2.2 under the heading *microscopic approach* [55].

The macroscopic approach under the steady-state assumption provides estimates for the fraction of dose absorbed F_a for the three cases, which refer to the magnitude of c_0 and c_{out} in Figure 6.2 relative to drug solubility c_s, namely:

1. Case I: $c_0 \leq c_s$ and $c_{out} \leq c_s$ (the drug is in solution throughout the transit process);

2. Case II: $c_0 > c_s$ and $c_{out} \leq c_s$ (solid drug at inlet; concentration reaches solubility at a certain point and diminishes thereafter);

3. Case III: $c_0 > c_s$ and $c_{out} > c_s$ (solid drug exists at both ends of the tube).

Irrespective of the specific case considered, the general mass balance relationship for the system depicted in Figure 6.2 under the steady-state assumption is

$$Q\left(c_0 - c_{out}\right) = 2\pi R P_{eff} \int_0^L c\left(z\right) \mathrm{d}z,$$

where P_{eff} is the effective permeability of drug and $\mathrm{d}z$ the infinitesimal axial length. The fraction of dose absorbed F_a can be expressed in terms of c_0 and c_{out} using the previous equation:

$$\begin{aligned} F_a &= 1 - \frac{c_{out}}{c_0} = \frac{2\pi R P_{eff}}{Q c_0} \int_0^L c(z)\,\mathrm{d}z \\ &= \frac{2\pi R P_{eff} L}{Q} \int_0^1 c^*(z^*)\,\mathrm{d}z^* = 2A_n \int_0^1 c^*(z^*)\,\mathrm{d}z^*. \end{aligned} \tag{6.5}$$

The last two integrals of the previous equation are expressed in dimensionless variables, $c^* = c/c_0$, $z^* = z/L$ with normalized limits $(0, 1)$, while the symbol A_n is the *absorption number* of the drug:

$$A_n \triangleq \frac{\pi R L}{Q} P_{eff} = \langle T_{si} \rangle \frac{P_{eff}}{R}. \tag{6.6}$$

The first fraction of the previous equation shows that A_n is exclusively determined by the effective permeability P_{eff} of drug since all other variables are species-dependent physiological parameters. In terms of characteristic times, the A_n of a drug can also be defined as the ratio of the mean small intestinal transit time $\langle T_{si} \rangle$, to its absorption time R/P_{eff}.

For the calculation of F_a, one should first express the dimensionless concentration $c^*(z^*)$ as a function of z^*, for each one of the three cases considered above, and then integrate (6.5).

1. For case I, the concentration profile $c^*(z^*)$ diminishes exponentially as a function of distance z^* assuming the complete radial mixing model [163] in the tube,

$$c^*(z^*) = \frac{c_{out}}{c_0} = \exp(-2A_n z^*),$$

   and for the fraction of dose absorbed,

$$F_a = 1 - \exp(-2A_n).$$

   This last equation shows that when the drug is in solution throughout the transit process and $c_0 \leq c_s$ and $c_{out} \leq c_s$, then F_a is dependent exclusively and exponentially on A_n. According to this equation, large values of A_n ensure complete absorption for this type of drugs.

2. For case III, the concentration c_{out} can be considered equal to the solubility since $c_0 > c_s$ and $c_{out} > c_s$; therefore

$$c^*(z^*) = \frac{c_s}{c_0} = \frac{1}{\theta}, \tag{6.7}$$

   and for the fraction of dose absorbed,

$$F_a = \frac{2A_n}{\theta}. \tag{6.8}$$

Although this equation indicates that F_a is proportional to A_n and inversely proportional to θ, this should be judged with caution since the conditions of case III, expressed in terms of concentration, are physically irrelevant ($c_0 > c_s$ and $c_{out} > c_s$). In addition, the use of (6.7) for the derivation of (6.8) assumes instantaneous dissolution in order to maintain the value of c_s constant throughout the transit process.

3. Case II can be viewed as a hybrid of cases I and III. As long as $c_0 > c_s$, the conditions assumed for case III are prevailing. Then, using a simple mass balance equation up to the temporal (spatial) point when c_0 reaches solubility ($c_0 = c_s$) and (6.7), the fraction absorbed F_{a1} can be calculated as

$$F_{a1} = \frac{c_0 V - c_s V}{c_0 V} = 1 - \frac{1}{\theta}.$$

Beyond this spatiotemporal point until the drug exits from the tube, the inequality $c_0 < c_s$ holds and therefore the fraction absorbed F_{a2} in this region follows the results obtained for case I conditions:

$$F_{a2} = \frac{1}{\theta}\left[1 - \exp\left(-2A_n + \theta - 1\right)\right].$$

Consequently, the total fraction of dose absorbed F_a is the sum of F_{a1} and F_{a2}:

$$F_a = 1 - \frac{1}{\theta}\exp\left(-2A_n + \theta - 1\right).$$

The most significant result of the macroscopic approach was derived from the analysis of case I conditions. It was found that the absorption number A_n and in particular its major determinant, the effective permeability, control the intestinal absorption of drugs. This observation triggered a large number of studies, and in recent years several attempts have been made to model the fraction of dose absorbed, F_a, with experimental in situ and in vitro models such as cell cultures (Caco-2, HT-29, and MDCK) [164–166] and artificial membranes (IAM, PAMPA) [167]. The aim of these studies is to find a correlation between the apparent permeability estimates P_{app} measured in these systems and the experimental F_a values. The most popular among these systems is the in vitro Caco-2 monolayer system [168], which is a donor–receptor compartment apparatus separated by a cell monolayer grown on a porous polycarbonate filter and is used to estimate the apparent permeability of compounds. In reality, an estimate for P_{app} is obtained from the experimental permeation data using (6.1) and solving it in terms of P; the flux rate $\dot{q}(t)$ is obtained from the slope of the receptor chamber solute mass vs. time plot, while $\mathcal{A}$ is the cross-sectional area of cell surface and $c(t) = c_0$ is the initial solute concentration in the donor compartment. Extensive research in the passive transport mechanisms of a great number of compounds in cell culture monolayers indicates that an apparent permeability estimate in the range of 2×10^{-6}–$10^{-5}\,\mathrm{cm\,s^{-1}}$ [168–170] ensures complete absorption of the solute provided that absorption is not solubility- and/or dissolution-limited, Figure 6.3.

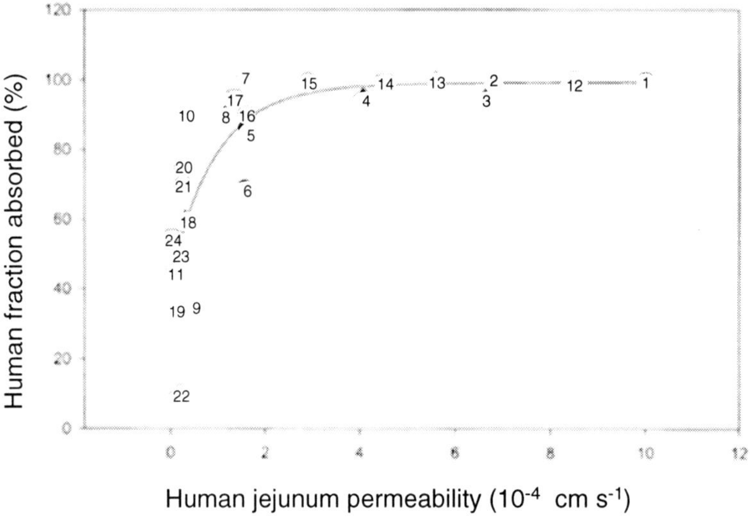

Figure 6.3: Plot of the fraction of dose absorbed (in %) of various drugs as a function of the permeability estimates in the Caco-2 system. Key: 1 D-glucose; 2 verapamil; 3 piroxicam; 4 phenylalanine; 5 cyclosporin; 6 enalapril; 7 cephalexim; 8 losartan; 9 lisinopril; 10 amoxicillin; 11 methyldopa; 12 naproxen; 13 antipyrine; 14 desipramine; 15 propanolol; 16 amiloride; 17 metoprolol; 18 terbutaline; 19 mannitol; 20 cimetidine; 21 ranitidine; 22 enalaprilate; 23 atenolol; 24 hydrochlorothiazide.

6.2.2 Microscopic Approach

This approach deals with the analysis of intestinal absorption of poorly soluble drugs, administered as suspensions, assuming that drug particles are spheres of the same initial radius size ρ_0. The resulting mathematical model [55] assumes complete radial mixing, takes into account drug dissolution, transit, and uptake, and relies on the homogeneous cylindrical intestinal tube depicted in Figure 6.2. Under the steady-state assumption, mass balance relationships for the drug processes in both solid and solution phase are considered in a volume element of the intestine of axial length $\mathrm{d}z$. Two differential equations expressed in dimensionless variables govern the reduction of the radius $\rho(z)$ of the particles from their initial value ρ_0 and the change of the luminal concentration of the drug $c(z)$:

$$\begin{aligned} \frac{\mathrm{d}\rho^*(z^*)}{\mathrm{d}z^*} &= -\frac{D_n}{3}\frac{1-c^*(z^*)}{\rho^*(z^*)} && \rho^*(0)=1, \\ \frac{\mathrm{d}c^*(z^*)}{\mathrm{d}z^*} &= \theta D_n \rho^*(z^*)\left[1-c^*(z^*)\right] - 2A_n c^*(z^*) && c^*(0)=0, \end{aligned} \tag{6.9}$$

where $z^* = z/L$, $c^*(z^*) = c(z^*)/c_s$, $\rho^*(z^*) = \rho(z^*)/\rho_0$, and D_n is the dissolution number defined by the following equation:

$$D_n = \frac{(\mathcal{D}/\rho_0)\, c_s \left(4\pi\rho_0^2\right)\left(\pi R^2 L\right)}{Q\left(4\pi\rho_0^3\varrho\right)},$$

where $\mathcal{D}$ is the diffusivity and ϱ is the density of the drug. Using a mass balance relationship for the solid and solution phases at the outlet of the tube ($\rho^* = 1$), the following equation is obtained for the fraction of dose absorbed, F_a:

$$F_a = 1 - \left(\rho^*|_{z^*=1}\right)^3 - \frac{1}{\theta}\left(c^*|_{z^*=1}\right).$$

This equation can be used in conjunction with (6.9) for the estimation of F_a. The microscopic approach points out clearly that the key parameters controlling drug absorption are three dimensionless numbers, namely, absorption number A_n, dissolution number D_n, and θ. The first two numbers are the determinants of membrane permeation and drug dissolution, respectively, while θ reflects the ratio of the dose administered to the solubility of drug.

6.3 Dynamic Models

These models are dependent on the temporal variable associated with the drug transit along the small intestine. Drug absorption phenomena are assumed to take place in the time domain of the physiological mean transit time. For those dynamic models that rely on diffusion-dispersion principles both the spatial and temporal variables are important in order to simulate the spatiotemporal profile of the drug in the intestinal lumen.

6.3.1 Compartmental Models

The compartmental approach to the process of a drug passing through the gastrointestinal tract has been used to simulate and explain oral drug absorption. The simplest approach relies on a single mixing tank model of volume V where the drug has a uniform concentration while a flow rate Q is ascribed to the contents of the tank. Thus, the ratio V/Q corresponds to the time period beyond which drug dissolution and/or absorption is terminated. In other words, it is equivalent to the small-intestinal transit time for the homogeneous tube model. Similarly, the ratio Q/V indicates the first-order rate constant for drug removal from the absorption sites. One or two mixing tanks in series have been employed for the study of various oral drug absorption phenomena [53, 171, 172].

Mixing tanks in series with linear transfer kinetics from one to the next with the same transit rate constant k_t have been utilized to obtain the characteristics of flow in the human small intestine [173, 174]. The differential equations of mass transfer in a series of m compartments constituting the small intestine for a nonabsorbable and nondegradable compound are

$$\dot{q}_i(t) = k_t q_{i-1}(t) - k_t q_i(t), \qquad i = 1, \ldots, m, \tag{6.10}$$

where $q_i(t)$ is the amount of drug in the ith compartment. The rate of exit of the compound from the small intestine is

$$\dot{q_m}(t) = -k_t q_m(t). \tag{6.11}$$

Solving the system of (6.10) and (6.11) in terms of the fraction of dose absorbed, we obtain

$$F_a = \frac{q_m(t)}{q_0} = 1 - \exp(-k_t t)\left[1 + k_t t + \frac{(k_t t)^2}{2} + \cdots + \frac{(k_t t)^{m-1}}{(m-1)!}\right]. \tag{6.12}$$

Analysis of experimental human small-intestine transit time data collected from 400 studies revealed a mean small-intestinal transit time $\langle T_{si}\rangle = 199\,\text{min}$ [173]. Since the transit rate constant k_t is inversely proportional to $\langle T_{si}\rangle$, namely, $k_t = m/\langle T_{si}\rangle$, (6.12) was further fitted to the cumulative curve derived from the distribution frequency of the entire set of small-intestinal transit time data in order to estimate the optimal number of mixing tanks. The fitting results were in favor of seven compartments in series and this specific model, (6.10) and (6.11) with $m = 7$, was termed the *compartmental transit model.*

The incorporation of a passive absorption process in the compartmental transit model led to the development of the *compartmental absorption transit model* (CAT) [175]. The rate of drug absorption in terms of mass absorbed $q_a(t)$ from the small intestine of the compartmental transit model is

$$\dot{q_a}(t) = k_a \sum_{i=1}^{7} q_i(t),$$

where k_a is the first-order absorption rate constant. Then, the fraction of dose absorbed F_a, using the previous equation, is

$$F_a = \frac{q_a(t)}{q_0} = \frac{k_a}{q_0} \sum_{i=1}^{7} \int_0^{\infty} q_i(t)\,dt. \tag{6.13}$$

The solution of (6.12) and (6.13) yields

$$F_a = 1 - \left(1 + \frac{k_a}{k_t}\right)^{-7}.$$

Recall that k_t is equal to $7/\langle T_{si}\rangle$, while k_a can be expressed in terms of the effective permeability and the radius R of the small intestine [55]:

$$k_a = \frac{2P_{eff}}{R}. \tag{6.14}$$

The previous equation can be written as

$$F_a = 1 - \left(1 + \frac{2P_{eff}\langle T_{si}\rangle}{7R}\right)^{-7}.$$

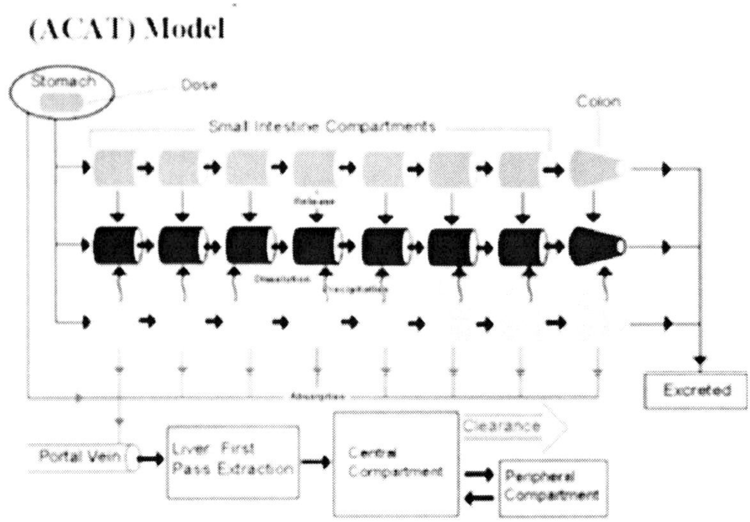

Figure 6.4: Schematic of the ACAT model. Reprinted from [176] with permission from Elsevier.

The CAT model presupposes that dissolution is instantaneous and therefore the kinetics of the permeation step control the gastrointestinal absorption of drug. This is reflected in the previous equation, which indicates that the effective permeability is the sole parameter controlling the intestinal absorption of highly soluble drugs.

Due to its compartmental nature, the CAT model can be easily coupled with the disposition of drug in the body using classical pharmacokinetic modeling. In this respect the CAT model has been used to interpret the saturable small-intestinal absorption of cefatrizine in humans [175].

The CAT model was further modified to include pH-dependent solubility, dissolution/precipitation, absorption in the stomach or colon, first-pass metabolism in gut or liver, and degradation in the lumen. Physiological and biochemical factors such as changes in absorption surface area, transporter, and efflux protein densities have also been incorporated. This advanced version of CAT, called ACAT [176], has been formulated in a commercially available simulation software product under the trademark name GastroPlus^{TM}. A set of differential equations, which is solved by numerical integration, is used to describe the various drug processes of ACAT as depicted in Figure 6.4.

6.3.2 Convection–Dispersion Models

The use of convection–dispersion models in oral drug absorption was first proposed in the early 1980s [177, 178]. The small intestine is considered a 1-

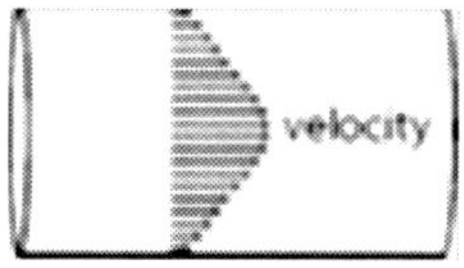

Figure 6.5: The velocity of the fluid inside the tube is larger near the axis and much smaller near the walls. This is considered to be the main factor for the dispersion of the distribution of the drug.

dimensional tube that is described by a spatial coordinate z that represents the axial distance from the stomach. In addition, the tube contents have constant axial velocity v and constant dispersion coefficient D, which arises from molecular diffusion, stirring due to the motility of the intestines, and Taylor dispersion due to the difference of the axial velocity at the center of the tube compared with the tube walls (Figure 6.5). The small-intestine transit flow for a nonabsorbable and nondegradable compound in this type of model is described by [173, 178]

$$\frac{\partial c(z,t)}{\partial t} = D\frac{\partial^2 c(z,t)}{\partial z^2} - v\frac{\partial c(z,t)}{\partial z}, \tag{6.15}$$

where $c(z,t)$ is the concentration. An analytical solution of this equation can be obtained if one assumes that the stomach operates as an infinite reservoir with constant output rate in terms of concentration and volume. Under these assumptions, the following analytical solution was obtained [178]:

$$\frac{c(z,t)}{c_0} = \frac{1}{2}\left[\operatorname{erf} c\left(\frac{z}{\sqrt{4Dt}} - \sqrt{\frac{v^2 t}{4D}}\right) + \exp\left(\frac{vz}{D}\right)\operatorname{erf} c\left(\frac{z}{\sqrt{4Dt}} + \sqrt{\frac{v^2 t}{4D}}\right)\right], \tag{6.16}$$

where $\operatorname{erf} c$ is the complementary error function defined by

$$\operatorname{erf} c(x) = 1 - \frac{2}{\sqrt{\pi}}\int_0^x \exp\left(-z^2\right) dz.$$

Equation (6.16) allows one to generate the axial profile of normalized concentration $c(z,t)/c_0$ at different times, Figure 6.6 A. The second term in the parentheses of (6.16) is relatively small compared to the first; therefore, (6.16) can be approximated by the following:

$$\frac{c(z,t)}{c_0} = \frac{1}{2}\operatorname{erf} c\left(\frac{z}{\sqrt{4Dt}} - \sqrt{\frac{v^2 t}{4D}}\right).$$

By replacing the spatial coordinate z with the length of the tube L in the previous equation, the fraction of dose exiting the small intestine as a function of time is obtained:

$$\frac{c(L,t)}{c_0} = \frac{1}{2}\operatorname{erf} c\left(\frac{L - vt}{\sqrt{4Dt}}\right).$$

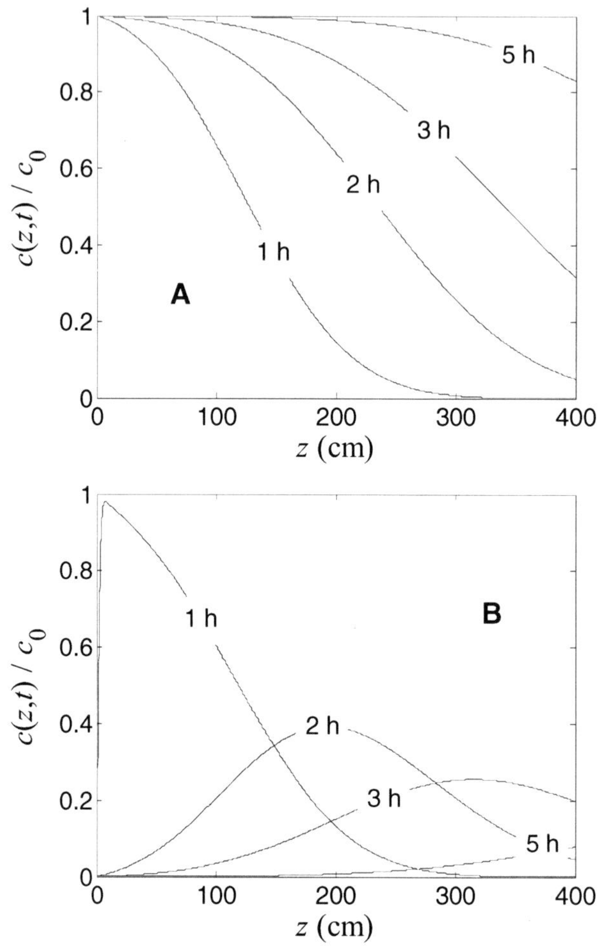

Figure 6.6: Axial profile snapshots of normalized concentration (with respect to the constant input concentration) inside the intestinal lumen, at various times. (**A**) (6.16) is used, with $D = 0.78\,\mathrm{cm}^2\,\mathrm{s}^{-1}$, $\upsilon = 1.76\,\mathrm{cm\,min}^{-1}$, and a constant-concentration infinite reservoir input. (**B**) the analytical solution of (6.17) with initial condition $c(z, 0) = 0$ is used, with $D = 0.78\,\mathrm{cm}^2\,\mathrm{s}^{-1}$, $\upsilon = 1.76\,\mathrm{cm\,min}^{-1}$, $k_a = 0.18\,\mathrm{h}^{-1}$, and a constant-concentration reservoir input, applied only for the first hour, $t_\circ = 1\,\mathrm{h}$.

This equation allows one to consider the cumulative distribution of small-intestinal transit time data with respect to the fraction of dose entering the colon as a function of time. In this context, this equation characterizes well the small-intestinal transit data [173, 174], while the optimum value for the dispersion coefficient D was found to be equal to $0.78\,\mathrm{cm}^2\,\mathrm{s}^{-1}$. This value is much greater than the classical order of magnitude $10^{-5}\,\mathrm{cm}^2\,\mathrm{s}^{-1}$ for molecular diffusion coefficients since it originates from Taylor dispersion due to the difference of the axial velocity at the center of the tube compared with the tube walls, as depicted in Figure 6.5.

For absorbable substances, a first-order absorption term can be coupled with the convection–dispersion (6.15) to model both the fluid flow and the absorption process:

$$\frac{\partial c(z,t)}{\partial t} = D\frac{\partial^2 c(z,t)}{\partial z^2} - v\frac{\partial c(z,t)}{\partial z} - k_a c(z,t), \tag{6.17}$$

where k_a is the first-order absorption rate constant. Although the previous equation is solved numerically, an analytical solution can be obtained [179] for appropriate initial and boundary conditions. More specifically, with a zero initial condition $c(z,0) = 0$ and boundary conditions that correspond to a constant reservoir for an initial period $t_\circ$ only,

$$c(0,t) = \begin{cases} c_0 & \text{for} \quad 0 < t \leq t_\circ, \\ 0 & \text{for} \quad t_\circ < t, \end{cases} \qquad \left.\frac{\partial c(z,t)}{\partial z}\right|_{z\to\infty,t} = 0,$$

the analytical solution of (6.17) is

$$c(z,t) = \begin{cases} c_0\Phi(z,t) & \text{for} \quad 0 < t \leq t_\circ, \\ c_0\Phi(z,t) - c_0\Phi(z,t-t_\circ) & \text{for} \quad t_\circ < t, \end{cases}$$

where

$$\Phi(z,t) = \frac{1}{2}\exp\left[\frac{(v-\alpha)z}{2D}\right]\operatorname{erf}c\left(\frac{z-vt}{\beta}\right) + \frac{1}{2}\exp\left[\frac{(v+\alpha)z}{2D}\right]\operatorname{erf}c\left(\frac{z+vt}{\beta}\right)$$

and

$$\alpha = v\sqrt{1+4k_a D v^{-2}}, \qquad \beta = 2\sqrt{Dt}.$$

Profiles of the analytical solution of (6.17) were plotted in Figure 6.6 B.

In this category of dispersion models, one can also classify a "continuous plug flow with dispersion" model for the simulation of gastrointestinal flow and drug absorption [180]. In this model, the drug is passively absorbed, while the intestinal transit is described via a Gaussian function. The drug solution moves in a concerted fashion along the intestines, but with an ever-widening distribution about the median location in contrast to the time-distribution theoretical profiles of classical dispersion–convection models shown in Figure 6.6. The model described nicely the dose-dependent absorption of chlorothiazide in rats [180], and it has been used for the development of a physiologically based model for gastrointestinal transit and absorption in humans [181].

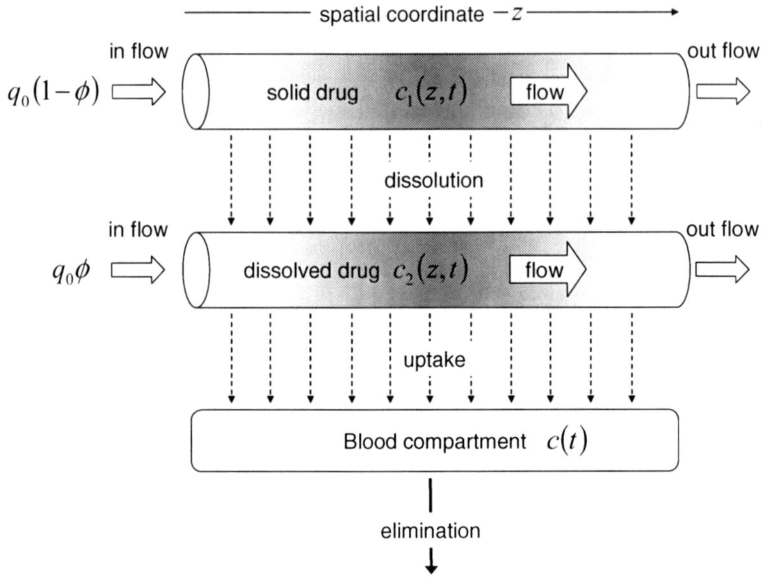

Figure 6.7: A dispersion model that incorporates spatial heterogeneity for the gastrointestinal absorption processes. q_0 denotes the administered dose and φ is the fraction of dose dissolved in the stomach.

Recently, a novel convection–dispersion model for the study of drug absorption in the gastrointestinal tract, incorporating spatial heterogeneity, was presented [182]. The intestinal lumen is modeled as a tube (Figure 6.7), where the concentration of the drug is described by a system of convection–dispersion partial differential equations. The model considers:

- two drug concentrations, for the dissolved and the undissolved drug species, and

- spatial heterogeneity along the axis of the tube for the various processes included, i.e., axial heterogeneity for the velocity of the intestinal fluids, the constants related to the dissolution of the solid drug, and the uptake of the dissolved drug from the intestinal wall.

The model includes more realistic features than previously published dispersion models for the gastrointestinal tract, but the penalty for that is that it can be solved only numerically and includes a large number of parameters that are difficult to be estimated based solely on blood data.

6.4 Heterogeneous Approaches

The approaches discussed in Sections 6.1, 6.2, and 6.3 were based on the concept of homogeneity. Hence, the analysis of drug dissolution, transit, and uptake in the gastrointestinal tract was accompanied by the assumption of perfect mixing in the compartment(s) or the assumption of homogeneous flow. In the same vein, the convection–dispersion models [173, 174, 177–180, 182] consider the small intestine as a uniform tube with constant axial velocity, constant dispersion behavior, and constant concentration profile across the tube diameter. The heterogeneous approaches attempt to incorporate the geometrically heterogeneous features of the internal structure of the intestinal tube, e.g., microvilli as well as the inhomogeneous flow of drug toward the lower end of the intestinal tube.

The assumptions of homogeneity and/or well-stirred media used in Sections 6.1 to 6.3 are not only not obvious, but they are in fact contrary to the evidence given the anatomical and physiological complexity of the gastrointestinal tract. Both in vivo drug dissolution and uptake are heterogeneous processes since they take place at interfaces of different phases, i.e., the liquid–solid and liquid–membrane boundaries, respectively. In addition, both processes occur in heterogeneous environments, i.e., variable stirring conditions in the lumen. The mathematical analysis of all models described previously relies furthermore on the assumption that an isotropic 3-dimensional space exists in order to facilitate the application of Fick's laws of diffusion. However, recent advances in physics and chemistry, as discussed in Chapter 2, have shown that the geometry of the environment in which the processes take place is of major importance for the treatment of heterogeneous processes. In media with topological constraints, well-stirred conditions cannot be postulated, while Fick's laws of diffusion are not valid in these spaces. Most of the arguments questioning the validity of the diffusion theory in a biological context seem to be equally applicable in the complex media of the gastrointestinal tract [183, 184]. However, advances in heterogeneous kinetics have led to the development of fractal-like kinetics that are suitable for processes taking place in heterogeneous media and/or involving complicated mechanisms. In the light of the above-mentioned gastrointestinal heterogeneity, the drug gastrointestinal processes are discussed below in terms of fractal concepts [185].

6.4.1 The Heterogeneous Character of GI Transit

Since gastrointestinal transit has a profound effect on drug absorption, numerous studies have focused on the gastric emptying and the intestinal transit of different pharmaceutical dosage forms. Gastric emptying is totally controlled by the two patterns of upper gastrointestinal motility, i.e., the interdigestive and the digestive motility pattern [186]. The interdigestive pattern dominates in the fasting state and is organized into alternating phases of activity and quiescence. Studies utilizing gamma scintigraphy have shown that gastric emptying is slower and more consistent in the presence of food [187, 188]. The transit through the small intestine, by contrast, is largely independent of the feeding conditions and physical properties of the system [187, 188], with an average transit time of ≈ 3 h [173]. Thus, normal transport seems to operate in the various segments of the small intestine and therefore a linear evolution in time of the mean position of the propagating packet of drug molecules or particles can be conceived.

Several studies with multiparticulate forms have indicated that the movement of pellets across the ileo–caecal junction involves an initial regrouping of pellets prior to their entrance and spreading in the colon [189–191]. According to Spiller et al. [192] the ileocolonic transit of 1 ml solution of a ^{99m}Tc-diethyltriamino-pentaacetic acid (DTPA) in humans is rapid postprandially and slow and erratic during fasting. Under fasting conditions the ileum is acting as a reservoir in several cases and the colonic filling curves of DTPA exhibit long plateaus and low slopes that are indicative of episodic colonic inflow and wide spreading of the marker in the colon [192]. Similarly, Krevsky et al. [193] have shown that an 8 ml bolus containing ^{111}In-DTPA installed into the cecum was fairly evenly distributed throughout all segments of the colon after 3 h. Finally, the colonic transit of different-sized tablets has also been shown to follow the same spreading pattern [194]. This type of marker movement is most likely due to the electrical activity of the proximal and distal parts of the colon [186]. The electrical waves in these regions are not phase locked and therefore random contractions of mixing and not propulsion of contents is observed. From a kinetic point of view, the wide spreading of the marker in the colon is reminiscent of what is known in physics as *dispersive transport* [195]. This conclusion can be derived if one compares time distribution analysis data of colonic transit (cf. for example the data of the first 3 hours in Figure 3 of [193]) with the general pattern of dispersive transport (Figure 4 in [195]). These observations substantiate the view that dispersive transport [195] operates in the large intestine and therefore the mean position of the propagating packet of drug particles is a sublinear function of time. However, dispersive transport is a scale-invariant process with no intrinsic transport coefficients; in other words, a mean transit time does not exist since transport coefficients become subject- and time-dependent [195]. These observations provide an explanation for the extremely variable whole-bowel transit, i.e., 0.5–5 d [194], since the greater part of the transit is attributable to residence time in the large intestine.

6.4.2 Is in Vivo Drug Dissolution a Fractal Process?

In the pharmaceutical literature there are several reports that demonstrate that flow conditions in the gastrointestinal tract do not conform to standard hydrodynamic models. Two investigations [196, 197] assessed the gastrointestinal hydrodynamic flow and the mechanical destructive forces around a dosage form by comparing the characteristics of in vitro and in vivo release of two different types of controlled-release paracetamol tablets. The results [196] indicate that the hydrodynamic flow around the dosage forms in the human gastrointestinal tract are very low, corresponding to a paddle speed of 10 rpm in the paddle method of dissolution or a velocity of about 1 cm min^{-1} (1–2 ml min^{-1} flow rate) in the flow-through cell method. In parallel, low and high in vitro destructive forces were found to be physiologically meaningful and essential for establishing a useful in vitro dissolution testing system [196, 197].

Furthermore, data from gastrointestinal physiology have long since shown the heterogeneous picture of the gastrointestinal contents as well as the importance of mechanical factors in the gastrointestinal processes [186]. It is very well established that the gastric contents are viscous, while shearing forces in the chyme break up friable masses of food. Since chyme moves slowly down the intestine by segmentation and short, weak propulsive movements, the flow is governed by resistance as well as by pressure generated by contraction [186]. Thus, there is a progressive reduction of the transit rate from duodenum to the large intestine [198, 199].

All the above observations [186–199] substantiate the view that the flow is forced in the narrow and understirred spaces of the colloidal contents of the lower part of the gastrointestinal tract. Consequently, friction becomes progressively more important than intermolecular diffusion in controlling the flow as the drug moves down the intestine. The characteristics of this type of flow have been studied [200, 201] with Hele–Shaw channels ensuring a quasi 2-dimensional space using miscible fluids of different viscosities. These studies revealed that when a less-viscous fluid moves toward a fluid with higher viscosity (polymer solution or colloidal suspension), the interface ripples and very soon becomes extremely meandering (fractal). These viscous, fractal fingers have been observed in experiments mimicking the secretion of HCl and its transport through the mucus layer over the surface epithelium [202]. Confirmation of this type of morphology (channel geometry) in the mucus layer has been provided by an in vivo microscopic study of the acid transport at the gastric surface [203]. The results obtained with the dyes Congo red and acridine strongly suggest that secreted acid (and pepsin) moves from the gastric crypts across the surface mucus layer into the luminal bulk solution only at restricted sites [203].

In the light of these observations one can argue that the dissolution of sparingly soluble drugs should be performed in topologically constrained media since the drug particles traverse the larger part or even the entire length of the intestines and attrition is a significant factor for their dissolution. However, one can anticipate poor reproducibility of dissolution results in topologically constrained media [204, 205] since the dissolution of particles will be inherently linked with the fractal fingering phenomenon, Figure 6.8:

1. The square in Figure 6.8 A represents geometrically all currently used well-stirred dissolution media, which ensure at any time a homogeneous concentration of drug throughout their volume. Due to homogeneity a sample taken from a well-stirred dissolution medium can provide the amount of drug dissolved (white squares) after separation of the undissolved drug (black squares).

2. Dissolution in topologically constrained media gives rise to fractal fingering, Figure 6.8 B (cf. also figures in [201, 204, 205]). The tree-like structure shown here indicates the flow of liquid where dissolution takes place.

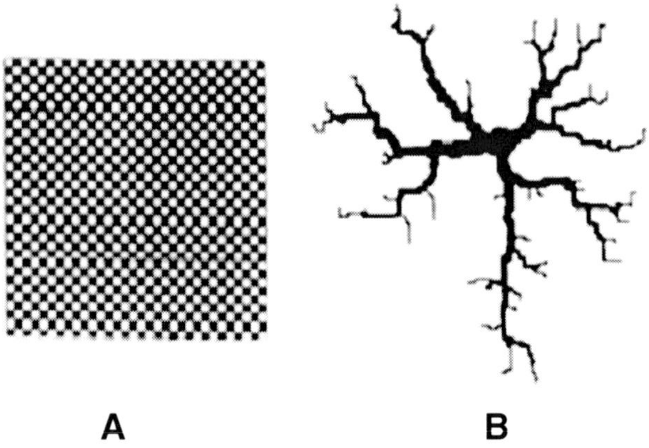

Figure 6.8: Geometric representation of dissolution under (**A**) homogeneous and (**B**) heterogeneous conditions at a given time t. Reprinted from [185] with permission from Springer.

> This structure is generated via the modified diffusion-limited aggregation (DLA) algorithm of [205] using the law $\rho = \alpha\,(m/N)^{\beta}$. Here, $N = 2{,}000$ (the number of particles of the DLA clusters), $\alpha = 10$ and $\beta = 0.5$ are constants that determine the shape of the cluster, ρ is the radius of the circle in which the cluster is embedded, $\rho_c = 0.1$ is the lower limit of ρ (always $\rho_c < \rho$), and m is the number of particles sticking to the downstream portion of the cluster. This example corresponds to a radial Hele–Shaw cell where water has been injected radially from the central hole. Due to heterogeneity a sample cannot be used to calculate the dissolved amount at any time, i.e., an average value for the percent dissolved amount at any time does not exist. This property is characteristic of fractal objects and processes.

According to van Damme [201], fractal fingering is in many respects a chaotic phenomenon because it exhibits a sensitive dependence on the initial conditions. Although this kind of performance for a dissolution system is currently unacceptable, it might mirror more realistically the erratic dissolution of drugs with very low extent of absorption.

6.4.3 Fractal-like Kinetics in Gastrointestinal Absorption

Derivation of the equations used in linear compartmental modeling relies on the hypothesis that absorption takes place from a homogeneous drug solution in the gastrointestinal fluids and proceeds uniformly throughout the gastrointestinal tract. Homogeneous gastrointestinal absorption is routinely described by the following equation [206]:

$$\dot{q_a}(t) = F_a q_0 k_a \exp(-k_a t),$$

where F_a is the fraction of dose (q_0) absorbed, and k_a is the first-order absorption rate constant. Nevertheless, the maximum initial absorption rate ($F_a q_0 k_a$) associated with the previous equation is not in accord with stochastic principles applied to the transport of drug molecules in the absorption process [206]. Theoretically, the absorption rate must be zero initially and increase to reach a maximum over a finite period of time. This type of time dependency for the input rate has been verified in deconvolution and maximum entropy studies of rapid-release dosage forms [206–208]. To overcome the discrepancies between the above equation and the actual input rates observed in deconvolution studies, investigators working in this field have utilized a cube-root-law input [209], polynomials [210], splines [208], and multiexponential [211] functions of time. In the same vein, but from a pharmacokinetic perspective, Higaki et al. [212] have considered models for time-dependent rate "constants" in oral absorption. Although these approaches [206, 208–212] are purely empirical, their capability in approximating the real input function indicates that power functions of time can be of value in describing the gastrointestinal drug absorption.

A more realistic approach to modeling drug absorption from the gastrointestinal tract should take into account the geometric constraints imposed by the heterogeneous structure and function of the medium. A diffusion process under such conditions is highly influenced, drastically changing its properties. For example, for a random walk in disordered media, the mean square displacement $\langle z^2(t) \rangle$ of the walker is given by (2.10):

$$\langle z^2(t) \rangle \propto t^{2/d_w},$$

where d_w is the random-walk dimension (cf. Section 2.2). The value of d_w is larger than 2, typically $d_w = 2.8$ (2 dimensions), and $d_w = 3.5$ (3 dimensions), so the overall exponent is smaller than 1. Furthermore, in understirred media, where reactions or processes take place in a low-dimensional space, the rate "constant" is in fact time-dependent at all times (cf. Section 2.5). Hence, the transit, dissolution, and uptake of drug under the heterogeneous gastrointestinal conditions can obey the principles of fractal kinetics [16, 213], where rate "constants" depend on time. For these heterogeneous processes, the time dependency of the rate coefficient k is expressed by

$$k = k_\circ t^\lambda,$$

where $k_\circ$ is a constant, while the exponent λ is different from zero and is the outcome of two different phenomena: the heterogeneity (geometric disorder of the medium) and the imperfect mixing (diffusion-limit) condition. Therefore, k depends on time since $\lambda \neq 0$ in inhomogeneous spaces while in 3-dimensional homogeneous spaces $\lambda = 0$ and therefore $k = k_\circ$, i.e., classical kinetics prevail and the rate constant does not depend on time. For "ideal" drugs having high solubility and permeability the homogeneous assumption ($\lambda = 0$, gastrointestinal absorption proceeds uniformly from a homogeneous solution) seems to be reasonable. In contrast, this assumption cannot be valid for the majority of drugs and in particular for drugs having low solubility and/or permeability. For these drugs a suitable way to model their gastrointestinal absorption kinetics under the inhomogeneous gastrointestinal conditions is to consider a time-dependent absorption rate coefficient k_a,

$$k_a = k_1 t^{\alpha},$$

and a time-dependent dissolution rate coefficient k_d,

$$k_d = k_2 t^{\beta}.$$

In reality, the exponents α and β determine how sensitive k_a and k_d are in temporal scale and the kinetic constants k_1 and k_2, determine whether the processes happen slowly or rapidly. The dimensions of k_1 and k_2 are time$^{-(1+\alpha)}$ and time$^{-(1+\beta)}$, respectively. Thus, the absorption rate $\dot{q_a}(t)$ is

$$\dot{q_a}(t) = k_a q_a(t) = k_1 t^{\alpha} q_a(t),$$

where $q_a(t)$ is the dissolved quantity of drug in the gastrointestinal tract. Since the change of $q_a(t)$ is the result of dissolution and uptake, which are both taking place under heterogeneous conditions ($\alpha \neq 0$ and/or $\beta \neq 0$), the previous equation exhibits a nonclassical time dependency for the input rate. Consequently, this equation provides a theoretical basis for the empirical power functions of time utilized in deconvolution studies [206, 208–211].

The values of the parameters α and β for drugs exhibiting heterogeneous absorption kinetics are inherently linked with the physicochemical properties of the drug, the formulation, the topology of the medium (gastrointestinal contents), and the initial distribution of drug particles in it [16]. It is worthy of mention that the initial conditions (the initial random distribution of the reactants: solid drug particles and gastrointestinal contents) are very important in fractal kinetics [16]. For all these reasons, population parameters for drugs having $\alpha \neq 0$ and/or $\beta \neq 0$ are unlikely since the topology of the medium and the initial conditions are by no means consistent or controlled, being dependent on subject and time of day. For the sake of completion, one should add that under homogeneous conditions ($\alpha = \beta = 0$) both k_a and k_d are independent of time and therefore classical kinetics can be applied.

6.4.4 The Fractal Nature of Absorption Processes

Relying on the above considerations one can argue that drugs can be classified with respect to their gastrointestinal absorption characteristics into two broad categories, i.e., homogeneous and heterogeneous. Homogeneous drugs have satisfactory solubility and permeability, and are dissolved and absorbed mostly prior to their arrival to the large intestine. It seems likely that the gastrointestinal absorption characteristics of the homogeneous group of drugs are adequately described or modeled with the homogeneous approach, i.e., well-stirred in vitro dissolution systems and classical absorption kinetics. In contrast, drugs with low solubility and permeability can be termed heterogeneous, since they traverse the entire gastrointestinal tract, and are most likely to exhibit heterogeneous transit, dissolution, and uptake and therefore heterogeneous absorption kinetics. In this context, the following remarks can be made for the heterogeneous drugs that exhibit limited bioavailability and high variability, and most of them can be classified in categories II and IV of the biopharmaceutics classification system [153] (cf. also Section 6.6.1):

- Mean or median values should not be given for the whole bowel transit since most of the dissolved and/or undissolved drug traverses the entire gastrointestinal tract. The complex nature of transit involving normal and dispersive transport [195] as well as periods of stasis would be better expressed by reporting the range of the experimental values.

- Dissolution testing with the officially used in vitro systems ensuring homogeneous stirring conditions, should be solely viewed as a quality control procedure and not as a surrogate for bioequivalence testing. According to the current view [153], limited or no in vitro–in vivo correlations are expected using conventional dissolution tests for the category IV drugs and the drugs of category II used in high doses. Since this unpredictability is routinely linked with our inability to adequately mimic the in vivo conditions, one should also consider whether the chaotic character of in vivo dissolution is a valid hypothesis for the failure of the in vitro tests. It is advisable, therefore, to perform physiologically designed dissolution experiments in topologically constrained media [201, 204, 205] for drugs of categories II and IV [153] in order to determine potential cutoffs for dose and solubility values as well as flow characteristics for drug classifications (homogeneous and heterogeneous drugs). Further, these cutoffs could be used for setting standards for in vitro drug dissolution methodologies of drugs classified as heterogeneous.

- A notion that routinely accompanies oral absorption studies is that the mathematical properties of the underlying processes have a Gaussian distribution where the moments, such as the mean and variance, have well-defined values. Relying on this notion, drugs and/or formulations are categorized as low or highly variable. Thus, any drug that generates an intraindividual coefficient of variation greater than 30% as measured by the residual coefficient of variation (from analysis of variance) is arbitrarily characterized as highly variable. The use of a statistical measure of dispersion for drug classification is based on the law of large numbers, which dictates that the sample means for peak blood concentration, $c_{\max}$, and the area under the blood time–concentration curve, AUC, converge to fixed values while the variances decrease to nonzero finite values as the number used in averaging is increased. The conventional assessment of bioequivalence relies on the analysis of variance to get an estimate for the intraindividual variability prior to the construction of the 90% confidence interval between 80 and 125% for AUC and $c_{\max}$. The basic premise of this approach is that errors are normally distributed around the estimated mean values and two one-sided t-tests can be performed. Although the validity of this assumption seems to be reasonable for drugs following classical kinetics, concern is arising for the parameters $c_{\max}$ and $t_{\max}$ (time corresponding to $c_{\max}$) when fractal-like kinetics govern absorption since for many fractal time-dependent processes [4, 195] the mean and the variance may not exist. Under heterogeneous conditions, both $c_{\max}$ and $t_{\max}$ will depend on α and β, and therefore mean values for these parameters cannot be justified when fractal kinetics are operating. Apparently, a significant portion of variability with the heterogeneous drugs can be mistaken as randomness and can be caused by the time dependency of the rate coefficients of the in vivo drug processes. These observations provide a plausible explanation for the high variability in $c_{\max}$ values and the erroneous results obtained in bioequivalence studies [214]. From the above it appears that is inappropriate to apply rigorous statistical tests in bioequivalence studies for heterogeneous drugs using parameter estimates for $c_{\max}$ and $t_{\max}$ that do not actually represent sample means. The suggested [215] comparison of the time–concentration curve profiles of test and reference products in bioequivalence studies seems to be in accord with the reservations pointed out regarding use of specific parameters for the assessment of the absorption rate.

6.4.5 Modeling Drug Transit in the Intestines

The small-intestinal transit flow is a fundamental process for all gastrointestinal absorption phenomena. However, the structure of the gastrointestinal tract is highly complex and it is practically impossible to explicitly write and solve the equations of motion for the drug flow. Instead, numerical computer simulation techniques that incorporate the heterogeneous features of the gastrointestinal wall structure and of the drug flow are used in this section to characterize the intestinal transit process in humans.

An algorithm is built from first principles, where the system structure is recreated and subsequently the drug flow is simulated via Monte Carlo techniques [216]. This technique, based on principles of statistical physics, generates a microscopic picture of the intestinal tube. The desired features of the complexity are built in, in a random fashion. During the calculation all such features are kept frozen in the computer memory (in the form of arrays), and are utilized accordingly. The principal characteristic of the method is that if a very large number of such units is built, then the average behavior of all these will approach the true system behavior.

Construction of the Heterogeneous Tube

The model is based on a cylinder whose length is several orders of magnitude larger than its radius. Thus, any entanglements that are present are ignored,

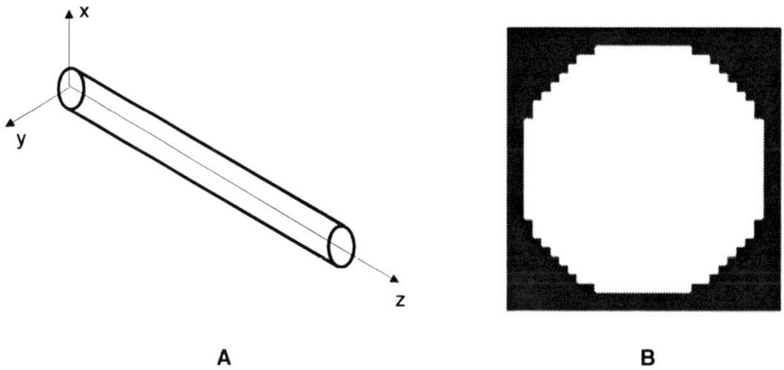

Figure 6.9: (**A**) The cylinder used for the tube construction. (**B**) Cross section of the tube. Reprinted from [216] with permission from Springer.

since they do not influence the dynamics of the phenomena. Initially, a 3-dimensional parallelepiped with a square cross section, of size $x : y : z$ equal to $31 : 31 : 3000$ is constructed, Figure 6.9 A. Inside it a cylinder with a radius of 14 units is built, a cross section of which appears in Figure 6.9 B. Hence, the quotient of [radius/length]$= R/L = 14/3000$ in the tube model is quite similar to the ratio of physiological data $1.3\,\text{cm}/3\,\text{m}$ for the human small intestine.

For convenience in the calculations, an underlying lattice of discrete spacing forming in effect a 3-dimensional grid is used. This grid covers the entire cylinder, while for all spatial considerations the grid sites are utilized. The interior of the cylinder has a finite concentration of villi attached to the cylinder wall, which have the property that they may absorb the dissolved drug particles flowing through the cylinder. The villi have the usual random dendritic structure, and they are formed by the DLA method [205]. The absorption of the drug particles in the model takes place when a flowing particle happens to have a position right next to the villi coordinates, implying that when a particle comes in contact with a villi structure it can be absorbed. The probability for absorption by the villi or walls is p_a. Since the present model focuses on the tube structure and the characteristics of flow, $p_a = 0$, while the case of $p_a \neq 0$ is treated in the following section.

The villi have a random dendritic-type structure, and they are formed initially by use of an algorithm based on the well-known DLA [205] model from solid-state physics. At random positions, $2z$ seed particles (z the cylinder length, Figure 6.9 A) are placed on the cylinder surface by positioning 2 particles on each z value. Following the DLA model, another particle, starting at a random point of each cross section, makes a 3-dimensional random walk (diffusion) inside the cylinder. The walk stops when the moving particle visits any of the neighbor sites of the original seed particles. At this point it stops and becomes

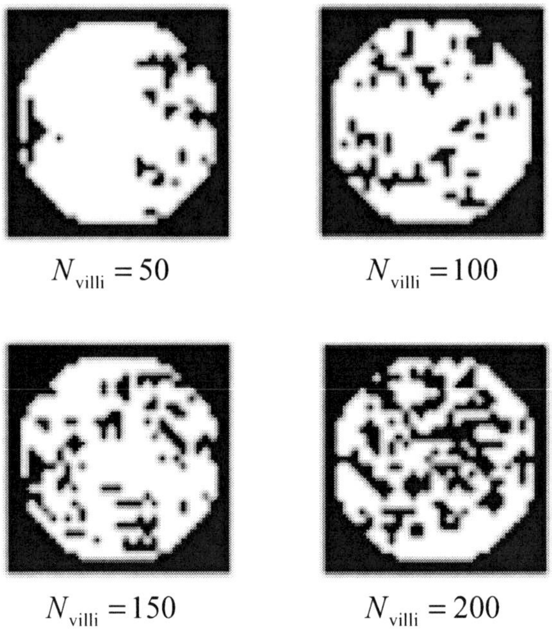

Figure 6.10: Cross sections of the tube at random positions for various concentrations of villi, $N_{\text{villi}} = 50, 100, 150, 200$. Reprinted from [216] with permission from Springer.

attached to the neighboring seed particle. The particle is constrained to move inside the cylinder. Then a second particle starts a random walk, until it meets either one of the seeds or the already *frozen* particle. The process continues and the internal structure of the tube, which can be of varying complexity, is built using a total of N_{villi} particles per unit length. The size of each villi cluster is limited to the value $1.5N_{\text{villi}}$. This is done in order to achieve a uniform distribution of villi cluster sizes. The higher the N_{villi} value, the more ramified is the ensuing structure. Some examples for various values of N_{villi} are shown in Figure 6.10. This figure shows typical 2-dimensional cross-sections of the cylinder, for four different N_{villi} values, $N_{\text{villi}} = 50, 100, 150$, and 200, at random places. It is clearly seen how the villi complexity is built up with increasing N_{villi}. Some squares appear not to be connected to any others in these pictures. In fact, these are indeed connected to adjacent (first neighbor) squares in the next or previous cross section of the tube (i.e., with $z' = z + 1$ or $z' = z - 1$), which are not shown in Figure 6.10.

Dynamics

The dynamics of the system are also followed utilizing the Monte Carlo technique. This includes motion of the particles through the tube, dissolution in the solvent flow, and absorption by the villi or the tube walls. Time is incremented by arbitrary time units, the MCS, which is the time it takes for a particle to move to one of its neighbor positions. A "tablet" can be inserted in one end of the tube (input end) at predefined time increments expressed in MCS. The "tablet" is modeled as an aggregate of drug particles of mass $q_0 = 100$. This means that one "tablet" can later be broken down successively into 100 units, which represent the solid drug particles. These can be further dissolved in the encompassing solution. But as long as the "tablet" has a mass larger than one it cannot be dissolved in the solution. All diffusing species (dissolved and undissolved) flow through the cylinder from the input end toward the direction of the other end (output end). This is accomplished by using a diffusion model of a biased random walk that simulates the fluid flow.

A simple random walk is the prototype model of the regular Brownian motion. Such a model is modified here, by including a bias factor, which makes the motion ballistic rather than simply stochastic. This bias factor, ε, increases the probability for motion in the z-direction, i.e., toward the output end, as compared to the probabilities in all other directions. This makes the flow of the particles and the dissolved drug molecules possible. If $\varepsilon = 0$, there is a motion but it is rather stationary and in all possible directions. If $\varepsilon > 0$, this makes the flow possible. The rate of the flow is also directly affected by the numerical value of ε, with increasing ε values resulting in increasing flow rates. With this statistical model the diffusing species can momentarily go against the flow, or sideways. This is a realistic feature, but it occurs with reduced probability.

Two different models of the biased random walk were envisaged. In model I the three directions of space, x, y, and z, are all equally probable, but in the z direction, the probability toward the output end (z_+) is now $(1/z) + \varepsilon$, while the corresponding probability toward the input end (z_-) is $(1/z) - \varepsilon$ (where z is the coordination number of the underlying space, e.g., $z = 6$ in a 3-dimensional space). This model has the characteristic that diffusion is equally probable in all possible directions, the species spending equal times in all of them, but due to the ε factor, when the z direction is chosen a positive flow drives the solution to the output end.

In a second model II, more emphasis is given to the motion toward the output and less to the other directions. The probabilities for motion in the different directions are now defined differently. While in the simple random walk the probability for motion in a specific direction is $1/z$, here the probability for motion in the output direction is $(1/z) + \varepsilon$, while the probability in any of the other five directions is

$$\frac{1 - \left(\frac{1}{z} + \varepsilon\right)}{z - 1}.$$

Thus, the values that ε can take are in the range

$$0 < \varepsilon < 1 - \frac{1}{z},$$

while the overall forward probability p_f, i.e., the probability toward the output end, is in the range

$$\frac{1}{z} < p_f < 1.$$

At each time step there is a probability p_d for the "tablet" to dissolve, i.e., $0 < p_d < 1$. In the Monte Carlo method the "tablet" is tested at every step to determine whether a fragment (one new particle) is to be released. When this happens a fragment of the "tablet" with mass $\psi = 1$ breaks off, and gets separated from the larger mass. It is understood that this $\psi = 1$ particle is immediately dissolved, and it is never reattached to the original mass. This dissolved particle now performs a random walk of its own, with the same characteristics (bias) as the main "tablet." The mass q_0 of the "tablet" is then reduced by ψ. The virtual experiment of the flow starts when a large number of drug particles (e.g., $10,000$) with mass $\psi = 1$ are inserted simultaneously at time $t = 0$ in the tube and are allowed to diffuse. To concentrate on the transit process exclusively, dissolution is considered instantaneous and p_d is set equal to 1, while absorption is not allowed by setting $p_a = 0$. When the fragments of the "tablet" reach the end of the tube, they are discarded. At the end of the simulation time the mass that has exited from the end of the tube is computed. The mean transit time is also computed by keeping track of the time it took for the particles to reach the end of the tube.

When the diffusing species come in contact with a closed site (such as the villi sites of the model) they have two options. In the first option, the particle does not "feel" the presence of the closed site, and it may attempt, unsuccessfully, to go to it. This model is called the *blind ant model*. In the second model, the particle feels the presence of the closed site, and thus it never attempts to land on it. This is called the *myopic ant model*. The difference between these two models is that the blind ant consumes long times in unsuccessful attempts, and thus its motion is slower than the myopic ant case.

Simulated vs. Experimental Data

The details of the flow of particles in the heterogeneous tube were studied using a model II biased random walk. In Figure 6.11, the mean transit time of the drug particles vs. the forward probability p_f (i.e., the probability toward the output along the z-axis) is plotted for various villi concentrations, for the two cases of the blind ant (part A), and the myopic ant (part B). For no villi structures, $N_{\text{villi}} = 0$, and for $N_{\text{villi}} = 50$ we observe that for larger p_f values the transit times of the particles were shorter, as one would expect. For larger villi concentrations the transit time became longer as p_f was increased. This behavior may seem inconsistent, but can be easily explained if we consider that when a drug fragment meets an obstacle (villi) then its forward motion is hampered,

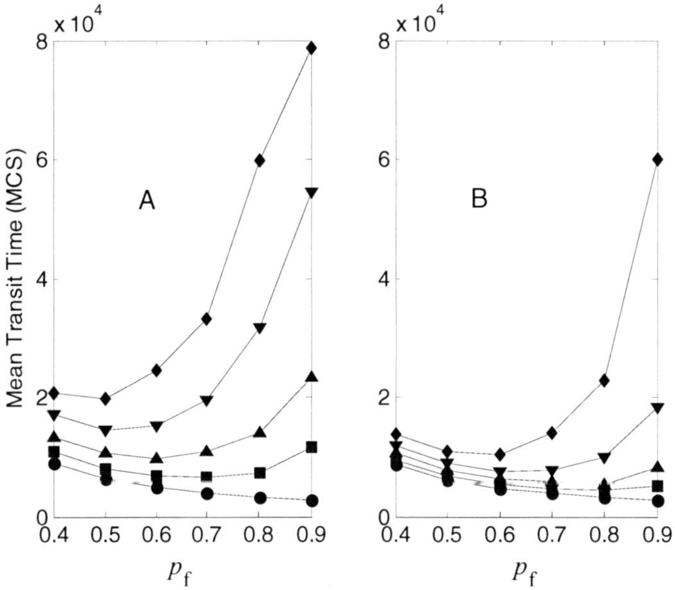

Figure 6.11: Mean transit times vs. the forward probability for various concentrations of villi, (**A**) blind ant model; (**B**) myopic ant model. Key (N_{villi}): ● 0; ■ 50; ▲ 100; ▼ 150; ♦ 200.

and it must move in the x or y direction (sideways) in order to circumvent it and continue moving toward the end of the tube. What happens is that when p_f values are large, then the probability for movement along the x- or y-axis is reduced. This does not give the particle the freedom to easily pass the obstacle, so it wastes time trying to move in the z direction. This explains the rise in the transit times, which is larger for larger villi concentrations. This qualitative picture is valid for both models in parts (A) and (B) of Figure 6.11. Plausibly, in comparing the two figures, the transit times are always longer in the blind ant case, for any villi concentration. The system behavior as shown in Figure 6.11 implies that the interplay of these two factors, namely the villi structure and the bias probability (flow rate), is important in determining the dynamics of the flow.

The frequency of transit times that result from the simulations for various values of villi and forward probability p_f are also compared to experimental data [173]. Model I consistently produces narrower frequencies than do model II and the experiments. This is because in model I, motion in the preferred z direction occurs with the same frequency as motion in the other directions. The effect of the flow along the tube length is downplayed, as opposed to the other model (II), in which it is emphasized. In Figure 6.12 the results for model I

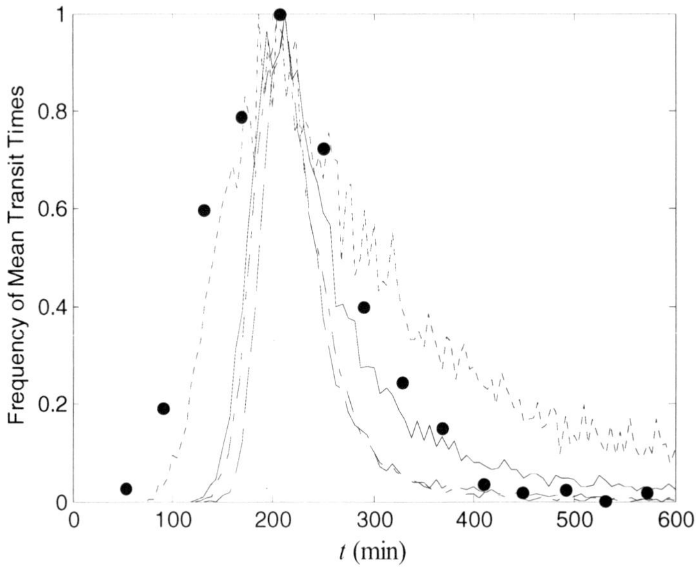

Figure 6.12: Frequency of mean transit times vs. time (min) using the diffusion model II for the blind ant model positions for various concentrations of villi and forward probabilities p_f values. Key: • experimental data; *solid line*, $N_{\text{villi}} = 200$ and $p_f = 0.6$; *dashed line*, $N_{\text{villi}} = 200$ and $p_f = 0.5$; *dotted line*, $N_{\text{villi}} = 180$ and $p_f = 0.7$; *dashed-dotted line*, $N_{\text{villi}} = 180$ and $p_f = 0.5$.

of the biased diffusion, together with the experimental data are presented. A wide range of variation for the two parameters, i.e., the bias factor ε and the villi concentration N_{villi}, was used, and the best resemblance between simulation and experimental data was achieved for the values of $N_{\text{villi}} = 190$ and forward probability $p_f = 0.65$, Figure 6.12. The x-axis here is in units of minutes. This is done by establishing a correspondence of $1\,\text{s} = 1.5$ MCS, since this is the value that produces the best possible fit.

Overall, the biased random walk, which places more emphasis on the motion toward the output end and less on the other directions, mimics more closely the transit profile of the experimental data. Both diffusion models, i.e., the blind and the myopic ant models, can reproduce the basic features of the real small-intestinal transit profile.

6.4.6 Probabilistic Model for Drug Absorption

The probabilistic absorption model described herein [217] was based on the cylinder built in [216] that incorporates all the random heterogeneities that make up the gastrointestinal tube. The optimal heterogeneous characteristics found in [216] were assigned to the number of villi and the type of the biased random walk. Thus, the parameter number of villi N_{villi} was set equal to 190, while the blind ant model for the biased random walk with forward probability $p_f = 0.65$ was used to simulate the motion of the dissolved and undissolved drug species. The dissolved species are tagged and continue the random walk and can be absorbed by the cylinder wall structure, or exit the tube if they reach its end. The quantities input and exiting through the tube, their transit time, and the fraction of the species absorbed and dissolved during the flow are monitored.

Simulation of Dissolution and Uptake Processes

A "tablet," which is modeled as an aggregate of drug particles of mass q_0, is inserted in one end of the tube (input end). At each time step a portion of the mass of the "tablet" can be dissolved. The rate of dissolution is considered to be dependent on three factors, which are all expressed in probability values.

1. The first factor, k_d, mimics the conventional dissolution rate constant; it is inherent for every drug and takes values in the range $0 < k_d < 1$. A value close to unity denotes a drug with rapid dissolution characteristics. Thus, a specific k_d value is conceived for a given drug under certain experimental conditions. As a probability value, k_d corresponds to p_d and it expresses the number of events occurring in a time unit. Consequently, k_d has dimension of time^{-1}.

2. The second factor, k_c, is related to the first-order concentration dependence of the dissolution rate. As dissolution proceeds the amount of drug in solution increases exponentially and therefore the value of k_c is reduced exponentially. This reduction is controlled by the relative amount dissolved, $q(t)/q_s$, as defined in Section 5.1.4, at each time point:

$$k_c \to k_c(t) = \exp\left[-\ln(10)\, q(t)/q_s\right],$$

   where $q(t)$ is the mass of the dissolved drug at any moment during the simulation and q_s is the dissolved mass at saturation. q_s is computed by multiplying the minimum physiologic solubility $c_{s,\min}$ of the drug by the luminal volume, which is assumed to be 250 ml. The $\ln(10)$ factor was chosen so that the magnitude of k_c, when the dissolved mass was equal to the dissolved mass at saturation, should arbitrarily be one-tenth of the value of k_c when the dissolved mass is equal to zero. Thus, k_c is reduced exponentially as dissolution proceeds. Of course, at saturation $(q(t) = q_s)$ no more material is allowed to dissolve.

3. The third factor, k_s, depends on the surface area of the drug particles. It is known that the reduction of the surface area is related nonlinearly to the reduction of mass as dissolution proceeds. Since the nonlinear

relationship between the undissolved mass, $q_0 - q(t)$, and surface area is dependent on the geometric characteristics of the drug particles, the value of k_s is considered to decrease proportionally to $\exp\{[q_0 - q(t)]/q_0\} = \exp[1 - \varphi(t)]$ in order to avoid any shape assumptions. Therefore, k_s is not computed directly in the simulation, but is calculated from the undissolved drug mass at any moment during the simulation. The exact equation that gives k_s is

$$k_s \rightarrow k_s(t) = 0.01 \exp\{4.5[1 - \varphi(t)]\}.$$

The constants in the last equation are chosen so that k_s arbitrarily equals 0.9 when $q(t)$ is close to zero and $k_s = 0.01$ when $q(t)$ equals q_0. In essence, the probability factor k_s is related to the diminution of the surface area of drug particles during the dissolution process.

The quantities k_c and k_s in the last two equations result from a calculation of an exponential, and thus have no physical dimensions. The effective dissolution probability rate "constant" $k_{d,eff}$ is calculated by multiplying the above three factors, so that $k_{d,eff} = k_d k_c k_s$. Thus, $k_{d,eff}$ has dimension of time^{-1} and denotes the fraction of the total number of drug particles that can be dissolved per MCS. The mass of the "tablet" that will break off at any moment is given by multiplying the value of $k_{d,eff}$ by the undissolved mass of the tablet. If $q_d(t)$ is this mass, then $q_d(t) = [q_0 - q(t)] k_{d,eff}$ and $q_d(t)/\psi$ particles of the "tablet" with mass ψ will break off, and will get separated from the larger mass. The dissolved particles now flow on their own, with the same characteristics (forward probability) as the undissolved particles. The mass $q_0 - q(t)$ of the undissolved drug is then reduced by $q_d(t)$.

Dissolved particles are tagged in the calculation at all times, so their location relative to all other particles and the tube walls is known. When one of the dissolved particles comes "in contact" (when it is in a lattice site adjacent to villi or tube wall) with the tube walls or the villi there is a probability k'_a that it will be absorbed. It is obvious that the higher the value of k'_a, the higher the probability of a dissolved particle of being absorbed. This proportionality implies that only passive mechanisms are considered. If a dissolved particle is absorbed it is immediately removed from the system. If it is not absorbed, it remains on its site and continues the flow. When a dissolved or undissolved particle reaches the end of the tube, it is discarded.

At the end of the simulation time, the mass that was absorbed and the mass that has exited from the end of the tube can be computed. Further, the dimensionless absorption number A_n can be computed [153] from

$$A_n = \frac{1}{2} \langle T_{si} \rangle k_a$$

using (6.6) and (6.14). In this relation $\langle T_{si} \rangle$ is equal to 24,500 MCS, i.e., the mean intestinal transit time found in [216]. It must be noted that k_a as it appears above is not identical to the one used as a parameter in the simulation. While they both describe probabilities, k_a is a first-order macroscopic rate constant expressed in dimension of time^{-1}, while the k'_a in the simulations describes the microscopic probabilistic events of the simulation model.

Absorption of Freely Soluble Drugs

The absorption of freely soluble drugs having various values of k'_a was studied. Initially, the relationship between the simulated k'_a values and the corresponding conventional k_a values, which are computed from the simulation assuming first-order absorption, was explored. An amount of instantly dissolved mass of $q_0 = 20,000$ was inserted in the input end of the tube and both profiles of the fraction of the mass that was absorbed and exited the tube were recorded. To find out the relationship between k'_a and k_a, the following exponential equation was used to fit the simulated data of the fraction of dose absorbed F_a vs. time:

$$F_a = 1 - \exp\left(-k_a t\right),$$

where the fitting parameter is k_a in MCS^{-1} units, and time t is also expressed in MCS. Focusing on k'_a values, which ensure that most of the drug is absorbed and does not exit the tube, the following relation between k'_a and k_a was found:

$$k_a = 0.885 k'_a.$$

This relationship shows the proportionality between the first-order macroscopic rate constant k_a and the k'_a that describes the microscopic probabilistic events (the "successful" visits of the dissolved species to the villi). Similar simulations for instantly dissolved $20,000$ drug particles were carried out using various values of k'_a, and the fraction of the drug dose absorbed, F_a, at $24,500$ MCS was calculated. The k'_a values were then translated to MCS^{-1} values using the last equation, and the absorption number A_n was computed as delineated above. The fraction of the dose that was absorbed vs. the absorption number A_n is shown in Figure 6.13. The symbols represent the experimental data of various drugs [55], while the line gives the simulation results obtained from the model by adjusting the intestinal transit time to $24,500$ MCS. From the different intestinal transit times evaluated it was found that $24,500$ MCS gave the best description of the experimental data. Using the correspondence between MCS and real time units [216], the $24,500$ MCS are $16,333$ s or 4.5 h. The duration of 4.5 h is physiologically sound as an effective intestinal transit time to study gastrointestinal drug absorption in the model.

Absorption of Sparingly Soluble Drugs

The model was also applied to the study of low-solubility drugs. Numerical results of the system of differential equations reported in [55] were compared to the simulations based on the heterogeneous tube. In the simulations the z^* variable is computed using the mean transit time of the particles, $\langle T_{si} \rangle = 24,500$ MCS, and $z^* = t / \langle T_{si} \rangle$, expressing both t and $\langle T_{si} \rangle$ in MCS. The "tablet" was

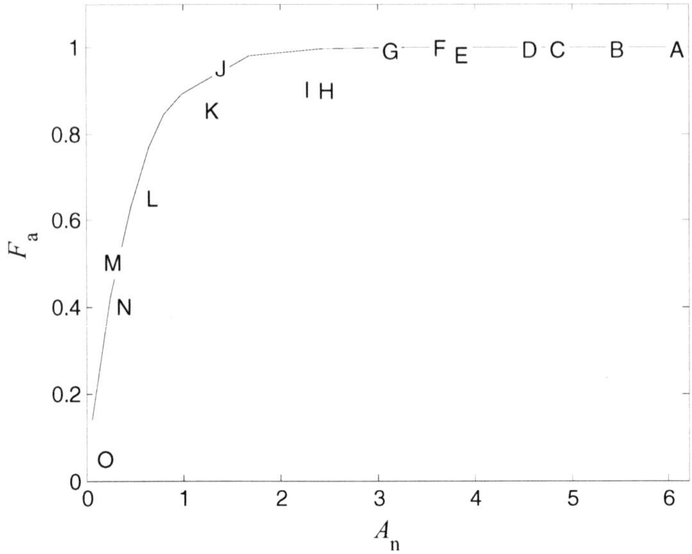

Figure 6.13: Fraction of dose absorbed vs. A_n. The solid line represents results for 24,500 MCS and the points the experimental data. Key: A D-glucose; B ketoprofen; C naproxen; D antipyrine; E piroxicam; F L-leucine; G phenylalanine; H beserazide; I L-dopa; J propranolol; K metoprolol; L terbutaline; M furosemide; N atenolol; O enalaprilate.

inserted in the tube entrance as a bolus of a given weight q_0 (e.g., 200 or 500 mg) and it was arbitrarily set that the bolus may break up eventually into a large number of particles, each weighing 0.01 mg. Thus, each "tablet" of mass q_0 can be finally broken down to $q_0/0.01$ particles. The values of k_c and k_s were continuously computed during the simulation-fitting procedure. Various values of the parameter k_d were used to get a good matching of the simulation and the theoretical curves obtained from the solution of equations [55] for the normalized concentration profile in the tube.

Finally, a 3-dimensional plot of the fraction of dose absorbed F_a at 24,500 MCS for various values of the parameters k_a' and k_d is shown in Figure 6.14 using values for dose and $c_{s,\min}$ corresponding to those of digoxin and griseofulvin. The plots of Figure 6.14 are indicative of the effect of dose on the fraction of dose absorbed for sparingly soluble drugs. For example, for a highly permeable drug ($k_a' \approx 0.5$) given in a large dose (500 mg) and having the dissolution characteristics of griseofulvin, $\approx 25\%$ of the administered dose will be absorbed according to Figure 6.14 B. In contrast, a drug like digoxin, which exhibits the same permeability and dissolution characteristics as griseofulvin, given at a low

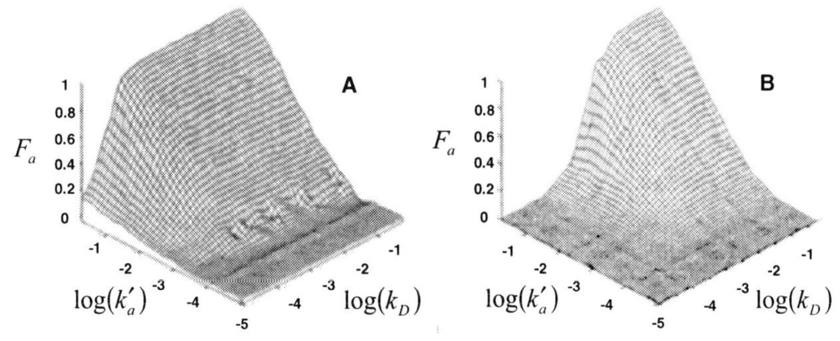

Figure 6.14: Three-dimensional graph of fraction dose absorbed vs. k'_a and k_d. Dose and $c_{s,\min}$ values [153] correspond to those of digoxin (**A**) and griseofulvin (**B**).

dose (0.5 mg) will be almost completely absorbed, Figure 6.14 A.

6.5 Absorption Models Based on Structure

The ability to predict the fraction of dose absorbed F_a and/or bioavailability is a primary goal in the design, optimization, and selection of potential candidates in the development of oral drugs. Although new and effective experimental techniques have resulted in a vast increase in the number of pharmacologically interesting compounds, the number of new drugs undergoing clinical trial has not increased at the same pace. This has been attributed in part to the poor absorption of the compounds. Thus, computer-based models based on calculated molecular descriptors have been developed to predict the extent of absorption from chemical structure in order to facilitate the lead optimization in the drug discovery process. Basically, the physicochemical descriptors of drug molecules can be useful for predicting absorption for passively absorbed drugs. Since dissolution is the rate-limiting step for sparingly soluble drugs, while permeability becomes rate-controlling if the drug is polar, computer-based models are based on molecular descriptors related to the important drug properties solubility and permeability across the intestinal epithelium.

A rapid popular screen for compounds likely to be poorly absorbed is Lipinski's [218] "rule of 5," which states that poor absorption of a compound is more likely when its structure is characterized by:

- molecular weight > 500,
- $\log P > 5$,
- more than 5 H-bond donors expressed as the sum of NHs and OHs, and
- more than 10 H-bond acceptors expressed as the sum of Ns and Os.

However, compounds that are substrates for biological transporters are exceptions to the rule. Based on the analysis of 2, 200 compounds in the World Drug Index that survived Phase I testing and were scheduled for Phase II evaluation, Lipinski's "rule of 5" revealed that less than 10% of the compounds showed a combination of any two of the four parameters outside the desirable range. Accordingly, the "rule of 5" is currently implemented in the form "if two parameters are out of range, a poor absorption is possible." However, compounds that pass this test do not necessarily show acceptable absorption.

Although various computational approaches for the prediction of intestinal drug permeability and solubility have been reported [219], recent computer-based absorption models utilize a large number of topological, electronic, and geometric descriptors in an effort to take both aqueous drug solubility and permeability into account. Thus, descriptors of "partitioned total surface areas" [168], Abraham molecular descriptors [220,221], and a variety of structural descriptors in combination with neural networks [222] have been shown to be determinants of oral drug absorption.

Overall, the development of a robust predictor of the extent of absorption requires a careful screening of a large number of drugs that undergo passive transport to construct well populated training and external validation test sets. The involvement in the data sets of compounds with paracellular, active transport, carrier-mediated transport mechanisms, or removal via efflux transporters can complicate the problem of in silico prediction of the extent of absorption. Another problem arises from the fact that published drug data for F_a or bioavailability are skewed toward high values (≈ 1), while the compounds in the training and external validation data sets should evenly distributed across the complete range of oral absorption.

6.6 Regulatory Aspects

Over the past fifteen years the advances described in the previous sections of this chapter have enhanced our understanding of the role of:

- the physicochemical drug properties,
- the physiological variables, and
- the formulation factors in oral drug absorption.

As a result, the way in which regulatory agencies are viewing bioavailability and bioequivalence issues has undergone change. In this section, we discuss the scientific basis of the regulatory aspects of oral drug absorption

6.6.1 Biopharmaceutics Classification of Drugs

As mentioned in Section 6.1.2, the first attempts to quantitatively correlate the physicochemical properties of drugs with the fraction of dose absorbed were

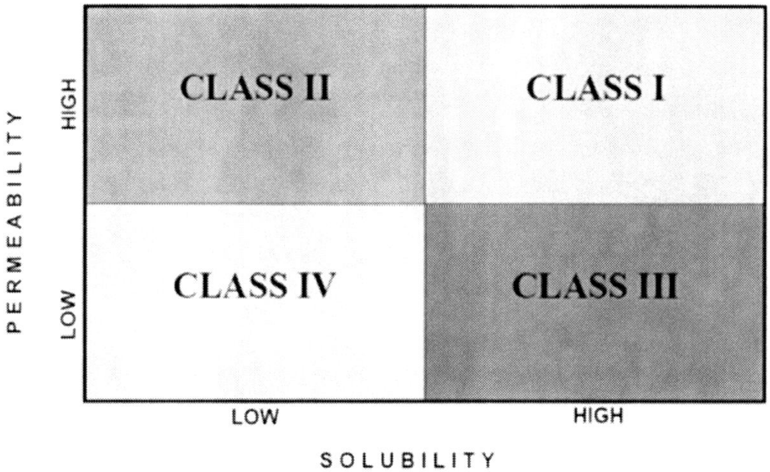

Figure 6.15: The Biopharmaceutics Classification System (BCS).

based on the absorption potential concept in the late 1980s [156, 158]. The elegant analysis of drug absorption by Amidon's group in 1993 based on a microscopic model [55] using mass balance approaches enabled Amidon and his colleagues [153] to introduce a Biopharmaceutics Classification System (BCS) in 1995. According to BCS a substance is classified on the basis of its aqueous solubility and intestinal permeability, and four drug classes were defined as shown in Figure 6.15. The properties of drug substance were also combined with the dissolution characteristics of the drug product, and predictions with regard to the in vitro–in vivo correlations for each of the drug classes were pointed out.

This important achievement affected many industrial, regulatory, and scientific aspects of drug development and research. In this context, the FDA guidance [223] on BCS issued in 2000 provides regulatory benefit for highly permeable drugs that are formulated in rapidly dissolving solid immediate release formulations. The guidance [223] defines a substance to be highly permeable when the extent of absorption in humans is 90% or more based on determination of the mass balance or in comparison to an intravenous reference dose. In parallel, the guidance [223] classifies a substance to be highly soluble when the highest dose strength is soluble in 250 ml or less of aqueous media over the pH range 1–7.5, while a drug product is defined as rapidly dissolving when no less than 85% of the dose dissolves in 30 min using USP Apparatus 1 at 100 rpm in a volume of 900 ml in 0.1N HCl, as well as in pH 4.5 and pH 6.8 buffers.

It has been argued [224] that the use of a single solubility value in the original BCS article [153], Figure 6.15, for solubility classification is inadequate since

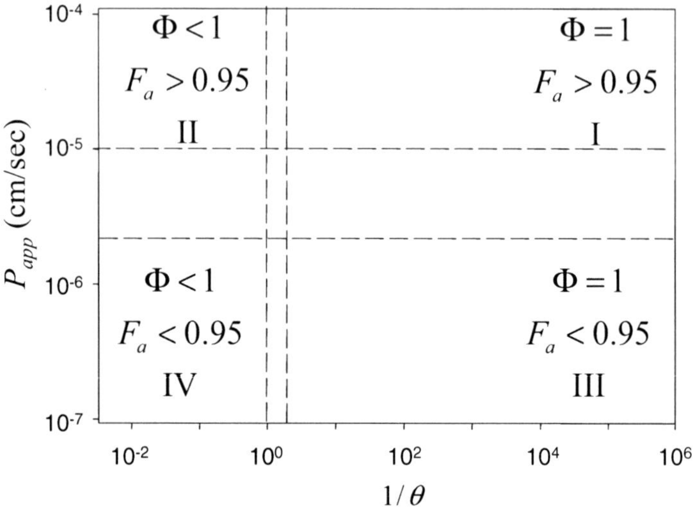

Figure 6.16: The Quantitative Biopharmaceutics Classification System (QBCS) utilizes specific cutoff points for drug classification in the solubility–dose ratio $(1/\theta)$, apparent permeability (P_{app}) plane. Each class of the QBCS can be characterized on the basis of the anticipated values for the fraction of dose absorbed, F_a and the fraction of dose dissolved, Φ at the end of the dissolution process assuming no interplay between dissolution and uptake. In essence the classification system is static in nature.

drugs are administered in various doses. Moreover, solubility is a static equilibrium parameter and cannot describe the dynamic character of the dissolution process. Both aspects are treated in the guidance on biowaivers [223]; solubility is related to dose, while dissolution criteria are specified. However, the reference of the FDA guidance exclusively to "the highest dose strength" for the definition of highly soluble drugs implies that a drug is always classified in only one class regardless of possible variance in performance with respect to solubility of smaller doses used in actual practice. This is not in accord with the dose dependency (non-Michaelian type) of oral drug absorption, which consistently has been demonstrated in early [156, 158] and recent studies [160, 161] related to the absorption potential concept and its variants as well as in the dynamic absorption models [55, 180, 181]. Moreover, the dissolution criteria of the FDA guidance [223], which unavoidably refer to a percentage of dose dissolved within a specific time interval:

- are not used as primary determinants of drug classification,
- have been characterized as conservative [225],
- have had pointed out suggestions for broadening them [226], and

- suffer from a lack of any scientific rationale.

In parallel, the current dissolution specifications [223] are not correlated with the drug's dimensionless solubility–dose ratio $1/\theta$, which has been shown [90] to control both the extent of dissolution as well as the mean dissolution time, MDT, which is a global kinetic parameter of drug dissolution.

The latter finding prompted the development of the Quantitative Biopharmaceutics Classification System (QBCS) [224] in which specific cutoff points are used for drug classification in the solubility–dose ratio permeability plane, Figure 6.16. Unity was chosen as the critical parameter for the dimensionless solubility–dose ratio axis because of the clear distinction between the two cases of complete dissolution (when $(1/\theta) \geq 1$) and incomplete dissolution (when $(1/\theta) < 1$) [90]. To account for variability related to the volume content, a boundary region of 250 to 500 ml was assumed and thus a boundary region for $1/\theta$ was set from 1 to 2. The boundary region of highly permeable drugs, P_{app} values in the range 2×10^{-6}–10^{-5} cm s^{-1} on the y-axis of Figure 6.16, can ensure complete absorption. It was based on experimental results [168–170], which indicate that drug absorption in Caco-2 monolayers can model drug transport in vivo.

In full analogy with BCS [153], the QBCS [224] classifies drugs into four categories based on their permeability (P_{app}) and solubility–dose ratio $1/\theta$ values defining appropriate cutoff points. For category I (high P_{app}, high $1/\theta$), complete absorption is anticipated, whereas categories II (high P_{app}, low $1/\theta$) and III (low P_{app}, high $1/\theta$) exhibit dose–solubility ratio- and permeability-limited absorption, respectively. For category IV (low P_{app}, low $1/\theta$), both permeability and solubility–dose ratio are controlling drug absorption. A set of 42 drugs was classified into the four categories of QBCS [224] and the predictions of their intestinal drug absorption were in accord with the experimental observations, Figure 6.17. However, some of the drugs classified in category II of the QBCS (or equivalently Class II of the BCS) exhibit a greater extent of absorption than the theoretically anticipated value based on a relevant semiquantitative analysis of drug absorption [224].

6.6.2 The Problem with the Biowaivers

According to the FDA guidance [223], petitioners may request biowaivers for high solubility-high permeability substances (Class I of BCS) formulated in immediate release dosage forms that exhibit rapid in vitro dissolution as specified above. The scientific aspects of the guidance as well as issues related to the extension of biowaivers using the guidance have been the subjects of extensive discussion [225, 226]. Furthermore, Yazdanian et al. [227] suggested that the high solubility definition of the FDA guidance on BCS is too strict for acidic drugs. Their recommendation was based on the fact that several nonsteroidal anti-inflammatory drugs (NSAID) exhibit extensive absorption and, according to the current definition of the FDA guidance, are classified in Class II (low soluble–high permeable) of the BCS. An important concluding remark of this

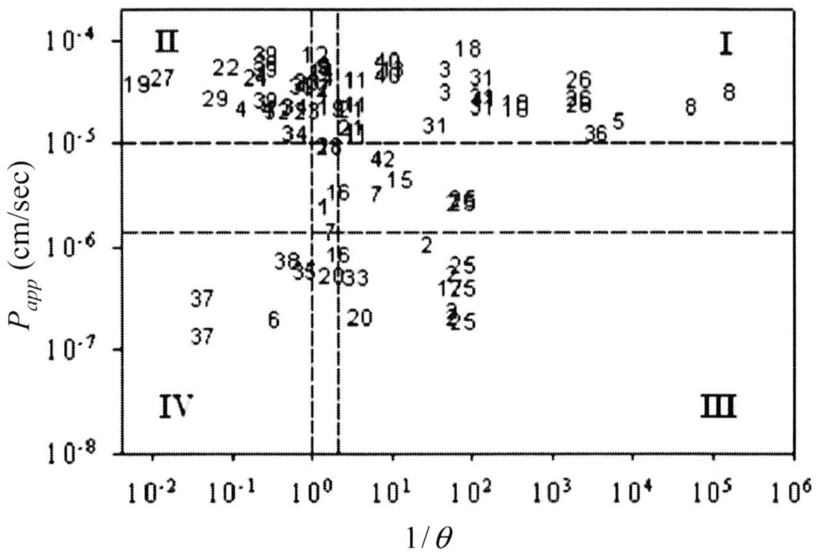

Figure 6.17: The classification of 42 drugs in the (solubility-dose ratio, apparent permeability) plane of the QBCS. The intersection of the dashed lines drawn at the cutoff points form the region of the borderline drugs. Key: 1 acetyl salicylic acid; 2 atenolol; 3 caffeine; 4 carbamazepine; 5 chlorpheniramine; 6 chlorothiazide; 7 cimetidine; 8 clonidine; 9 corticosterone; 10 desipramine; 11 dexamethasone; 12 diazepam; 13 digoxin; 14 diltiazem; 15 disopyramide; 16 furosemide; 17 gancidovir; 18 glycine; 19 grizeofulvin; 20 hydrochlorothiazide; 21 hydrocortisone; 22 ibuprofen; 23 indomethacine; 24 ketoprofen; 25 mannitol; 26 metoprolol; 27 naproxen; 28 panadiplon; 29 phenytoin; 30 piroxicam; 31 propanolol; 32 quinidine; 33 ranitidine; 34 salicylic acid; 35 saquinavir; 36 scopolamine; 37 sulfasalazine; 38 sulpiride; 39 testosterone; 40 theophylline; 41 verapamil HCl; 42 zidovudine.

study [227] is, "an inherent limitation in the solubility classification is that it relies on equilibrium solubility determination, which is static and does not take into account the dynamic nature of absorption." Moreover, the measurement of intrinsic dissolution rates [228] or the use of dissolution–absorption in vitro systems [229] appears more relevant than solubility to the in vivo drug dissolution dynamics for regulatory classification purposes. Also, the development of QBCS [224] is based on the key role of the solubility–dose ratio for solubility classification, since it is inextricably linked to the dynamic characteristics of the dissolution process [90]. All these observations point to the need for involvement of the dynamics of dissolution and uptake processes for the regulatory aspects of biopharmaceutical drug classification.

Recently, this type of analysis was attempted [230] for several nonsteroidal

Table 6.1: Dose and human bioavailability data of NSAIDs [227].

n°	Drug	Highest Dose (mg)	Bioavailability (%)
1	Diclofenac	50	54
2	Etodolac	400	> 80
3	Indomethacin	50	98
4	Ketorolac	20	100
5	Sulindac	200	88
6	Tolmetin	600	> 90
7	Fenoprofen	600	85
8	Flurbiprofen	100	92
9	Ibuprofen	800	> 80
10	Ketoprofen	75	100
11	Naproxen	500	99
12	Oxaprozin	600	95–100
13	Mefenamic acid	250	Rapidly absorbed
14	Acetylsalicylic acid	975	68 (unchanged drug)
15	Diflunisal	500	90
16	Salicylic acid	750	100
17	Meloxicam	15	89
18	Piroxicam	20	Rapidly absorbed
19	Celecoxib	200	-
20	Rofecoxib	25	93

anti-inflammatory drugs listed in Table 6.1, which are currently classified as Class II drugs. The dynamics of the two consecutive drug processes, dissolution and wall permeation, were considered in the time domain of the physiologic transit time using a tube model that considers constant permeability along the intestines, a plug flow fluid with the suspended particles moving with the fluid, and dissolution in the small-particle limit. The radius of the spherical drug particles, ρ, and the concentration of dissolved drug in the intestinal tract, $c(z)$, are modeled as suggested by Oh et al. [55] for the development of BCS [153] by a system of differential equations, with independent variable the axial intestinal distance z, which is considered to be proportional to time, since the fluid flow rate is constant:

$$\begin{aligned} \frac{\mathrm{d}\rho(z)}{\mathrm{d}z} &= -\frac{\mathcal{D}\pi R^2}{Q\varrho}\frac{c_s - c(z)}{\rho(z)}, & \rho(0) &= \rho_0, \\ \frac{\mathrm{d}c(z)}{\mathrm{d}z} &= \frac{\mathcal{D}(n/V)4\pi^2 R^2}{Q}\rho(z)\left[c_s - c(z)\right] - \frac{2P_{eff}\pi R}{Q}c(z), & c(0) &= 0, \end{aligned}$$

where $\mathcal{D}$ is the diffusion coefficient of the drug, ϱ is the density of the solid drug, R is the radius of the intestinal lumen, c_s is the solubility of the drug, Q is the volumetric flow rate, n is the number of drug particles in the dose, V is the luminal volume and P_{eff} is the effective permeability of the drug.

These equations can be rewritten with respect to time if one multiplies both sides by $L/MITT$ (where L is the length of the tube and $MITT$ is the mean intestinal transit time) and simplifies:

$$\begin{aligned} \dot{\rho}(t) &= -\frac{\mathcal{D}}{\varrho}\frac{c_s - c(t)}{\rho(t)}, \\ \dot{c}(t) &= \frac{3\mathcal{D}}{\varrho V}\frac{q_0}{\rho_0^3}\rho(t)\left[c_s - c(t)\right] - \frac{2P_{eff}}{R}c(t), \end{aligned}$$

where q_0 is the dose and ρ_0 is the initial radius of the drug particles.

Both sides of the last two equations are divided by q_0/V, and $c(t)$ and c_s are substituted by the fraction $\varphi(t)$ of dose dissolved and the dimensionless dose–solubility ratio θ, respectively, yielding

$$\begin{aligned} \dot{\rho}(t) &= \begin{cases} -\frac{\mathcal{D}}{\varrho}\frac{q_0}{V\rho(t)}\left[\frac{1}{\theta} - \varphi(t)\right] & \text{if} \quad \rho(t) > 0, \\ 0 & \text{if} \quad \rho(t) = 0, \end{cases} & \rho(0) = \rho_0, \\ \dot{\varphi}(t) &= \frac{3\mathcal{D}}{\varrho V}\frac{q_0}{\rho_0^3}\rho(t)\left[\frac{1}{\theta} - \varphi(t)\right] - \frac{2P_{eff}}{R}\varphi(t), & \varphi(0) = 0. \end{aligned} \tag{6.18}$$

The mass balance equation for the fraction F_a of dose absorbed at the end of the tube is

$$F_a = \frac{1}{q_0}\left[q_0 - q_{solid} - q_{dissolv}\right],$$

where q_{solid} and $q_{dissolv}$ denote the mass of the undissolved and dissolved drug, respectively, at the end of the intestine. This equation simplifies to the following:

$$F_a = 1 - \left[\frac{\rho(MITT)}{\rho(0)}\right]^3 \Phi, \tag{6.19}$$

where $\rho(MITT)$, and Φ refer to their values at $t = MITT = 199\,\text{min}$ [173].

The system of (6.18) and (6.19) describes the intestinal drug absorption as a function of four fundamental drug/formulation properties: dose q_0, dose–solubility ratio θ, initial radius of the particles ρ_0, and effective permeability P_{eff}. Typical values can be used for the constants $\mathcal{D}$ ($10^{-4}\,\text{cm}^2\,\text{min}^{-1}$), ϱ ($1000\,\text{mg}\,\text{ml}^{-1}$), V (250 ml), and R (1 cm) [55]. Thus, one can assess, using (6.18) and (6.19), whether practically complete absorption ($F_a = 0.90$) of category II drugs of the QBCS is feasible by setting the permeability in (6.18) equal to $P_{eff} = 1.2 \times 10^{-2}\,\text{cm}\,\text{min}^{-1}$, which is equivalent [170] to the upper boundary limit $P_{app} = 10^{-5}\,\text{cm}\,\text{s}^{-1}$ of the apparent permeability borderline region of QBCS [224], Figure 6.16. The correlations developed [170] between effective permeability P_{eff}, values determined in humans and the Caco-2 system allowed the conversion of the Caco-2 to P_{eff} estimates. Figure 6.18 shows the simulation results in a graph of q_0 vs. $1/\theta$ for the three particle sizes $\rho_0 = 10$, 25, and 50 μm. The areas above the lines, for each of the particle sizes considered, correspond to drug/formulation properties q_0, $1/\theta$, ensuring complete absorption, i.e., $F_a > 0.90$ for drugs classified in category II of the QBCS [224]. It is worth noting that for a given value of $1/\theta$, a higher fraction of dose is absorbed from a larger rather than a smaller dose. This finding is reasonable since the common $1/\theta$ value ensures higher solubility for the drug administered in a larger dose.

The underlying reason for a region of fully absorbed drugs in category II of the QBCS, shown in Figure 6.18, is the dynamic character of the dissolution-uptake processes. A global measure of the interplay between dissolution and

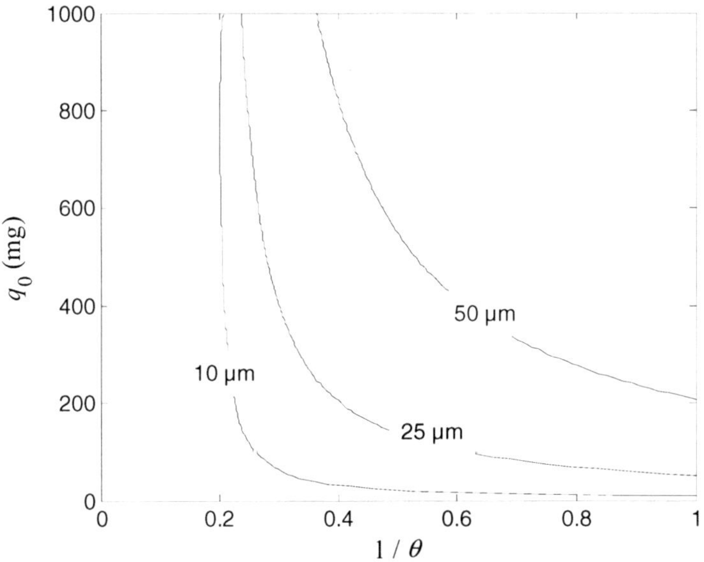

Figure 6.18: Plot of dose q_0 vs. the dimensionless solubility–dose ratio $1/\theta$. The curves indicate 90% absorption for three radius sizes 10, 25, and 50 μm assuming $P_{eff} = 1.2 \times 10^{-2}$ cm min^{-1}. Since the value assigned to P_{eff} corresponds to the upper boundary limit (expressed in apparent permeability values, [170]) of the borderline permeability region of QBCS [224], compounds of category II of QBCS exhibiting complete absorption are located above the curves.

uptake can be seen in Figure 6.19, which shows the mean dissolution time, MDT, in the intestines as a function of the effective permeability for a Class II drug ($1/\theta = 0.2$). Clearly, the MDT value is reduced as effective permeability increases. Needless to say that the MDT would be infinite for this particular drug ($1/\theta = 0.2$) if dissolution were considered in a closed system ($P_{eff} = 0$) [90]. The plot of Figure 6.19 verifies this observation since $MDT \rightarrow \infty$ as $P_{eff} \rightarrow 0$.

According to Yazdanian et al. [227] most of the NSAIDs listed in Table 6.1 are classified in Class II based on their solubility data at pH 1.2, 5.0, and fed state simulated intestinal fluid at pH 5.0. A series of simulations based on (6.18) and (6.19) revealed that the extensive absorption (Table 6.1) of the NSAIDs can be explained using the solubility–dose ratio values in buffer or fed state simulated intestinal fluid, both at pH 5.0, Figure 6.20. This plot shows the experimental data along with the curves generated from (6.18) and (6.19) assuming $F_a = 0.90$, radius sizes 10 and 25 μm, and assigning $P_{eff} = 2 \times 10^{-2}$ cm min^{-1}, which corresponds [170] to the mean of the apparent permeability values of the NSAIDs

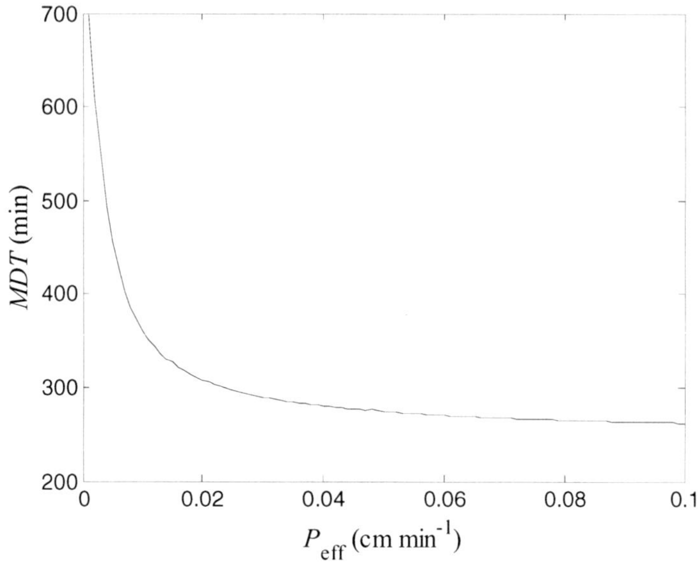

Figure 6.19: The mean dissolution time MDT in the intestines as a function of P_{eff} for parameter values $q_0 = 10\,\text{mg}$, $(1/\theta) = 0.2$, and $\rho(0) = 10\,\mu\text{m}$. MDT is calculated as the area under the curve of the undissolved fraction of dose using the integral $MTD = \int_0^\infty \left[\frac{\rho(t)}{\rho(0)}\right]^3 dt$ in conjunction with (6.18).

($P_{app} = 1.68 \times 10^{-5}\,\text{cm}\,\text{s}^{-1}$) [227]. Visual inspection of the plot based on the solubility at pH 5.0, Figure 6.20 A, reveals that only the absorption of sulindac (n°5, $F_a = 0.88$) can be explained by the generated curve adhering to $25\,\mu\text{m}$, while flurbiprofen (n°8, $F_a = 0.92$) lies very close to the theoretical line of $10\,\mu\text{m}$.

In contrast, the extensive absorption of tolmetin (n°6, $F_a > 0.90$), sulindac (n°5, $F_a = 0.88$), etodolac (n°2, $F_a > 0.80$), diflunisal (n°15, $F_a = 0.90$), ibuprofen (n°9, $F_a > 0.80$), using the corresponding doses listed in Table 6.1, can be explained on the basis of the solubility data in fed state simulated intestinal fluid at pH 5.0, in conjunction with the generated curve assigning $\rho(0) = 25\,\mu\text{m}$, Figure 6.20 B. Also, the curve generated from $\rho(0) = 10\,\mu\text{m}$ and the solubility in the biorelevant medium of indomethacin (n°3) and piroxicam (n°18) explain their extensive absorption. Although naproxen (n°11, $F_a = 0.99$) lies very close and meloxicam (n°17, $F_a = 0.89$) in the neighborhood of the theoretical line of $10\,\mu\text{m}$, oxaprozin (n°12, $F_a = 0.95$–1.00) is located far away from the simulated curve of $10\,\mu\text{m}$, Figure 6.20 B. Special caution is required in the interpretation for diclofenac (n°1, $F_a = 0.54$), which lies between the theoretical curves of 10 and $25\,\mu\text{m}$ in Figure 6.20 B. Some reports suggest that diclofenac undergoes first-

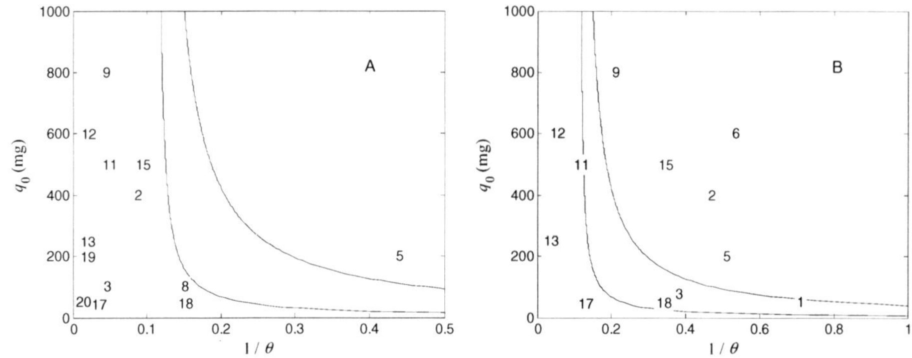

Figure 6.20: Plot of q_0 vs. $1/\theta$, for the experimental data of Table 6.1 classified in Class II. The curves denote 90% absorption for two particle sizes (from left to right 10 and 25 μm) assigning $P_{eff} = 2 \times 10^{-2}$ cm min^{-1}, which corresponds [170] to the mean, $P_{app} = 1.68 \times 10^{-5}$ cm s^{-1} of the Caco-2 permeability values of the data [90]. Drugs located above the curves are fully absorbed ($F_a > 0.90$) Class II drugs. Key (solubility values in): (**A**) buffer, pH 5.0; (**B**) fed state simulated intestinal fluid, pH 5.0.

pass metabolism ($F_a = 0.60$), while some others refer to absolute bioavailability 0.90 [231]. Explicit data for the extent of absorption of mefenamic acid (n°13), Figure 6.20 B, are not reported [227], while solubility data in the fed state simulated intestinal fluid (pH 5.0) for the two nonacidic NSAIDs, celecoxib (n°19) and rofecoxib (n°20), have not been measured [227].

These results point out the importance of the dynamic nature of the absorption processes for those drugs classified in Class II. It should also be noted that a conservative approach was utilized for the interpretation of the NSAIDs' extensive absorption, Table 6.1. In fact, only the highest doses of drugs were analyzed, while the duration of absorption was restricted to the mean intestinal transit time, 199 min [173], i.e., absorption from the stomach or the large intestine was not taken into account. Moreover, the lower value for the volume of the intestinal content, 250 ml [224–226], was used in the simulations. This means that drugs like naproxen (n°11) and meloxicam (n°17) in Figure 6.20 B would also have been explained if higher values of the two physiological parameters for time and volume had been used.

For the sake of completeness one should also add that Blume and Schug [232] suggested that Class III compounds (high solubility and low permeability) are better candidates for a waiver of bioavailability and bioequivalence studies since bioavailability is not so much dependent on the formulation characteristics as on the permeability of the compound. Finally, the recent extension [233] of BCS toward disposition principles underlines the importance of using the drug properties behind BCS toward a general biopharmaceutic-pharmacokinetic classification.

6.7 Randomness and Chaotic Behavior

Pharmacokinetic studies are in general less variable than pharmacodynamic studies. This is so since simpler dynamics are associated with pharmacokinetic processes. According to van Rossum and de Bie [234], the phase space of a pharmacokinetic system is dominated by a point attractor since the drug leaves the body, i.e., the plasma drug concentration tends to zero. Even when the system is as simple as that, tools from dynamic systems theory are still useful. When a system has only one variable a plot referred to as a *phase plane* can be used to study its behavior. The phase plane is constructed by plotting the variable against its derivative. The most classical, quoted even in textbooks, phase plane is the $\dot{c}(t)$ vs. $c(t)$ plot of the ubiquitous Michaelis–Menten kinetics. In the pharmaceutical literature the phase plane plot has been used by Dokoumetzidis and Macheras [235] for the discernment of absorption kinetics, Figure 6.21. The same type of plot has been used for the estimation of the elimination rate constant [236].

A topic in which there is a potential use of dynamic systems theory is the analysis of variability encountered in bioavailability and bioequivalence studies with highly variable orally administered formulations [237–239]. For example, the dissolution of a sparingly soluble drug takes place in the continuously changing environment of the gastrointestinal lumen. Due to the interactive character of the three principal physiological variables that affect drug dissolution, i.e., the motility of intestines, the composition and volume of gastrointestinal contents, a dynamic system of low dimension can be envisaged. If this is a valid hypothesis, a significant portion of the high variability encountered in the gastrointestinal absorption studies can be associated with the dynamics of the physiological variables controlling drug dissolution, transit, and uptake. However, the inaccessibility of the region and thus the difficulty of obtaining detailed information for the variables of interest compel one to infer that the observed variability originates exclusively from classical randomness.

Despite the hypothetical character of the previous paragraph, recent findings [240] have revealed the chaotic nature of the gastric myoelectrical complex. It seems likely that the frequently observed high variability in gastric emptying data should not be attributed exclusively to the classical randomness of rhythmic electrical oscillation in the stomach. Plausibly, one can argue that this will have an immediate impact on the absorption of highly soluble and permeable drugs from immediate release formulations since their absorption is controlled by the gastric emptying rate. Hence, the high variability of $c_{\max}$ values for this type of drugs originates from both classical experimental errors and the chaotic dynamics of the underlying processes.

Finally, the heterogeneous dynamic picture of the gastrointestinal tract becomes even more complicated by the coexistence of either locally or centrally

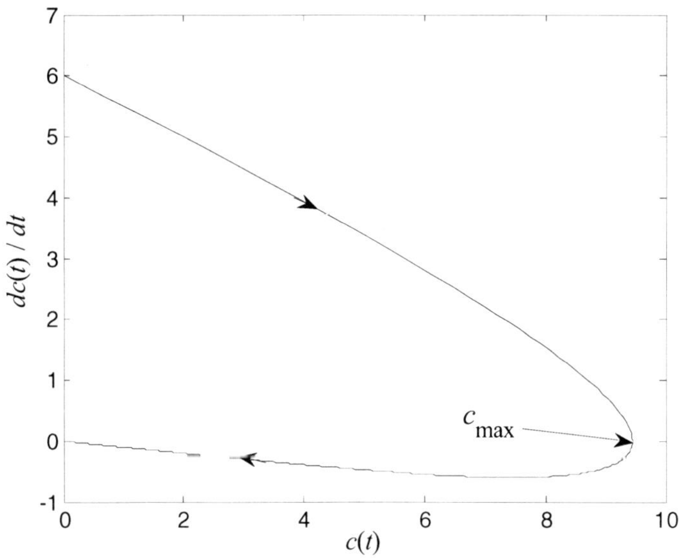

Figure 6.21: Phase plane plot for a drug obeying one-compartment model disposition with first-order absorption and elimination. Time indexes each point along the curve. The time flow is indicated by the arrows, while the x-axis intercept corresponds to $c_{\max}$.

driven feedback mechanisms, e.g., avitriptan controlling drug absorption. Experimental observations indicate [241] that when avitriptan blood levels exceed a certain threshold level, a centrally driven feedback mechanism that affects gastric emptying is initiated. Consequently, the presence or absence of double or multiple peaks of avitriptan blood levels is associated with the dynamic system describing the dissolution and uptake of drug as well as the feedback mechanism controlling the functioning of the pylorus.

It can be concluded that the use of nonlinear dynamics in gastrointestinal absorption studies can provide a tool for:

- the interpretation of variability and
- the understanding of unpredictability in situations in which double, or multiple peaks are observed and classical explanations, e.g., enterohepatic cycling, are not applicable.

Part III

MODELING IN PHARMACOKINETICS

The kinetic description of the overall disposition of drug in the body is based on experimental observations, most frequently time–drug plasma concentration data. In essence, the quantitative description of these kinetic patterns constitutes the discipline of pharmacokinetics.

Pharmacokinetic models are, or at least should be, created through a repeated *seesaw* process of monitoring against experimental evidence. Starting from a comparatively simple model, a comparison with experimental evidence will normally point to modifications of the original model, which in turn will call for new experimental evidence to assist a decision between alternative model modifications, and so on.

Mathematical models are typically classified as *phenomenological* or *empirical* [242, 243]. For the construction of phenomenological models, one must have some knowledge of the processes in terms of structural connectivity and functional mechanisms. On the other hand, when the underlying processes are unknown, one often relies on mathematical functional forms for the observed profile. These mathematical functions are empirical models. However, one can start with an empirical model and incorporate some mechanistic assumptions so that the model looks more "phenomenological." For this reason, a clear borderline cannot be drawn between phenomenological and empirical models. The full contrast between phenomenological and empirical models can be exemplified by the physiologically based and the input–output models. Neither of these extreme cases is suitable because:

- in the physiologically based models the compiled information is highly complex implying a decline in prediction performances and
- in the input–output models only a rough description of the external behavior of the process is established.

We believe that the current science is a dynamic process of knowledge requiring updating the formal tools of analysis that are our mathematical models. For example, the starting point may be a purely phenomenological or purely empirical model, but as the knowledge is accumulated, the phenomenological must become more empirical, or the empirical more phenomenological.

Most models should be considered "temporary." All require validation and many will be subject to change. Modeling may be viewed either as a screening process that employs variable selection methods to construct candidate models, or as a testing tool that validates a specific model. From a mathematical point of view, the art of good modeling relies on:

- a sound understanding and appreciation of the biological problem,
- a realistic mathematical representation of the important biological processes,
- finding useful solutions, preferably quantitative; and most importantly,
- a biological interpretation of the mathematical results in terms of insights and predictions.

In this part of the book, we first present in Chapter 7 the empirical pharmacokinetic models used to fit the observed kinetic data, placing particular emphasis on the relevance of power functions and the heterogeneous processes. The deterministic compartmental approach is described in Chapter 8 as the standard technique to analyze homogeneous processes. The last chapter of this part, Chapter 9, is devoted to the stochastic modeling techniques, a powerful tool in mathematical biology suitable for the analysis of both homogeneous and heterogeneous processes.

7

Empirical Models

It is through a few empirical functions that I am able to approach contemplation of the whole.

William A. Calder III (1934-2002)
Size, function and life history

In experimental or clinical pharmacokinetics, the simplest experiment consists in administering, in a rapid input, a large number of drug molecules having the same pharmacological properties and then in the subsequent time interval, sampling biological fluids in order to follow the decline in number of molecules or in drug concentration. The investigators are primarily interested in describing the observed decrease in time of the data by simple mathematical functions called *empirical models*. The most commonly employed model profiles are the negative exponential, the power-law, and the gamma profiles.

Exponential Profiles These have the form $c(t) = \gamma \exp(-\beta t)$. Differentiating with respect to time, one obtains

$$\dot{c}(t) = -\beta c(t), \qquad \text{or} \qquad \frac{[\mathrm{d}c(t)/c(t)]}{\mathrm{d}t} = \frac{\mathrm{d}\ln c}{\mathrm{d}t} = -\beta, \tag{7.1}$$

i.e., "the *relative* variation of the concentration c of the material divided by the *absolute* variation of time t is constant," which is the expression of Fick's law (cf. Section 2.3 and equation 2.14) under the assumption of constant volume of distribution V of the material in the medium. The constant β with dimension time^{-1} represents the ratio of the clearance CL to the volume V.

Power-Law Profiles These profiles follow the form $c(t) = \gamma t^{-\alpha}$. Differentiating with respect to time, one obtains

$$\dot{c}(t) = -\frac{\alpha}{t} c(t), \qquad \text{or} \qquad \frac{[\mathrm{d}c(t)/c(t)]}{[\mathrm{d}t/t]} = \frac{\mathrm{d}\ln c}{\mathrm{d}\ln t} = -\alpha, \tag{7.2}$$

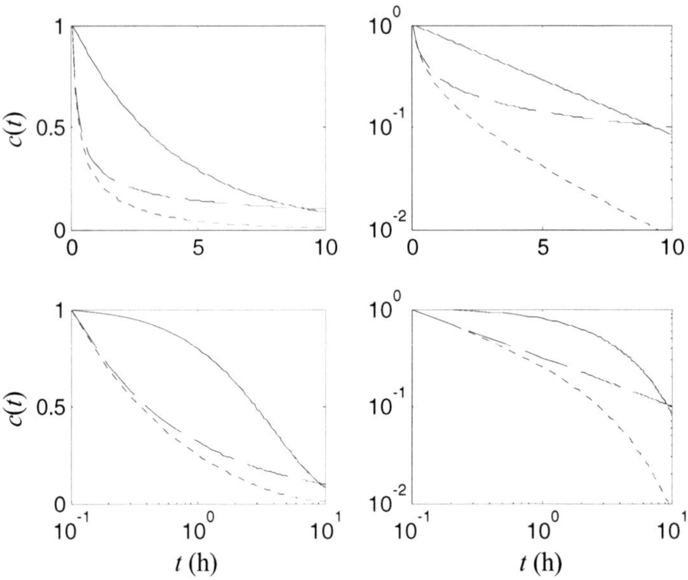

Figure 7.1: Plots of the exponential, power-law, and gamma empirical models (*solid*, *dashed*, and *dotted lines*, respectively).

i.e., "the *relative* variation of the concentration c of the material divided by the *relative* variation of time t is constant." Similarly, we can argue that the dimensionless constant α relates to how many new molecules are eliminated from the experimental medium or from the body by a mechanism similar to the overall process as the time resolution becomes finer. Attention will be given below to clarifying the power law.

Gamma Profiles These profiles follow the form $c(t) = \gamma t^{-\alpha} \exp(-\beta t)$, which is reported in the literature as the gamma-function model [244]. This model was used to fit pharmacokinetic data empirically [245, 246]. Differentiating with respect to time, we obtain

$$\dot{c}(t) = -\left(\frac{\alpha}{t} + \beta\right) c, \tag{7.3}$$

i.e., the gamma profiles might be considered as the mixed exponential and power-law profiles; the general expression for the behavior of the process in specific cases becomes either exponential or power-law.

In the three profiles above, the coefficient γ is set according to the initial conditions. For instance, if $c(t_0) = c_0$ at $t_0 \neq 0$, γ is equal to

$$c_0 \exp(-\beta t_0) \qquad \text{or} \qquad c_0 (t_0)^{\alpha} \qquad \text{or} \qquad c_0 (t_0)^{\alpha} \exp(-\beta t_0)$$

for the exponential, power-law, or gamma model, respectively. Figure 7.1 illustrates, in linear, semilogarithmic, and logarithmic scales, the behavior of these basic profiles with $\alpha = 0.5$, $\beta = 0.25$, and $c(0.1) = 1$. From these plots, we can decide in practice which empirical model we need to use:

- The y-semilogarithmic plot distinguishes the exponential model, which is depicted as one straight-line profile.
- The log-log plot distinguishes the power-law model, which is depicted as one straight-line profile.
- Both y-semilogarithmic and log-log plots are needed to decide for the gamma profile. It behaves like a power-law model in the early times (cf. the log-log plot) and as an exponential model in the later times (cf. the y-semilogarithmic plot).

The linear and x-semilogarithmic plots are uninformative for such decisions.

7.1 Power Functions and Heterogeneity

In a more realistic context, the observed data usually decay according to a sum of m negative exponentials

$$c(t) = \sum_{i=1}^{m} B_i \exp(-b_i t),$$

which correspond to a series of well-stirred tanks where drug administration is in the first tank and the concentration is computed for the mth tank.

In many cases, it was observed that when the fit of data improves as m increases, they would also be well fitted by a function of a negative power of time. It does seem extraordinary that the power function, with only two adjustable parameters, fits the data nearly as well as the sum of three or more exponential functions [244]. In fact, the scheme of the series of tanks corresponds to the states of a random walk that describes the retention of the molecules by movement of elements between nearest-neighbor sites from the administration to the sampling site. For large m, this random walk can be thought of as approximating a diffusion in a single heterogeneous site that is fitted by the empirical power-law model.

When the real process generates power-law data, alternatively a sum of exponentials and power function models may be used. But:

- power functions are defined by fewer parameters than the sums of exponentials;
- power functions seem to yield better long-term predictions;
- furthermore, the exponential parameters have little or no physiological meaning, under inhomogeneous conditions.

Overall, a large number of drugs that exhibit apparently multiexponential kinetics obey power-law kinetics. The cogent question is why many of the observed time–concentration profiles exhibit power function properties. Although the origin of the power function remains unclear, some empirical explanations could elucidate its origin:

1. A power function can be related to the sum of an infinitely large number of exponential functions:

$$t^{-\alpha} = \frac{1}{\Gamma(\alpha)} \int_0^{\infty} u^{\alpha-1} \exp(-ut)\, du, \qquad \alpha > 0.$$

   Therefore, within a given range of time, the power functions can always be fitted by sums of negative exponentials within limits that are typical for experimental error. But the converse is not true: one cannot fit power functions to data generated by sums of negative exponentials.

2. Beard and Bassingthwaighte [247] showed that a power function can be represented as the sum of a finite number of scaled basis functions. Any probability density function may serve as a basis function. They considered as basis function a density corresponding to the passage time of a molecule through two identical well-stirred tanks in series. The weighted sum of such m models leads to the power function

$$t^{-\alpha} \propto \sum_{i=1}^{m} k_i^{\alpha+1} t \exp(-k_i t), \qquad \alpha > 0.$$

   This sum can be viewed also as the parallel combination of m pathways, each characterized by a different rate constant and a uniform distribution of flow in the input of these pathways. Then, the negative power function behavior can be attributed to the heterogeneity of the flow in the system.

3. Power functions can arise if the administered molecules undergo random walks with drift, as in the well-known Wiener process [248]. The concept of random walk in series can be expressed in terms of compartments in series that have one-way entrances and exits. Each series of compartments constitutes one region, and according to the inhomogeneous assumption the administered molecules move through such a region, while according to the homogeneous assumption they move randomly within it. The inhomogeneous process could be related to active transport, i.e., through membranes.

Therefore, it seems that when the response can be fitted by power-law empirical models, the underlying process is rather heterogeneous. This probably occurs because of inhomogeneous initial mixing and transport of the molecules by bloodstream that is understirred [249], or because of elimination of molecules by organs with structural heterogeneity. Perhaps the most obvious origin of the simple power function is a diffusion process that constitutes a rate-limiting step for removal of certain substances from the circulation [4]. Moreover, drug molecules can differ in their kinetic behavior because of inherent variability in their characteristics such as molecular weight, chemical composition, or hepatic clearance involving a large number of metabolites. All these features introduce *functional heterogeneity*. Overall, homogeneity and heterogeneity can originate respectively when:

- Most substances intermix rapidly within their distribution spaces, and the rate-limiting step in their removal from the system is biochemical transformation or renal excretion. Substances of this nature are best described by compartmental models and exponential functions.

- Conversely, some substances are transported relatively slowly to their site of degradation, transformation, or excretion, so that the rate of diffusion limits their rate of removal from the system. Substances of this nature are best described by noncompartmental models and power functions.

7.2 Heterogeneous Processes

Description of distribution and elimination under homogeneous conditions can be done using classical kinetics, while fractal kinetics should be applied to describe distribution and elimination mechanisms under heterogeneous conditions. Classical transport theories, and the resulting mass-action kinetics, applicable to Euclidean structures do not apply to transport phenomena in complex and disordered media. The geometric constraints imposed by the heterogeneous fractal-like structure of the blood vessel network and the liver strongly modify drug dynamics [250]. Topological properties like connectivity and the presence of loops or dead ends play an important role. Hence, it is to be expected that media having different dimensions or even the same fractal dimension, but different spectral dimensions, could exhibit deviating behavior from that described by classical kinetics.

7.2.1 Distribution, Blood Vessels Network

According to Mandelbrot [251], fractal bifurcating networks mimic the vascular tree. Based on this observation, van Beek et al. [252] developed dichotomous branching fractal network models to explain the regional myocardium flow heterogeneity. Even though the developed models give overly simple descriptions of the fractal network, they describe adequately the dependence of the relative dispersion of flow distribution on the size of the supplied region of myocardium. These findings allow us to infer that such fractal approaches would be useful in describing other systems with heterogeneous flow distributions.

From a drug's site of administration, the blood is the predominant medium of transport of the molecules through the body to the drug's final destination. Conventionally, the blood is treated as a simple compartment, although the vascular system is highly complex and consists of an estimated 96,000 km of

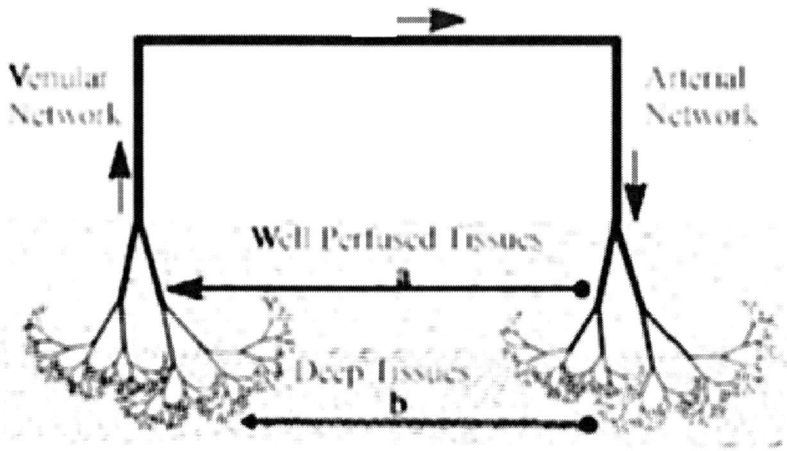

Figure 7.2: A complete vascular dichotomous network used to describe the distribution of drug in the body. The black circle represents the drug molecules. (**a**) The distribution of drug in well-perfused tissues takes place under homogeneous (well stirred) conditions. (**b**) The distribution of drug in deep tissues takes place under heterogeneous (understirred) conditions. Reprinted from [256] with permission from Springer.

vessels [253]. The key feature of the network is the continuous bifurcation of the parent vessels for many generations of branching. The vessels of one generation bifurcate to form vessels of the next generation in a continuous process toward smaller and smaller vessels. Some studies [254, 255] of the microvascular system have shown that the dimensions for vessel radii, branch length, and wall thickness in the mesenteric and renal arterial beds have fractal properties. The discovery of the fractal nature of the blood vessels, however, indicated that the distribution of flow within an organ might be fractal as well.

Building on the work of van Beek et al. [252], a dichotomous branching network of vessels representing the arterial tree connected to a similar venous network can be used to describe the distribution of the drug in the body, Figure 7.2. Thus, the general pattern of distribution of flow can also be assumed for the complete vascular system of Figure 7.2, envisaged for the distribution of drugs in the body. The flow will diverge in the arterial tree and converge in the venous tree, while at the ends of the arterial and venular networks the local flow will be slow and heterogeneous.

In the light of these network flow considerations, the distribution of drugs in the body can be classified into two broad categories. The distribution process of the drugs of the first category takes place under homogeneous (well-stirred) conditions. For the second category of drugs a significant part of the distribution process operates under heterogeneous (understirred) conditions.

- Drugs of the first category have physicochemical properties and permeability characteristics that allow them to leave the arteriole network and diffuse to the adjacent tissues under conditions of flow that ensure complete mixing (Figure 7.2 a). These drugs reach only the well perfused tissues and return rapidly to the venular draining network. The disposition of this category of drugs can be modeled with the "homogeneous model," which is identical mathematically to what we call the "one-compartment model." Obviously, the drug molecules obeying the homogeneous model permeate the walls of vessels prior to their arrival at the hugely dense ending of the networks; thus, the upper part of the vascular system and the well-perfused adjacent tissues comprise a homogeneous well-stirred "compartment."

- Based on the considerations of flow in the network, it is reasonable to argue that in close proximity with the terminal arteriolar ending, the blood flow and drug diffusion in the adjacent deep tissues will be so slow that the principle of the well-mixed system will no longer hold. Consequently, if a large portion of drug is still confined in the arterial system near its ending, the drug diffusion in the deep tissues will operate under heterogeneous (understirred) conditions (Figure 7.2 b). Transport limitations of drug in tissues have been dealt with so far with the flow- or membrane-limited physiological models [257] that maintain compartmental and homogeneity concepts. Albeit not specifying transport limitations, the previously developed description relies on the more realistic heterogeneous conditions of drug diffusion.

7.2.2 Elimination, Liver Structure

The liver is the major site of drug biotransformation in the body [258]. It is the largest composite gland of the body and weighs about $15\,\mathrm{g\,kg}^{-1}$ body weight. The physical structure of the liver exhibits unusual microcirculatory pathways [259]. Circulation in the liver can be divided into macrocirculation and microcirculation. The former comprises the portal vein, hepatic artery, and hepatic veins, while the latter consists of hepatic arterioles and sinusoids [259]. The sinusoids are the specialized capillaries of the liver that form an uninterrupted 3-dimensional network and are fully permeable by substances. This macrocirculation spans the axes of the liver while branching into successively smaller vessels. At the anatomical level, there exist small histological units, called lobules, made up of an interlacing channel network of sinusoids supplied with blood and drug by the terminal ends of the portal venules and hepatic arterioles. Between the individual sinusoids of the interior of a lobule, one-cell-thick sheets of hepatocytes are interspersed [260, 261].

In Vitro–in Vivo Correlations in Liver Metabolism

The in vitro studies in this field of research attempt to assess the rate of metabolism at an early stage of drug development in order to:

- identify problematic substances and
- allow extrapolation of the in vitro findings to in vivo conditions.

The driving force for the execution of these studies is the reduction of cost, which is related to expensive animal testing. However, replacement of in vivo testing with in vitro approaches presupposes well-based understanding of the scaling factors associating the in vitro with the in vivo measurements. The establishment of relationships between in vitro and in vivo data are known as in vitro–in vivo correlations.

Both isolated rat hepatocytes and rat liver microsomes [262–264] have been advocated for the determination of the kinetic parameters $V_{\max}$ and k_M (cf. equation 2.20) under in vitro conditions. The development of in vitro–in vivo correlations is based on two essential steps. Initially, the units of the in vitro intrinsic clearance CL_{int} (μl min^{-1} per 10^6 liver cells or μl min^{-1} per mg microsomal protein) are converted to ml min^{-1} per standard rat weight of 250 g using scaling factors reported in the literature [265]. Next, a liver model that incorporates physiological processes such as hepatic blood flow, Q, and plasma protein binding is used to provide the hepatic clearance CL_h. Therefore, the liver modeling step of the in vitro–in vivo correlations is crucial in the scaling process from the in vitro to the in vivo estimates of clearances.

Due to its mathematical simplicity, most in vitro–in vivo correlations are based on a homogeneous, "well-stirred" model for the liver such that all metabolic enzymes in the liver are exposed to the same drug concentration [266]. Under steady-state conditions, the predicted hepatic clearance CL_h for this model is

$$CL_h = \frac{Q f_u CL_{int}}{Q + f_u CL_{int}},$$

where f_u is the blood unbound fraction. Alternatively, liver has also been viewed as a parallel tube model [267]. In this case, the liver is considered as an organ receiving a series of parallel blood flows carrying the drug in identical parallel tubes representing the sinusoids. Here, the hepatic clearance assuming linear kinetics and steady-state conditions is

$$CL_h = Q \left[1 - \exp \left(- \frac{f_u CL_{int}}{Q} \right) \right].$$

However, these two models assume either perfect mixing conditions (well-stirred model) or no mixing at all (parallel tube model) and cannot explain several experimental observations. Therefore, other approaches such as the distributed model [268], the dispersion model [269], and the interconnected tubes model [270,271] attempt to capture the heterogeneities in flow and an intermediate level of mixing or dispersion. Despite numerous comparisons [264,265,272–

274] of the use of various liver models [266–271] for predicting the in vivo drug clearance from in vitro measurements, there is still controversy regarding the most suitable liver modeling approach. This is so since drug-specific factors, like high- or low-cleared drugs, seem to have a major impact on the quality of the in vitro–in vivo correlations. For example, low-clearance drugs are rather independent of blood-flow characteristics, while drugs with relatively higher clearance values show a more pronounced dependence on blood-flow properties.

Fractal Considerations in Liver Metabolism

Observations of the liver reveal an anatomically unique and complicated structure, over a range of length scales, dominating the space where metabolism takes place. Consequently, the liver was considered as a fractal object by several authors [4, 248] because of its self-similar structure. In fact, Javanaud [275], using ultrasonic wave scattering, has measured the fractal dimension of the liver as approximately $d_f \approx 2$ over a wavelength domain of 0.15–1.5 mm.

While there is no performance advantage over a well-stirred classical compartment, one with a rate constant due to a uniformly random distribution of drug and enzyme, such a compartment may well be impossible to achieve under biological designs, and the implied comparison is therefore an ill-posed one [276]. It may be that the fractal liver design is the best design possible, so that comparisons against nonideal theoretical models, like a poorly stirred sphere with enzymes adhered along the inner wall, are favorable. For example, the fractal structure, with many layers of membrane at its interface, allows the organ to possess a high number (concentration) of enzymes, thus giving it a high reaction rate despite time-dependent (decay) fractal kinetics. Indeed, the intricate interlacing of a stationary, catalytic phase of hepatocytes with a liquid phase of blood along a fractal border is what reduces the required diffusional distances for reactions to take place with any appreciable celerity. Moreover, the complicated structure of the liver, which provides for a huge interface between drug and hepatocytes, may be generated simply during the growth of the liver. The fractal form may be parsimoniously encoded in the DNA, indirectly specified by means of a simple recursive algorithm that instructs the biological machinery on how to construct the liver. In this way, a vascular system made up of fine tubing with an effective topological dimension of one may fill the 3-dimensional embedding space of the liver. These possibilities suggest that the structure of the liver may be that of a fractal.

In this context, Berry [277] studied the enzyme reaction using Monte Carlo simulations in 2-dimensional lattices with varying obstacle densities as models of biological membranes. That author found that the fractal characteristics of the kinetics are increasingly pronounced as obstacle density and initial concentration increase. In addition, the rate constant controlling the rate of the complex formation was found to be, in essence, a time-dependent coefficient since segregation effects arise due to the fractal structure of the reaction medium. In a similar vein, Fuite et al. [278] proposed that the fractal structure of the liver with attendant kinetic properties of drug elimination can explain the unusual nonlinear pharmacokinetics of mibefradil [279, 280]. These authors utilized a simple flow-limited physiologically based pharmacokinetic model where clearance of the drug occurs in the liver by fractal kinetics [278]. The analytical solution of the proposed model was fitted to experimental dog data and the estimates for the spectral dimension d_s of the dog liver were found to be in the range 1.78–1.91. This range of values is consistent with the value found in ultrasound experiments on the liver, $d_f \approx 2$ [275]. Furthermore, special attention was given to mibefradil pharmacokinetics by studying the effect of species segregation on the kinetics of the enzyme reaction in fractal media using a microscopic pharmacokinetic model mimicking the intravenous and oral administration of the substrate [281]. This mathematical model coupled with Monte Carlo simulations of the enzyme reaction in a 2-dimensional square lattice reproduced the classical Michaelis–Menten kinetics in homogeneous media as well as unusual kinetics in fractal media. Based on these findings, a time-dependent version of the classic Michaelis–Menten equation was developed for the rate of change of the substrate concentration in disordered media. This equation was successfully used to describe the experimental time–concentration data of mibefradil and to derive estimates for the model parameters.

7.3 Fractal Time and Fractal Processes

The concept of fractals may be used for modeling certain aspects of dynamics, i.e., temporal evolution of spatially extended dynamic systems in nature. Such systems exhibit fractal geometry and may maintain dynamic processes on all time scales. For example, the fractal geometry of the global cloud cover pattern is associated with fluctuations of meteorological parameters on all time scales from seconds to years. The temporal fluctuations exhibit structure over multiple orders of *temporal magnitude* in the same way that fractal forms exhibit details over several orders of *spatial magnitude*. Power-law behavior has been documented in the functioning of physiological systems [282, 283]. Long-range spatial correlations have also been identified at DNA level [284, 285]. Long-range correlations over time and space for geophysical records have also been investigated by Mandelbrot and Wallis [286] and, more recently, by Tang and Bak [287]. Recent studies have identified power laws that govern epidemiological phenomena [288]. All the reported long-range temporal correlations signify persistence or *memory*.

A major feature of this correlation is that the amplitudes of short-term and long-term fluctuations are related to each other by the scale factor alone, independent of details of growth mechanisms from smaller to larger scales. The macroscopic pattern, consisting of a multitude of subunits, functions as a unified whole independent of details of dynamic processes governing its individual subunits [289]. Such a concept, whereby physical systems consisting of a large number of interacting subunits obey universal laws that are independent of the microscopic details, is acknowledged as a breakthrough in statistical physics. The variability of individual elements in a system acts cooperatively to establish regularity and stability in the system as a whole [290]. Scale invariance implies that knowledge of the properties of a model system at short times or short length scales can be used to predict the behavior of a real system at large times and large length scales [291].

The spatiotemporal evolution of dynamic systems was not investigated as a unified whole, and fractal geometry of spatial patterns and fractal fluctuations in time of dynamic processes were investigated as two separate multidisciplinary areas of research till as late as 1987. In that year, Bak et al. [292, 293] postulated that fractal geometry in spatial patterns, as well as the associated fractal fluctuations of dynamic processes in time, are signatures of self-organized phase transition in the spatiotemporal evolution of dynamic systems. The relation between spatial and temporal power-law behavior was recognized much earlier in condensed-matter physics where long-range spatiotemporal correlations appear spontaneously at the critical point for continuous phase transitions. The amplitude of large- and small-scale fluctuations are obtained from the same mathematical function using an appropriate scale factor, i.e., ratio of the scale lengths.

Conversely, the relationship (7.2) expresses a time-scale invariance (self-similarity or fractal scaling property) of the power-law function. Mathematically, it has the same structure as (1.7), defining the capacity dimension d_c of a fractal object. Thus, α is the capacity dimension of the profiles following the power-law form that obeys the fundamental property of a fractal self-similarity. A fractal decay process is therefore one for which the rate of decay decreases by some exact proportion for some chosen proportional increase in time: the self-similarity requirement is fulfilled whenever the exact proportion, α, remains unchanged, independent of the moment of the segment of the data set selected to measure the proportionality constant.

Therefore, the power-law behavior itself is a self-similar phenomenon, i.e., doubling of the time is matched by a specific fractional reduction of the function, which is independent of the chosen starting time: self-similarity, independent of scale is equivalent to a statement that the process is fractal. Although not all power-law relationships are due to fractals, the existence of such a relationship should alert the observer to seriously consider whether the system is self-similar. The dimensionless character of α is unique. It might be a reflection of the fractal nature of the body (both in terms of structure and function) and it can also be linked with "species invariance." This means that α can be found to be "similar" in various species. Moreover, α could also be thought of as the reflection of a combination of structure of the body (capillaries plus eliminating organs) and function (diffusion characteristics plus clearance concepts).

7.4 Modeling Heterogeneity

From a kinetic viewpoint, the distribution of drugs operating under homogeneous conditions can be described with classical kinetics. When the distribution processes are heterogeneous, the rate constant of drug movement in the tissues is not linearly proportional to the diffusion coefficient of the drug. Then, modeling of the heterogeneity features should be based on fractal kinetics concepts [4,9,16].

7.4.1 Fractal Concepts

A better description of transport limitations can be based on the principles of diffusion in disordered media [294]. It has been shown [295] that in disordered media the value of the first-order rate constant is related to the geometry of the medium. In these media the diffusional propagation is hindered by its geometric heterogeneity, which can be expressed in terms of the fractal and spectral dimensions. For our purposes, the propagation of the drug's diffusion front in the heterogeneous space of tissues can be viewed as a diffusion process in a disordered medium. Both the diffusion coefficient of the drug and the rate constant are dependent on the position of the radial coordinate of the diffusion front, and therefore both parameters are time-dependent. In these lower-dimensional systems, diffusion is inhibited because molecules cannot move in all directions and are constrained to locally available sites.

The description of these phenomena in complex media can be performed by means of fractal geometry, using the spectral dimension d_s. To express the kinetic behavior in a fractal object, the diffusion on a microscopic scale of an exploration volume is analyzed [278]. A random walker (drug molecule), migrating within the fractal, will visit $n(t)$ distinct sites in time t proportional to the number of random walk steps. According to the relation (2.9), $n(t)$ is proportional to $t^{d_s/2}$, so that diffusion is related to the spectral dimension.

The case $d_s = 2$ is found to be a critical dimension value in the phenomena of self-organization of the reactants:

- For $d_s > 2$, a random walker has a finite escape probability-microscopic behavior conducive to re-randomize the distribution of reactants around a trap and deplete the supply of reactive pairs, and thus a stable macroscopic reactivity as attested by the classical rate constant [296, 297]. The scale of the self-organization is microscopic and independent of time, such that $n(t) \propto t$ (is linear) and $k = \dot{n}(t)$ is a constant, so the reaction kinetics are classical.

- For $d_s \leq 2$, a random walker (drug) is likely to stay at its original vicinity and will eventually recross its starting point, a microscopic behavior conducive to producing mesoscopic depletion zones around traps, e.g., enzymes. The compactness of the low-dimensional random walk implies ineffective diffusion, relevant mesoscopic density fluctuations of the drug, and an entailing aberrant macroscopic rate coefficient. Subsequently, the macroscopic reaction rate, which is given by the time derivative of $n(t)$, sometimes described as the efficiency of the diffusing, reacting random walker, will be

$$k(t) \propto \dot{n}(t) \propto t^{-(1-d_s/2)} = t^{-\lambda} \tag{7.4}$$

for transient reactions [278]. Since $0 < d_s \leq 2$, the parameter λ has values in the range $0 \leq \lambda < 1$. The minus sign in (7.4) is used to mimic the decrease of k with time as the walker (drug) has progressively less successful visits. This time-dependent rate "constant" in the form of a power law is the manifestation of the anomalous microscopic diffusion in a dimensionally restricted environment leading to anomalous macroscopic kinetics [278].

The kinetic consequences that are associated with the time-dependency of the rate "constant" are delineated in Section 2.5 under the heading, coined by Kopelman [9, 16], *fractal-like kinetics*.

7.4.2 Empirical Concepts

Heterogeneity could also be expressed and described by elementary operations with empirical models. The only difference between (7.1) and (7.2) lies in the coefficient of $c(t)$ on the right-hand side of the differential equations. This allows someone to infer empirically that these equations could be unified as

$$\dot{c}(t) = -\beta \left(\frac{\beta t}{\alpha} \right)^{-\lambda} c(t) \tag{7.5}$$

with initial condition $c(t_0) = c_0$ at $t_0 \neq 0$. The exponent λ takes integer 0 or 1 values corresponding to the exponential and power-law profiles, respectively, and α and β are as defined in (7.1) and (7.2). Since the gamma profile (7.3) is presented as the additive mixture of the previous ones, one wonders whether λ is allowed to attain fractional values between 0 and 1. Indeed, the previous equation could also be considered as a generalization of (7.1) and (7.2) assuming a fractional time exponent λ $(0 \leq \lambda \leq 1)$. Under this assumption, (7.5) is similar to what we reported previously (equations 5 and 7 in [256]), obtained from the classical first-order rate kinetics assuming that the rate coefficient is a time-varying rate coefficient.

The solution of (7.5) is

$$c(t) = c_0 \exp \left\{ -\frac{\alpha}{1-\lambda} \left[\left(\frac{\beta t}{\alpha} \right)^{1-\lambda} - \left(\frac{\beta t_0}{\alpha} \right)^{1-\lambda} \right] \right\} \tag{7.6}$$

for $\lambda \neq 1$ and

$$c(t) = c_0 \exp \left[-\alpha \ln \left(\frac{t}{t_0} \right) \right]$$

for $\lambda = 1$. Then, with fractional λ, the transition in output response is continuous between a homogeneous process $(\lambda = 0)$ and a heterogeneous one $(\lambda \neq 1)$ (or equivalently, how to generate multiexponential behavior starting from a monoexponential one). Inversely, after fitting observed data by empirical models such as (7.6), the estimated value of λ might help us classify drugs in two large groups:

- *Homogeneous drugs* with $\lambda \approx 0$: their kinetics can be described homogeneously with what we will call compartmental models. These drugs are characterized by small or medium volumes of distribution.
- *Heterogeneous drugs* with $\lambda \neq 0$: their kinetics are described with noncompartmental modeling, and in reality they approximate the true heterogeneous disposition, i.e., the time-dependent character of diffusion (flow). These drugs are characterized by high volumes of distribution.

Moreover, combinations of these models can also be used to roughly describe physiological considerations. For instance, if the drug is metabolized by the liver and simultaneously eliminated by the kidney, a gamma profile is obtained as solution of (7.3), where the α/t term expresses the structural heterogeneity of the liver, and the term β, the homogeneous elimination process from the kidney.

7.5 Heterogeneity and Time Dependence

It has been stated that heterogeneous reactions taking place at interfaces, membrane boundaries, or within a complex medium like a fractal, when the reactants are spatially constrained on the microscopic level, culminate in deviant reaction rate coefficients that appear to have a sort of *temporal memory*. Fractal kinetic theory suggested the adoption of a time-dependent rate "constant", with power-law form, determined by the spectral dimension. This time-dependency could also be revealed from empirical models.

In fact, the empirical models involve parameters without any physiological meaning. To obtain sound biological information from the observed data, these models should be converted to some more phenomenological ones, parametrized by volume of distribution, clearance, elimination rate constant, etc. In their simplest form, the phenomenological models are based on Fick's first law (2.14), where the concentration gradient is the force acting to diffuse the material q through a membrane:

$$\dot{q}(t) = -CLc(t), \tag{7.7}$$

where CL is the clearance. Concentration and amount of material are also linked via the well-known relationship

$$q(t) = Vc(t), \tag{7.8}$$

where V is the volume of distribution of the material. We also explicitly denote the time dependency in each parameter, $CL(t)$ and $V(t)$, and define the rate constant $k(t)$ as

$$\frac{CL(t)}{V(t)} \triangleq k(t). \tag{7.9}$$

Differentiating (7.8) with respect to t and using expressions (7.7) and (7.9) to substitute $\dot{q}(t)$ and $CL(t)$, we obtain

$$c(t)\dot{V}(t) = -k(t)c(t)V(t) - V(t)\dot{c}(t). \tag{7.10}$$

According to the exponential, power-law, or gamma empirical model, $\dot{c}(t)$ may take the form of relation (7.1), (7.2), or (7.3), respectively. By introducing these relations in (7.10) we get, respectively,

$$\dot{V}(t) = [\beta - k(t)] V(t) \tag{7.11}$$

or

$$\dot{V}(t) = \left[\frac{\alpha}{t} - k(t)\right] V(t) \tag{7.12}$$

or

$$\dot{V}(t) = \left[\frac{\alpha}{t} + \beta - k(t)\right] V(t). \tag{7.13}$$

A time-invariant process has time-independent parameters. Therefore, a time-invariant process is that for which both V and k are invariant in time. From the three previous relationships, the only time-independent situation occurs in the exponential empirical model when $k(t) = \beta$. In this case, from (7.11) one has $V(t) = V_0$, a time-invariant volume. The processes fitted by the power-law and gamma empirical models are necessarily time-varying processes, because when either V or k is kept constant, the other becomes time-varying.

In these cases, two extreme situations may occur:

- k is time-invariant. If we assume $k(t) = \beta$ in (7.12) and (7.13), the time courses of the volume are

  $$V(t) = V_0 t^{\alpha} \exp(-\beta t) \qquad \text{and} \qquad V(t) = V_0 t^{\alpha},$$

  respectively, where V_0 is set according to the initial conditions. Taking into account this time dependence of volume, a unique form of the amount profile is obtained, $q(t) = Q_0 \exp(-\beta t)$, irrespective of the exponential, power-law, or gamma concentration profiles.

- V is time-invariant. From (7.12) and (7.13) one obtains the time course of the rate constant:

  $$k(t) = \frac{\alpha}{t} \qquad \text{and} \qquad k(t) = \frac{\alpha}{t} + \beta,$$

  respectively. With time-invariant V, the amount profiles $q(t)$ will be proportional to the concentration profiles $c(t)$.

Figure 7.3 illustrates the time courses of the reduced volume of distribution $V(t)/V_0$ and of the reduced rate constant $k(t)/\beta$ with $\alpha = 0.5$ and $\beta = 0.25$. Certainly, mixed situations where both $k(t)$ and $V(t)$ are time-varying can be thought of.

This preliminary analysis highlights the difference between regular and irregular profiles associated with time-invariant and time-varying physiological parameters, respectively. Some authors have attempted to associate a functional physiological meaning to the gamma empirical model [298,299] or to describe by stochastic modeling the real processes leading to power-law outputs [300,301].

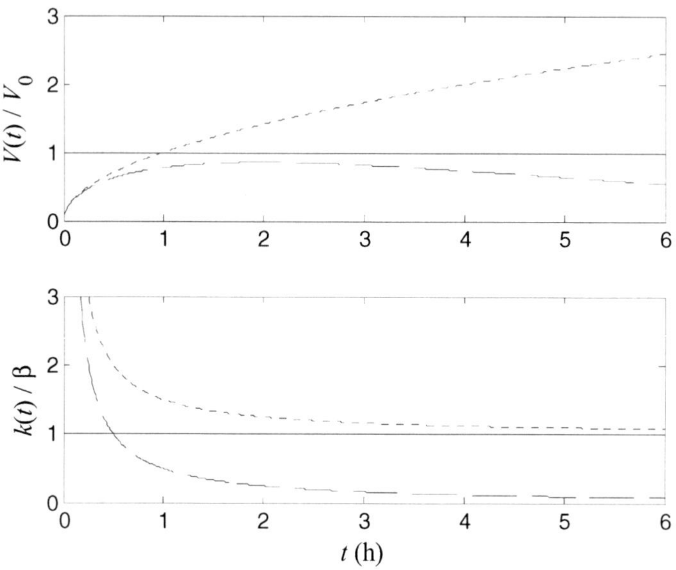

Figure 7.3: Time courses of $V(t)/V_0$ (*up*) and of $k(t)/\beta$ (*down*) associated with the exponential, power-law, and gamma empirical models (*solid*, *dashed*, and *dotted lines*, respectively).

In contrast, in the case of calcium pharmacokinetics [256], the possible mechanisms underlying (7.3), where the renal elimination of calcium was associated with the parameter β, and the other elimination mechanisms, with parameter α were discussed. Lastly, a simple approach for including, within a multicompartment model, time dependence of the transfer coefficients that vary continuously with the age of human patients was described by Eckerman et al. [302], but time dependence was over periods much greater than a single dose. This simplified the mathematics so that there was no time dependence of coefficients during the time course of a single dose. Within a physiological model, over a very long time scale of 98 days, Farris et al. [303] introduce time-dependent compartment volume changes due to growth in the studied rat model system.

Therefore, it is clear that when the outputs are optimally fitted by the power-law and gamma empirical models, the underlying processes are rather time-varying. The time-varying features of the observed processes are in fact the expression of functional or structural heterogeneities in the body.

7.6 Simulation with Empirical Models

The observed empirical models should now be employed to simulate and predict kinetic behaviors obtained with administration protocols other than that used for observation. Moreover, we must develop pharmacokinetics in a multicompartment system by including the presence of a fractal organ. We have argued that the liver, where most of the enzymatic processes of drug elimination take place, has a fractal structure. Hence, we expect transport processes as well as chemical reactions taking place in the liver to carry a signature of its fractality.

Little has so far been done to predict the effect of different modes of administration, according to inhomogeneous conditions, on the observed $c(t)$ when this contains a power function. In fact, the availability of the drug in the process was simply expressed by an initial condition $c(t_0) = c_0$. Later on, exponential, power-law, or gamma profiles were observed according to the inherent heterogeneity of the process.

Empirical models helped us to recognize heterogeneity in the process and to simply express it by mathematical models with time-varying parameters V and k. Nevertheless, the time in such time-varying parameters can be conceived only as a maturation time or as an age a associated with each administered molecule, i.e., $V(a)$ and $k(a)$. This time a must be distinguished from the exogenous time t associated with the evolution of the overall process. Several hypotheses based on fractal principles were formulated to explain heterogeneity and time dependency, but conceptual difficulties persist in explaining the time profiles of $V(a)$ and $k(a)$. The volume may represent the maximal space visited by a molecule and the elimination constant, the fragility of a molecule while it remains in the process. These parameters are dependent on the age a of each molecule, and they must be independent of the drug administration protocol, e.g., the repeated dosages, which are scheduled with respect to the exogenous time t of the process. Therefore, the relation between a and t must be resolved before integrating in the model the usual routes of administration. The heterogeneous process observed in several circumstances and the resulting complexity of the molecular kinetic behaviors, with respect to the actual experiments, required new techniques as well as modifications of Fick's law in order to comply with observations. In this way, two operational procedures may be retained:

- First operate at a molecular level and establish a probabilistic model for the behavior and the time spent by each molecule in the process. Second, take statistically into account all the molecules in the process. This stochastic formulation would be the most appropriate for capturing the structural and functional heterogeneity in the biological media. The resulting models supply tractable forms involving the time-varying parameters $V(a)$ and $k(a)$ [304]. This issue was greatly addressed in biological systems and only recently in pharmacokinetics [305,306]. It will be developed, here, in Chapter 9.

- From a holistic point of view, the time-varying parameters $V(t)$ and $k(t)$ fitting the observed data could represent the dynamic behavior of a complex system involving feedback mechanisms implying the states $q(t)$. So, these parameters can be assumed to be complex functions of $q(t)$, namely $V(q)$ and $k(q)$, leading to nonlinear kinetics (e.g., logistic saturable [307]), with time-varying coefficients [256], etc. For decades, this approach has had numerous applications in pharmacokinetics, and it allows any complex function to be assumed as V and k. Time variation in the parameters is treated in Appendix C.

8

Deterministic Compartmental Models

This is Polyfemos the copper Cyclops whose body is full of water and someone has given him one eye, one mouth and one hand to each of which a tube is attached. Water appears to drip from his body and to gush from his mouth, all the tubes have regular flow. When the tube connected to his hand is opened his body will empty within 3 days, while the one from his eye will empty in one day and the one from his mouth in 2/5 of a day. Who can tell me how much time is needed to empty him when all three are opened together?

Metrodorus (331-278 BC)

Compartmental modeling is a broad modeling strategy that has been used in many different fields, though under varying denominations. Virtually all current applications and theoretical research in compartmental analysis are based on deterministic theory. In this chapter deterministic compartmental models will be presented. The concept of compartmental analysis assumes that a process may be divided though it were occurring in homogeneous components, or "compartments." Various characteristics of the process are determined by observing the movement of material. A compartmental system is a system that is made up of a finite number of compartments, each of which is homogeneous and well mixed, and the compartments interact by exchanging material. Compartmental systems have been found useful for the analysis of experiments in many branches of biology.

We assume that compartment i is occupied at time 0 by q_{i0} amount of material and we denote by $q_i(t)$ the amount in the compartment i at time t. We also assume that no material enters in the compartments from the outside of the compartmental system and we denote by $R_{i0}(t)$ the rate of elimination from compartment i to the exterior of the system. Let also $R_{ji}(t)$ be the transfer rate of material from the jth to ith compartment. Because the material

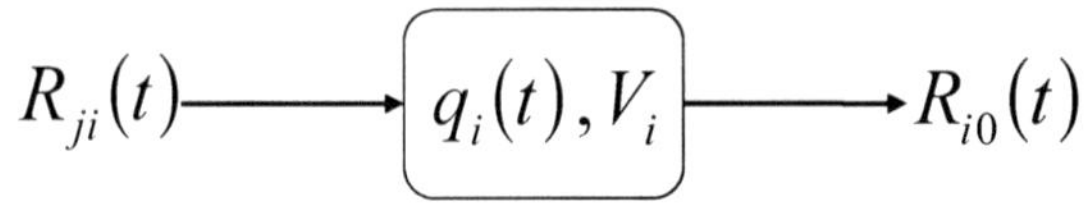

Figure 8.1: The rates of transfer of material for the ith compartment.

is distributed in each compartment at uniform concentration, we may assume that each compartment occupies a constant volume of distribution $V_{i.}$ The box in Figure 8.1 represents the ith compartment of a system of m compartments.

Mathematics is now called upon to describe the compartmental configurations and then to simulate their dynamic behavior. To build up mathematical equations expressing compartmental systems, one has to express the mass balance equations for each compartment i:

$$\dot{q}_i(t) = -R_{i0}(t) + \sum_{\substack{j=1 \\ j \neq i}}^{m} R_{ji}(t), \tag{8.1}$$

with initial condition $q_i(0) = q_{i0}$. Thus, we obtain m differential equations, one for each compartment i.

8.1 Linear Compartmental Models

Now, some fundamental hypotheses, or as commonly called laws, were employed to expand the transfer rates appearing in (8.1). Fick's law is largely used in current modeling (cf. Section 2.3 and equation 2.14). It assumes that the transfer rate of material by diffusion between regions l (left) and r (right) with concentrations c_l and c_r, respectively, is

$$R_{lr}(t) = -CL_{lr}(c_r - c_l). \tag{8.2}$$

This law may be applied to the transfer rates $R_{ji}(t)$ of the previous equation for all pairs j and i of compartments corresponding to l and r and for the elimination rate $R_{i0}(t)$, where the concentration is assumed nearly zero in the region outside the compartmental system. One has for the compartment i,

$$\dot{q}_i(t) = -CL_{i0}c_i(t) + \sum_{\substack{j=1 \\ j \neq i}}^{m} CL_{ji}[c_j(t) - c_i(t)],$$

where CL_{i0} is the *total clearance* from compartment i and CL_{ji} is the *intercompartmental clearance* between i and j. We recall that the clearance has a bidirectional property ($CL_{ji} = CL_{ij}$) and the subscript ij denotes simply the pair of compartments referenced. The initial condition associated with the previous differential equation is denoted by $q_i(0) = q_{i0}$. Using the volumes of distribution V_i and the well-known relationship $q_i(t) = V_i c_i(t)$, we substitute the concentrations by the corresponding amounts of material:

$$\dot{q}_i(t) = -k_{i0} q_i(t) + \sum_{\substack{j=1 \\ j \neq i}}^{m} k_{ji} q_j(t) - \sum_{\substack{j=1 \\ j \neq i}}^{m} k_{ij} q_i(t).$$

The constants k are called the *fractional flow rates*. They have the dimension of time^{-1} and they are defined as follows:

$$\frac{CL_{i0}}{V_i} \triangleq k_{i0}, \qquad \frac{CL_{ij}}{V_i} \triangleq k_{ij}, \qquad \frac{CL_{ij}}{V_j} \triangleq k_{ji}. \tag{8.3}$$

In contrast to the clearance, the fractional flow rates indicate the direction of the flow, i.e., $k_{ji} \neq k_{ij}$, the first subscript denoting the start compartment, and the second one, the ending compartment. The fractional flow rates and the volumes of distribution are usually called *microconstants.*

When the volume of the compartment being cleared is constant, the assumption that the fractional flow rate is constant is equivalent to assuming that the clearance is constant. But in the general case, in which the volume of distribution cannot be assumed constant, the use of the fractional flow rates k is unsuitable, because the magnitude of k depends as much upon the volume of the compartment as it does upon the effectiveness of the process of removal. In contrast, the clearance depends only upon the overall effectiveness of removal, and can be used to characterize any process of removal whether it be constant or changing, capacity-limited or supply-limited [308].

Through the following procedure the equations for a deterministic model can be obtained:

1. Represent the underlying mechanistic model with the desired physiological structure through a set of phenomenological compartments with their interconnections.

2. For each compartment in the configuration, apply the mass-balance law to obtain the differential equation expressing the variation of amount per unit of time. In these expressions, constant or variable fractional flow rates k can be used.

3. Solve the system of differential equations obtained for all the compartments by using classical techniques or numerical integration (e.g., Runge–Kutta) [309].

Therefore, Fick's law, when applied to all elements of the compartmental structure, leads to a system of linear differential equations. There are as many equations as compartments in the configuration. If we set

$$k_{ii} = k_{i0} + \sum_{\substack{j=1 \\ j \neq i}}^{m} k_{ij},$$

the equation for the ith compartment is

$$\dot{q}_i(t) = -k_{ii}q_i(t) + \sum_{\substack{j=1 \\ j\neq i}}^{m} k_{ji}q_j(t), \tag{8.4}$$

associated with initial conditions q_{i0}. In the previous equation, the $q_i(t)$ and q_{i0} amounts of material can be compiled in vector forms as $\underline{q}(t)$ and $\underline{q}_0$, respectively. In the same way, the fractional flow rates k_{ij} may be considered as the (i, j)th elements of the $m \times m$ fractional flow rates matrix $\mathbf{K}$. Thus, the set of linear differential equations can be expressed as

$$\dot{\underline{q}}^T(t) = \underline{q}^T(t)\mathbf{K},$$

and it has the following solution:

$$\underline{q}^T(t) = \underline{q}_0^T \exp(\mathbf{K}t), \tag{8.5}$$

where the initial conditions are *postmultiplied* by $\exp(\mathbf{K}t)$, which is defined as

$$\exp(\mathbf{K}t) = \mathbf{I} + \sum_{i=1}^{\infty} \frac{\mathbf{K}^i t^i}{i!}.$$

In most pharmacokinetic applications, one can assume that the system is open and at least weakly connected. This is the case of mammillary compartmental models, where the compartment n° 1 is referred to as the *central compartment* and the other compartments are referred to as the *distribution compartments*, characterized by $k_{i0} = 0$ and $k_{ij} = 0$ for $i, j = 2, \ldots, m$. For open mammillary compartmental configurations, the eigenvalues of $\mathbf{K}$ are distinct, real, and negative, implying that

$$q_i(t) = \sum_{j=1}^{m} B_{ij} \exp(-b_j t),$$

the so-called formula of *sum of exponentials*, which is common in pharmacokinetics. The B_{ij} and positive b_j are often called *macroconstants*, and they are functions of the microconstants. The equations relating these formulations are given explicitly for the common 2- and 3-compartment models in many texts [307, 310]. It should be noted, however, that the addition of a few more compartments usually complicates the analysis considerably.

8.2 Routes of Administration

In practice, it is unlikely to have compartmental models with initial conditions unless there are residual concentrations obtained from previous administrations. Drugs are administered either by extravascular, or intravascular in single or repeated experiments. Extravascular routes are oral, or intramuscular routes, and intravascular are the constant-rate short- and long-duration infusions.

- For the extravascular route, the rate of administration is

$$r_{ev}(t) = q_0 k_a \exp(-k_a t),$$

where q_0 is the amount of material initially given to the extravascular site of administration and k_a is the fractional flow rate for the passage of material from the site of administration toward the recipient compartment; k_a is the absorption rate constant.

- For the intravascular route with constant rate, we have

$$r_{iv}(t) = \frac{q_0}{T_E - T_S}\left[u(t - T_S) - u(t - T_E)\right],$$

where q_0 is the amount of material given at a constant rate in the venous compartment between the starting time T_S and the ending time T_E. Here, $u(t)$ is the step Heaviside function.

Extravascular and intravascular routes can be conceived as concomitant or repeated, e.g., delayed oral intake with respect to an intramuscular administration, or piecewise constant rate infusions, etc. Applying the superposition principle, the contribution of all administration routes in the same recipient compartment is given by the following input function:

$$r(t) = \sum_{i=1}^{m_{ev}} q_{0i} k_{ai} \exp\left[-k_{ai}(t - T_i)\right] + \sum_{i=1}^{m_{iv}} \frac{q_{0i}}{T_{E_i} - T_{S_i}}\left[u(t - T_{S_i}) - u(t - T_{E_i})\right],$$

where the m_{ev} and m_{iv} administrations preceding the time t are associated with the q_{0i} amounts of material. T_i is the time of the ith extravascular administration, and T_{S_i} and T_{E_i} are the starting and ending times in the ith intravascular administration. The contribution of the input function $r(t)$ in the mass-balance differential equation for the recipient compartment is represented by an additive term in the right-hand side of (8.1).

8.3 Time–Concentration Profiles

In (8.4), by dividing the amounts $q_i(t)$ by non-time-dependent volumes of distribution V_i, one obtains the differential equations for the concentrations $c_i(t)$:

$$\dot{c}_i(t) = -k_{ii} c_i(t) + \sum_{\substack{j=1 \\ j \neq i}}^{m} \frac{V_j}{V_i} k_{ji} c_j(t). \tag{8.6}$$

Additional assumptions further reduce the complexity of these equations. One such assumption is the incompressibility of the volumes of distribution or, as usually known, the *flow conservation*. This assumption applied to compartment j leads to

$$\sum_{\substack{i=1 \\ i \neq j}}^{m} V_i k_{ij} = V_j \sum_{\substack{i=1 \\ i \neq j}}^{m} k_{ji}.$$

In the special case of a mammillary compartmental configuration, the above relation allows one to express the volume of distribution in peripheral compartments as functions of the fractional flow rates and the volume of distribution of the central compartment $V_j = [k_{1j}/k_{j1}] V_1$ for $j = 2, \ldots, m$. Substituting this relationship in (8.6), we obtain

$$\dot{c}_i(t) = -k_{ii} c_i(t) + \sum_{\substack{j=1 \\ j \neq i}}^{m} k_{ij} c_j(t).$$

This set of linear differential equations can be expressed as $\dot{\underline{c}}(t) = \mathbf{K}\underline{c}(t)$, and it has the following solution:

$$\underline{c}(t) = \exp(\mathbf{K}t)\, \underline{c}_0,$$

where the initial conditions are *premultiplied* by $\exp(\mathbf{K}t)$ (instead of the *post-multiplication* in the case of amounts; cf. equation 8.5).

These equations are widely used to simulate simple or complex compartmental systems and currently to identify pharmacokinetic systems from observed time–concentration data. However, it is not always possible to write the equations in terms of concentrations that represent true physical blood or plasma levels. In practice, it may occur that some, say two, compartments exchange so rapidly on the time scale of an experiment that they are not distinguishable but merge kinetically into one compartment. If the two compartments represent material that exists at different concentrations in two different spaces, or two forms of a compound in one space, the calculated concentration may not correspond to any actual measurable concentration and so may be misleading. For this reason the development of differential equations in terms of compartment amounts $q_i(t)$ is more general. If these equations are available, it is not difficult to convert to concentrations $c_i(t)$ by assuming that V_i is a proportionality constant, called the *apparent volume of distribution*, and to solve the equations as long as the volumes are constant in time [311]. If the volumes are changing the problem becomes more difficult.

8.4 Random Fractional Flow Rates

The deterministic model with random fractional flow rates may be conceived on the basis of a deterministic transfer mechanism. In this formulation, a given replicate of the experiment is based on a particular realization of the random fractional flow rates and/or initial amounts $\underline{\Theta}$. Once the realization is determined, the behavior of the system is deterministic. In principle, to obtain from the assumed distribution of $\underline{\Theta}$ the distribution of $q_i(t)$, $i = 1, \ldots, m$, the common approach is to use the classical procedures for transformation of variables. When the model is expressed by a system of differential equations, the solution can be obtained through the theory of random differential equations [312–314].

However, in practice, one can find the moments directly using conditional expectations (cf. Appendix D):

$$\begin{aligned} E\left[q_i\left(t\right)\right] &= E_{\underline{\Theta}}\left[q_i\left(t \mid \underline{\Theta}\right)\right], \\ Var\left[q_i\left(t\right)\right] &= Var_{\underline{\Theta}}\left[q_i\left(t \mid \underline{\Theta}\right)\right]. \end{aligned}$$

Besides the deterministic context, the predicted amount of material is subjected now to a variability expressed by the second equation. This expresses the random character of the fractional flow rate, and it is known as *process uncertainty*. Extensive discussion of these aspects will be given in Chapter 9.

Example 4 *One-Compartment Model*

As an illustration of the procedure, consider the one-compartment model $q\left(t\right) = q_0 \exp\left(-kt\right)$. Assuming that k has a gamma distribution $k \sim$Gam(λ, μ), one has the solutions

$$\begin{aligned} E\left[q\left(t\right)\right] &= q_0 E\left[\exp\left(-kt\right)\right] = q_0\left(1 + t/\lambda\right)^{-\mu}, \\ Var\left[q\left(t\right)\right] &= q_0^2 Var\left[\exp\left(-kt\right)\right] = q_0^2\left[\left(1 + 2t/\lambda\right)^{-\mu} - \left(1 + t/\lambda\right)^{-2\mu}\right]. \end{aligned}$$

Figure 8.2 shows $E\left[q\left(t\right)\right]$ and $E\left[q\left(t\right)\right] \pm \sqrt{Var\left[q\left(t\right)\right]}$ with $q_0 = 1$ and $k \sim$Gam$(2, 2)$. Noteworthy is that confidence intervals are present due to the variability of the fractional flow elimination rate k. This variability is inherent to the process and completely different from that introduced by the measurement devices. ■

8.5 Nonlinear Compartmental Models

Many systems of interest are actually nonlinear:

- A first formulation considers the transfer rates of material from compartment i to j as functions of the amounts in all compartments $\underline{q}\left(t\right)$ and of time t, i.e., $R_{ij}\left[\underline{q}\left(t\right), t\right]$. In this case, $R_{ij}\left(t\right)$ in (8.1) should be substituted by $R_{ij}\left[\underline{q}\left(t\right), t\right]$. If we expand the $R_{ij}\left[\underline{q}\left(t\right), t\right]$ in a Taylor series of $\underline{q}\left(t\right)$ and retain only the linear terms, the nonlinear transfer rates take the form $k_{ij}\left(t\right) q_i\left(t\right)$ and one obtains a linear time-varying compartmental model.

- A second formulation considers the fractional flow rate of material as a function of $\underline{q}\left(t\right)$ and t, i.e., $k_{ij}\left[\underline{q}\left(t\right), t\right]$. In this case, k_{ij} in (8.4) should be substituted by $k_{ij}\left[\underline{q}\left(t\right), t\right]$.

Therefore, the transfer rates and the fractional flow rates are functions of the vector $\underline{q}\left(t\right)$ and t. The dependence on t may be considered as the exogenous environmental influence of some fluctuating processes. If no environmental dependence exists, it is more likely that the transfer rates and the fractional flow rates depend only on $\underline{q}\left(t\right)$. Nevertheless, since $\underline{q}\left(t\right)$ is a function of time, the

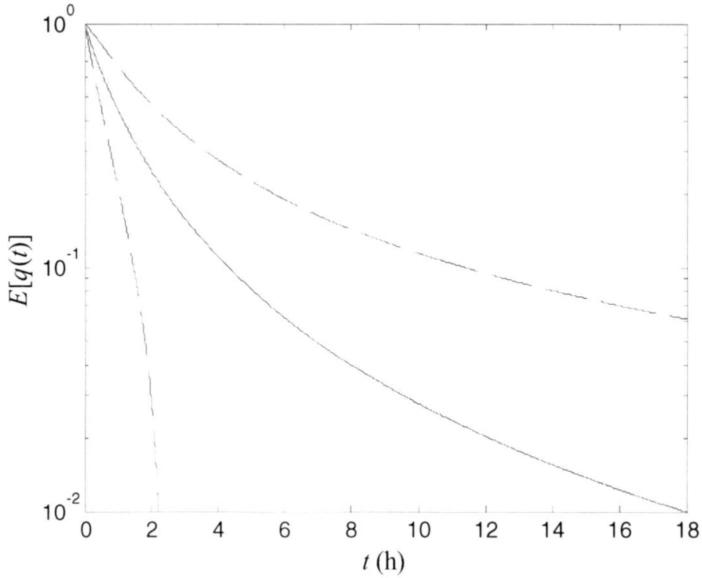

Figure 8.2: One-compartment model with gamma-distributed elimination flow rate $k \sim \mathrm{Gam}(2,2)$. The *solid line* represents the expected profile $E\left[q\left(t\right)\right]$, and *dashed lines*, the confidence intervals $E\left[q\left(t\right)\right] \pm \sqrt{Var\left[q\left(t\right)\right]}$.

observed data in the inverse problem can reveal only a time dependency of the transfer rate, i.e., $R_{ij}\left(t\right)$, or of the fractional flow rate, i.e., $k_{ij}\left(t\right)$. Hence, the dependency of $R_{ij}\left(t\right)$ and $k_{ij}\left(t\right)$ on $\underline{q}\left(t\right)$ is obscured, and a second-level modeling problem now arises, i.e., how to regress the observed dependency on the $\underline{q}\left(t\right)$ and t separately. This problem is mentioned in Appendix C.

Until now, the compartmental model was considered as consisting of compartments associated with several anatomical locations in the living system. The general definition of the compartment allows us to associate in the same location a different chemical form of the original molecule administered in the process. In other words, the compartmental analysis can include not only diffusion phenomena but also chemical reaction kinetics.

One source of nonlinear compartmental models is processes of enzyme-catalyzed reactions that occur in living cells. In such reactions, the reactant combines with an enzyme to form an enzyme–substrate complex, which can then break down to release the product of the reaction and free enzyme or can release the substrate unchanged as well as free enzyme. Traditional compartmental analysis cannot be applied to model enzymatic reactions, but the law of mass-balance allows us to obtain a set of differential equations describing mechanisms implied in such reactions. An important feature of such reactions is that the enzyme is sometimes present in extremely small amounts, the concentration of enzyme being orders of magnitude less than that of substrate.

8.5.1 The Enzymatic Reaction

The mathematical basis for enzymatic reactions stems from work done by Michaelis and Menten in 1913 [315]. They proposed a situation in which a substrate reacts with an enzyme to form a complex, one molecule of the enzyme combining with one molecule of the substrate to form one molecule of complex. The complex can dissociate into one molecule of each of the enzyme and substrate, or it can produce a product and a recycled enzyme. Schematically, this can be represented by

$$\begin{aligned}&[\text{substrate}] + [\text{enzyme}] \underset{k_{-1}}{\overset{k_{+1}}{\rightleftarrows}} [\text{complex}]\,,\\ &[\text{complex}] \overset{k_{+2}}{\rightarrow} [\text{product}] + [\text{enzyme}]\,.\end{aligned} \tag{8.7}$$

In this formulation k_{+1} is the rate parameter for the forward substrate–enzyme reaction, k_{-1} is the rate parameter for the backward reaction, and k_{+2} is the rate parameter for the creation of the product.

Let $s(t)$, $e(t)$, $c(t)$, and $w(t)$ be the amounts of the four species in the reaction (8.7), and s_0 and e_0 the initial amounts for substrate and enzyme, respectively. The differential equations describing the enzymatic reaction,

$$\begin{aligned}\dot{s}(t) &= -k_{+1}s(t)\left[e_0 - c(t)\right] + k_{-1}c(t)\,, & s(0) &= s_0,\\ \dot{c}(t) &= k_{+1}s(t)\left[e_0 - c(t)\right] - (k_{-1} + k_{+2})\,c(t)\,, & c(0) &= 0,\\ \dot{w}(t) &= k_{+2}c(t)\,, & w(0) &= 0,\end{aligned} \tag{8.8}$$

are obtained by applying the law of mass-balance for the rates of formation and/or decay, and the conservation law for the enzyme, $e_0 = e(t) + c(t)$.

Relying on a suggestion of Segel [316], we make the variables of the above equations dimensionless

$$\begin{aligned} x(\tau) &= \frac{s(t)}{s_0}, & y(\tau) &= \frac{c(t)}{e_0}, & z(\tau) &= \frac{w(t)}{s_0},\\ \lambda &= \frac{k_{+2}}{k_{+1}s_0}, & \kappa &= \frac{k_{-1}+k_{+2}}{k_{+1}s_0}, & \varepsilon &= \frac{s_0}{e_0},\end{aligned}$$

with $\tau = k_{+1}e_0t$ and $\kappa \geqslant \lambda$. The set of differential equations becomes

$$\begin{aligned}\dot{x}(\tau) &= -x(\tau)\left[1 - y(\tau)\right] + (\kappa - \lambda)\,y(\tau)\,, & x(0) &= 1,\\ \dot{y}(\tau) &= \varepsilon\left\{x(\tau)\left[1 - y(\tau)\right] - \kappa y(\tau)\right\}, & y(0) &= 0,\\ \dot{z}(\tau) &= \lambda y(\tau)\,, & z(0) &= 0.\end{aligned}$$

This system cannot be solved exactly, but numerical methods are easily able to generate good solutions. The time courses for all reactant species of reaction (8.7) generated from the previous equations with $(\kappa, \lambda) = (0.015, 0.010)$ and $\varepsilon = 2$ are shown in the semilogarithmic plot of Figure 8.3. We note that:

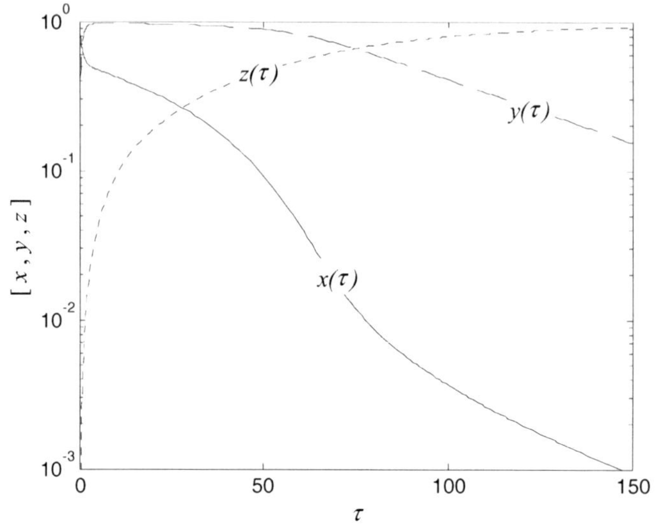

Figure 8.3: Profiles of dimensionless reactant amounts, substrate $x\,(\tau)$, complex $y\,(\tau)$, and product $z\,(\tau)$.

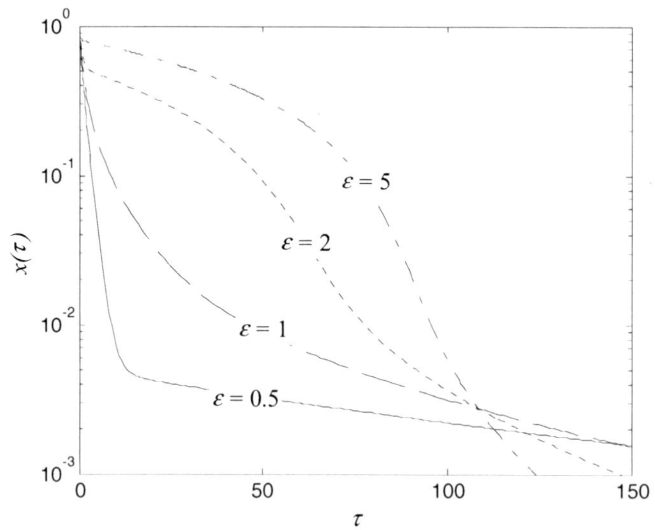

Figure 8.4: Influence of ε on the substrate $x\,(\tau)$ profiles with fixed $(\kappa, \lambda) = (0.015, 0.010)$ and $\varepsilon = (0.5, 1, 2, 5)$.

- The substrate $x(\tau)$ drops from its initial condition value, equal to 1, at a rapid rate, but quickly it decelerates. Progressively, and for $\tau > 50$, the substrate decreases rapidly in a first phase and then slowly, in a second phase. This irregular profile of substrate in the semilogarithmic plot is reflected as a *concavity* or *nonlinearity*, as it is usually called.

- The intermediate compound complex $y(\tau)$ reaches a maximum (called *quasi-steady state* in biology) that persists only for a time period and then decreases; this time period corresponds to the period of nonlinearity for the substrate time course. In fact, saturation of the complex form is responsible for the nonlinearity in the substrate time course. During this period, there is no free enzyme to catalyze the substrate conversion toward the product.

- The product $z(\tau)$ reaches the maximum plateau level asymptotically. In contrast to the substrate profile, the nonlinear behavior along the saturation of the complex is not easily defined on the product profile.

Figure 8.4 shows the influence of ε on the $x(\tau)$ shape. For fixed (κ, λ), we simulated the time courses for $\varepsilon = 0.5, 1, 2, 5$. It is noted that the shape of the substrate profiles varies remarkably with the values of ε; thus profiles of biphasic, power-law, and nonlinear type are observed. So, the sensitivity of the kinetic profile regarding the available substrate and enzyme amounts is studied by using several ε values: for low substrate or high enzyme amounts the process behaves according to two decaying convex phases, in the reverse situation the kinetic profile is concave, revealing nonlinear behavior.

Other processes that lead to nonlinear compartmental models are processes dealing with transport of materials across cell membranes that represent the transfers between compartments. The amounts of various metabolites in the extracellular and intracellular spaces separated by membranes may be sufficiently distinct kinetically to act like compartments. It should be mentioned here that Michaelis–Menten kinetics also apply to the transfer of many solutes across cell membranes. This transfer is called *facilitated diffusion* or in some cases *active transport* (cf. Chapter 2). In facilitated diffusion, the substrate combines with a membrane component called a carrier to form a carrier–substrate complex. The carrier–substrate complex undergoes a change in conformation that allows dissociation and release of the unchanged substrate on the opposite side of the membrane. In active transport processes not only is there a carrier to facilitate crossing of the membrane, but the carrier mechanism is somehow coupled to energy dissipation so as to move the transported material up its concentration gradient.

8.6 Complex Deterministic Models

The branching pattern of the vascular system and the blood flow through it has continued to be of interest to anatomists, physiologists, and theoreticians [4, 317, 318]. The studies focusing on the geometric properties such as lengths, diameters, generations, orders of branches in the pulmonary, venular, and arterial tree of mammals have uncovered the principles on which these properties are based. Vascular trees seem to display roughly the same dichotomous branching pattern at different levels of scale, a property found in fractal structures [319–321]. The hydrodynamics of blood flow in individual parts of the dichotomous branching network have been the subject of various studies. Recently, West et al. [322], relying on an elegant combination of the dynamics of energy transport and the mathematics of fractal geometry, developed a hydrodynamic model that describes how essential materials are transported through space-filling fractal networks of branching tubes.

Although these advances provide an analysis of the scaling relations for mammalian circulatory systems, models describing the transport of materials along the entire fractal network of the mammalian species are also needed. Pharmacokinetics and toxicokinetics, the fields in which this kind of modeling is of the greatest importance, are dominated by the concept of homogeneous compartments [323]. Physiologically based pharmacokinetic models have also been developed that define the disposition patterns in terms of physiological principles [257, 323, 324]. The development of models that study the heterogeneity of the flow and the materials distribution inside vascular networks and individual organs has also been fruitful in the past years [269, 325–327]. Herein, we present a simple model for the heterogeneous transport of materials in the circulatory system of mammals, based on a single-tube convection–dispersion system that is equivalent to the fractal network of the branching tubes.

8.6.1 Geometric Considerations

We consider a fractal arterial tree that consists of several branching levels where each level consists of parallel vessels, Figure 8.5 A. Each vessel is connected to m vessels of the consequent branching level [322]. We make the assumption that the vessel radii and lengths at each level k follow a distribution around the mean values ρ_k and μ_k, respectively. The variance of the vessel radii and lengths at each level produces heterogeneity in the velocities.

The total flow across a section of the entire tree is constant (conservation of mass). This allows us to replace the tree with a single 1-dimensional tube. Since the tree is not area-preserving and the area of the cross section of the tube is equal to the total area of the cross sections of each level of the tree, the total cross-sectional area of subsequent levels increases, i.e., the tube is not cylindrical (Figure 8.5 A-C).

Based on the scaling properties of the fractal tree, the noncylindrical tube is described in terms of a continuous spatial coordinate, z, which replaces the branching levels of the fractal tree from the aorta to the capillaries. As suggested by West [322], both the radii and the vessel lengths scale according to "cubic law" branching, i.e., $\rho_{k+1}/\rho_k = \mu_{k+1}/\mu_k = m^{-1/3}$. These assumptions allow us to obtain the expression for the area $\mathcal{A}(z)$ of the noncylindrical tube (Figure 8.5

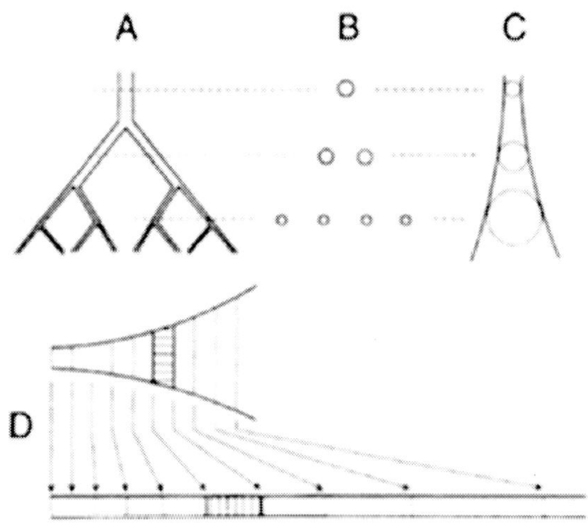

Figure 8.5: (**A**) Schematic representation of the dichotomous branching network. (**B**) Cross sections at each level. (**C**) Single tube with continuously increasing radius. (**D**) Volume-preserving transformation of the varying radius tube to a fixed radius tube. Reprinted from [328] with permission from Springer.

C) as a function of the coordinate z:

$$\mathcal{A}(z) = \frac{\pi \rho_0^2 \mu_0 \widetilde{m}}{z\,(1-\widetilde{m}) + \mu_0 \widetilde{m}}, \tag{8.9}$$

where ρ_0 and μ_0 are the radius and the length of aorta, respectively, and $\widetilde{m} = m^{1/3}$.

Further, a volume-preserving transformation allows the replacement of the varying radius tube with a tube of fixed radius ρ_0 and fixed area $\mathcal{A}_0 = \pi \rho_0^2$ (Figure 8.5 D). This is accomplished by replacing z with a new coordinate z^* with the condition that the constant total flow of the fluid across a section is kept invariant under the transformation:

$$z = \frac{\mu_0 \widetilde{m}}{\widetilde{m} - 1}\left\{1 - \exp\left[\frac{z^*\,(1-\widetilde{m})}{\mu_0 \widetilde{m}}\right]\right\}. \tag{8.10}$$

8.6.2 Tracer Washout Curve

The disposition of a solute in the fluid as it flows through the system is governed by convection and dispersion. The convection takes place with velocity

$$v\,(z) = \frac{\mathcal{A}_0}{\mathcal{A}(z)} v_0, \tag{8.11}$$

where v_0 is the velocity in the aorta and $\mathcal{A}(z)$ is given by (8.9). If molecular diffusion is considered negligible, dispersion is exclusively geometric and consists of two components originating from the variance of the path lengths and of the vessel radii. Because the components are independent of each other, the global form of the dispersion coefficient is

$$D(z) = \left[k_1 \sigma_{10}^2 + 2k_2 \sigma_{20}^2 \frac{\mu_0}{\rho_0}\right] \left[\frac{\mathcal{A}_0}{\mathcal{A}(z)}\right]^2 v_0, \tag{8.12}$$

where k_1 and k_2 are proportionality constants, and σ_{10}^2 and σ_{20}^2 are the variances of the radius and the length of aorta, respectively [326, 329, 330]. The equation that describes the concentration $c(z,t)$ of solute inside the tube is a convection–dispersion partial differential equation:

$$\frac{\partial c(z,t)}{\partial t} = \frac{\partial}{\partial z}\left[D(z) \frac{\partial c(z,t)}{\partial z}\right] - v(z) \frac{\partial c(z,t)}{\partial z}$$

with $D(z)$ and $v(z)$ given by (8.12) and (8.11), respectively. Applying the transformation (8.10), the previous equation becomes a simple convection–dispersion equation with constant coefficients:

$$\frac{\partial c(z^*,t)}{\partial t} = D_0^* \frac{\partial^2 c(z^*,t)}{\partial z^{*2}} - v_0^* \frac{\partial c(z^*,t)}{\partial z^*}, \tag{8.13}$$

where

$$D_0^* = k_0 v_0, \qquad v_0^* = \left(\widetilde{m}\frac{k_0}{\mu_0} + 1\right) v_0, \qquad k_0 = k_1 \sigma_{10}^2 + 2k_2 \sigma_{20}^2 \frac{\mu_0}{\rho_0}.$$

These forms relate the dependence on the system characteristics. Equation (8.13) describes the concentration $c(z^*,t)$ of a solute in a tree-like structure that corresponds to the arterial tree of a mammal. Considering also the corresponding venular tree situated next to the arterial tree and appropriate inflow and outflow boundary conditions, we are able to derive an expression for the spatiotemporal distribution of a tracer inside a tree-like transport network. We also make the assumption that the arterial and venular trees are symmetric, that is, have the same volume V; then, the total length is $L = V/\mathcal{A}_0$. The initial condition is $c(z^*,0) = 0$ and the boundary conditions are:

- Inflow at $z^* = 0$:

$$\left[-D_0^* \frac{\partial c(z^*,t)}{\partial z^*} + v_0^* c(z^*,t)\right]\bigg|_{z^*=0} = \frac{q_0}{a_0}\delta(t)$$

  where q_0 is the dose, and $\delta(t)$ is the Dirac delta function.

- Outflow at $z^* = L$:

$$\left.\frac{\partial c(z^*,t)}{\partial z^*}\right|_{z^*=L} = 0.$$

The outflow concentration $c(L,t)$ of the above model describes tracer washout curves from organs that have a tree-like network structure, and it is given by an analytic form reported in [328].

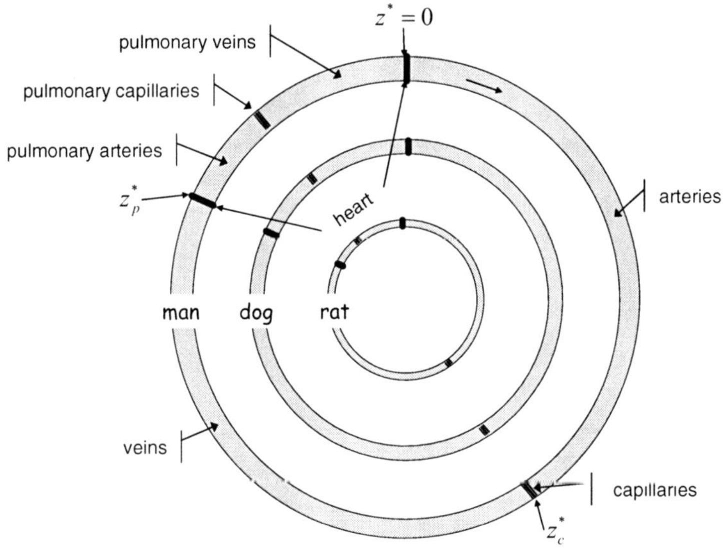

Figure 8.6: Schematic representation of the ring shaped tube that models the circulatory system of a mammal. Blood flows clockwise. The tube is divided into segments corresponding to the arterial, venular, pulmonary arterial, and pulmonary venular trees.

8.6.3 Model for the Circulatory System

Based on the above, an elementary pharmacokinetic model considering the entire circulatory system was constructed. Thus, apart from the arterial and venular trees, a second set of arterial and venular trees, corresponding to the pulmonary vasculature, must be considered as well. These trees follow the same principles of (8.10) and (8.13), i.e., tubes of radius ρ_0 are considered with appropriate length to accommodate the correct blood volume in each tree.

Structure

An overall tube of appropriate length L is considered and is divided into four sequential parts, characterized as arterial, venular, pulmonary arterial, and pulmonary venular, Figure 8.6.

We assign the first portion of the tube length from $z^* = 0$ to $z^* = z_c^*$ to the arterial tree, the next portion from $z^* = z_c^*$ to $z^* = z_p^*$ to the venular, and the rest from $z^* = z_p^*$ to $z^* = L$ to the two symmetrical trees of the lungs. We consider that the venular tree is a structure similar to the arterial tree, only of greater, but fixed, capacity. Also, the two ends of the tube are connected, to allow recirculation of the fluid. This is implemented by introducing a boundary condition, namely $c(0,t) = c(L,t)$, which makes the tube ring-shaped. The "heart" is located at two separate points. The left ventricle-left atrium is situated at $z^* = 0$, and the right ventricle-right atrium is situated at $z^* = z_p^*$, Figure 8.6.

Dispersion

Two separate values were used for the dispersion coefficient D_a for the arterial segment and D_p for the pulmonary segment. For the venular segment we consider that the dispersion coefficient has the value $D_a \left(z_p^* - z_c^*\right)/z_c^*$, meaning that it is proportional to the length of the segment. The flux preservation boundary condition,

$$D_p \left. \frac{\partial c\left(z^*, t\right)}{\partial z^*} \right|_{z^*=L} = D_a \left. \frac{\partial c\left(z^*, t\right)}{\partial z^*} \right|_{z^*=0},$$

must also be satisfied.

Elimination

The contribution of elimination of drugs is appreciable and is integrated into the model. A segment in the capillary region of the tube $(z^* \approx z_c^*)$ is assigned as the elimination site and a first-order elimination term $kc\left(z^*, t\right)$ is now introduced in (8.13). The length of the elimination segment is arbitrarily set to $0.02L$, which is in the order of magnitude of the capillary length. The position of the elimination site is imprecise in physiological terms, but it is the most reasonable choice in order to avoid further model complexity.

Drug Administration and Sampling

The necessary initial condition for the intravenous administration of an exogenous substance, $c\left(z^*, 0\right)$, which is the spatial profile of c at the time of administration, is determined by the initial dose and the type of administration. This profile may have the shape of a "thin" Gaussian function if an intravenous bolus administration is considered, or the shape of a "rectangular" gate for constant infusion. The reference location z_0^* of this profile for an intravenous administration must be set close to the heart. Similarly, when lung administration is considered, z_0^* should be set in the capillary area of the lungs. Due to the geometric character of the model, a sampling site z_s^* should be either specified, in simulation studies, or calculated when fitting is performed.

The final model can be summarized as follows:

$$\frac{\partial c\left(z^*, t\right)}{\partial t} = \frac{\partial}{\partial z^*}\left[D^*\left(z^*\right)\frac{\partial c\left(z^*, t\right)}{\partial z^*}\right] - v_0^* \frac{\partial c\left(z^*, t\right)}{\partial z^*} - W\left(z^*\right) kc\left(z^*, t\right),$$

where $W\left(z^*\right)$ is a combination of delayed in space Heaviside functions, i.e., $W\left(z^*\right) = u\left(z^* - z_c^* + 0.01L\right) - u\left(z^* - z_c^* - 0.01L\right)$, and

$$D^*\left(z^*\right) = \begin{cases} D_a & \text{for} \quad 0 < z^* \leq z_c^*, \\ D_a\left(z_p^* - z_c^*\right)/z_c^* & \text{for} \quad z_c^* < z^* \leq z_p^*, \\ D_p & \text{for} \quad z_p^* < z^* \leq L. \end{cases}$$

Boundary and initial conditions are considered as discussed above.

Example 5 *Indocyanine Green Injection*

The model was used to identify indocyanine green profile in man after a $q_0 = 10\,\text{mg}$ intravenous bolus injection. Both injection and sampling sites (z_0^* and z_s^*, respectively) were closely located on the ring-shaped tube. The model of drug administration was a "thin" Gaussian function:

$$c\left(z^*,0\right) = \frac{q_0}{V}\sqrt{\frac{b}{\pi}}\exp\left[-b\left(\frac{z^*}{L}-\frac{z_0^*}{L}\right)^2\right].$$

This administration corresponds to a bolus injection at the cephalic vein. The parameters set in the model were $m = 3$, $\mu_0 = 50\,\text{cm}$, $\mathcal{A}_0 = 3\,\text{cm}^2$, and $b = 10^5$. The estimated model parameters were:

- Structure: $z_c^*/L = 0.28$, $z_0^*/L = 0.83$, $z_p^*/L = 0.85$, and $V = 4.4\,\text{l}$. These values result in $L = 1470\,\text{cm}$.
- Dispersion and elimination: $D_a = 1826\,\text{cm}^2\,\text{s}^{-1}$, $D_p = 1015\,\text{cm}^2\,\text{s}^{-1}$, $v_0 = 44.98\,\text{cm}\,\text{s}^{-1}$, and $k = 1.13\,\text{s}^{-1}$.

Figure 8.7 depicts the fitted concentration profile of indocyanine green at the sampling site along with the experimental data. ∎

A 1-dimensional linear convection–dispersion equation was developed with constant coefficients that describes the disposition of a substance inside a tree-like fractal network of tubes that emulates the vascular tree. Based on that result, a simple model for the mammalian circulatory system is built in entirely physiological terms consisting of a ring shaped, 1-dimensional tube. The model takes into account dispersion, convection, and uptake, describing the initial mixing of intravascular tracers. This model opens new perspectives for studies dealing with the disposition of intravascular tracers used for various hemodynamic purposes, e.g., cardiac output measurements [331, 332], volume of circulating blood determination [331], and liver function quantification [333]. Most importantly, the model can be expanded and used for the study of xenobiotics that distribute beyond the intravascular space. In future developments of the model, the positioning of organs that play an important role in the disposition of substances can be implemented by adding parallel tubes at physiologically based sites to the present simple ring-shaped model. Consequently, applications can be envisaged in interspecies pharmacokinetic scaling and physiologically based pharmacokinetic–toxicokinetic modeling, since both fields require a realistic geometric substrate for hydrodynamic considerations.

8.7 Compartmental Models and Heterogeneity

Initially, the deterministic theory was applied to describe the movement of a population of tracer molecules. Briefly, a drug administered as a bolus input into an

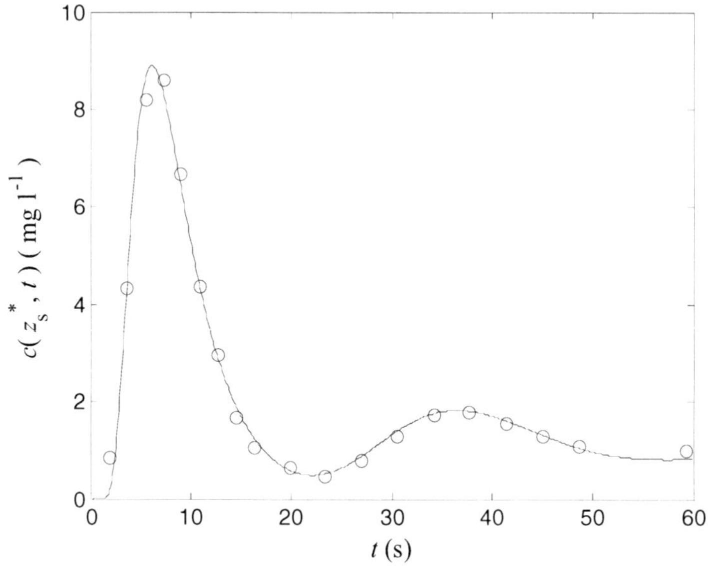

Figure 8.7: Indocyanine profile at the sampling location $z_s^* = 1220$ cm after intravenous bolus administration of 10 mg. The peaks correspond to successive passes of the drug bolus from the sampling site as a result of recirculation. The dots indicate the experimental data.

organ modeled by homogeneous compartments results in a time–concentration curve describing the amount of the drug remaining in the organ as a function of the elapsed time of the form of a sum of exponential terms. Possibly because the individual molecules are infinitesimal in size, in most of the literature the implicit assumption is made of deterministic flow patterns. So, compartmental analysis, grounded on deterministic theory, has provided a rich framework for quantitative modeling in the biomedical sciences with many applications to tracer kinetics in general [334, 335] and also to pharmacokinetics [310]. The linear combinations of exponential function forms have provided a very rich class of curves to fit to time–concentration data, and compartmental models turn out to be good approximations for many processes.

Thus, compartmental models have been used extensively in the pharmacokinetic literature for some time, but not without criticism. These criticisms were directed:

- First, at the compartmental approach per se grounded on the assumption of homogeneous compartments. Compartmental models are in fact appropriate when there is an obvious partitioning of the material in the process into discrete portions, the compartments that exchange amounts of materials. From a theoretical standpoint, there has always been a consensus that the notion of a homogeneous compartment is merely a simplified representation for different tissues that are pooled together [336, 337].

- Second, at the fact that the models obtained are not necessarily exact because mixing in a compartment is not instantaneous. How good a compartment model is depends on the relative rates of mixing within a compartment as compared to the transfer rates between the compartments. Mixing may occur by diffusion, various types of convection, and combinations of them, so it is difficult to come up with a uniform theory of mixing. Ideally, we should measure the concentration of material throughout the process and define mixing in terms of the time course of a ratio such as the standard deviation divided by the mean concentration.

- Third, at the ill-conditioning of numerical problems for parameter estimation with models involving a large number of exponential terms. Wise [299] has developed a class of powers of time models as alternatives to the sums of exponentials models and has validated these alternative models on many sets of experimental data. From an empirical standpoint, Wise [244] reported "1000 or more" published time–concentration curves where alternative models fit the data as well or better than the sums-of-exponentials models.

Moreover, it is clear that even the continuous models are often unreliable models. Matter is atomic, and at a fine enough partition, continuity is no longer an acceptable solution. Furthermore, living tissues are made up of cells, units of appreciable size that are the basic structural and functional units of living things. And cells are not uniform in their interiors; they contain smaller units, the cellular organelles. There is inhomogeneity at a level considerably above the molecular. All these facts enhanced the criticism against determinism and the use of homogeneous compartments. More realistic alternatives have aimed at removing the limiting assumption of homogeneity:

- The process was considered as continuous and compartmental models were used to approximate the continuous systems [335]. For such applications, there is no specific compartmental model that is the best; the approximation improves as the number of compartments is increased. It order to put compartmental models of continuous processes in perspective it may help to recall that the first step in obtaining the partial differential equation, descriptive of a process continuous in the space variables, is to discretize the space variables so as to give many *microcompartments*, each uniform in properties internally. The differential equation is then obtained as the limit of the equation for a microcompartment as its spatial dimensions go to zero. In approximation of continuous processes with compartmental models one does not go to the limit but approximates the process with a finite compartmental system. In that case, the partial differential equation is approximated by a set of simultaneous ordinary differential equations. In philosophy, compartmental modeling shares basic ideas with the finite element method, where the structure of the system is also used to define the elements of a partition of the system. But even if a finite compartmental approximation is used, how can we define the approximation error and its dependence on the size of the compartmental model? In addition, many compartmental models approximating continuous processes are so large that it may be difficult to deal with them and it may be useful or necessary to lump some of the compartments into one compartment. This raises a set of questions about the errors incurred in aggregation and about the optimal way of aggregating compartments.

- Noncompartmental models were introduced as models that allow for transport of material through regions of the body that are not necessarily well mixed or of uniform concentration [248]. For substances that are transported relatively slowly to their site of degradation, transformation, or excretion, so that the rate of diffusion limits their rate of removal from the system, the noncompartmental model may involve diffusion or other random walk processes, leading to the solution in terms of the partial differential equation of diffusion or in terms of probability distributions. A number of noncompartmental models deal with plasma time–concentration curves that are best described by power functions of time.

- Physiological and circulatory models have been developed, and they have provided information of physiological interest that was not available from compartmental analysis. Rapidly, physiological models turned to the modeling of complex compartmental structures. In contrast, circulatory models associated with a statistical framework have proved powerful in describing heterogeneity in the process [246, 338]. Recently, the above presented complex model for the entire circulatory system was built, describing the initial mixing following an intravascular administration in a treelike network by a relatively simple convection–dispersion equation [328, 339].

- Stochastic compartmental analysis assumes probabilistic behavior of the molecules in order to describe the heterogeneous character of the processes. This approach is against the unrealistic notion of the "well-stirred" system, and it is relatively simpler mathematically than homogeneous multicompartment models. At first glance, this seems to be a paradox since the conventional approaches rely on the simpler hypothesis of homogeneity. Plausibly, this paradox arises from the analytical power of stochastic approaches and the unrealistic hypothesis of homogeneity made by the compartmental analysis. Nevertheless with only few exceptions, stochastic modeling has been slow to develop in pharmacokinetics and only recently have some applications also included stochastic behavior in the models.

In conclusion, compartmental models are generally well determined if there is an obvious partitioning of the material in compartments, and if the mixing processes within these compartments are considerably faster than the exchanges between the compartments.

9

Stochastic Compartmental Models

Résumons nos conclusions... C'est donc en termes probabilistes que les lois de la dynamique doivent être formulées lorsqu'elles concernent des systèmes chaotiques.

Ilya Prigogine (1917-2003)
1977 Nobel Laureate in Chemistry
La fin des certitudes

The "real world" of compartmental systems has a strong stochastic component, so we will present a stochastic approach to compartmental modeling. In deterministic theory developed in Chapter 8, each compartment is treated as being both homogeneous and a continuum. But:

- Biological media are inhomogeneous, and the simplest way to capture structural and functional heterogeneity is to operate at a molecular level. First, one has to model the time spent by each particle in the process and second, to statistically compile the molecular behaviors. As will be shown in Section 9.3.4, this compilation generates a process uncertainty that did not exist in the deterministic model, and this uncertainty is the expression of process heterogeneity.

- Matter is atomic, not continuous, and cells and molecules come in discrete units. Thus, in compartmental models of chemical reactions and physiological processes, a compartment contains an integral number of units and in any transfer only an integral number of units can be transferred. Consequently, it is important to develop the theory for such systems in which transfers occur in discrete numbers of units, and that is done in terms of the probabilities of transfer of one unit from one compartment to another or to the outside.

As concluded in Chapter 7, the observed time-varying features of a process are expressions of structural and functional heterogeneity. Observations gathered from such processes were fitted by power-law and gamma-type functions. Marcus was the first to suggest stochastic modeling as an alternative working hypothesis to the empirical power-law or gamma-type functions [300]. At the same time, stochastic modeling began to provide applications in compartmental analysis either as multivariate Markov immigration–emigration models [340–342], or as random-walk models [298, 299], or as semi-Markov (Markov renewal) models [343, 344].

In deterministic theory we started with the definition of a compartment as a kinetically homogeneous amount of material. The equivalent definition in stochastic theory is that the probability of a unit participating in a particular transfer out of a compartment, at any time, is the same for all units in the compartment.

9.1 Probabilistic Transfer Models

The present stochastic model is the so-called *particle model*, where the substance of interest is viewed as a set of particles.[1] We begin consideration of stochastic modeling by describing Markov-process models, which loosely means that the probability of reaching a future state depends only on the present and not the past states. We assume that the material is composed of particles distributed in an m-compartment system and that the stochastic nature of material transfer relies on the independent random movement of particles according to a continuous-time *Markov process.*

9.1.1 Definitions

The development of probabilistic transfer models is based on two probabilities, a conditional probability and a marginal one, commonly stated as *transfer* and *state probabilities*, respectively.

- The transfer probability $p_{ij}(t_\circ, t)$ gives the conditional probability that "a given particle resident in compartment i at time $t_\circ$ will be in compartment j at time t." Because the particles move independently, the transfer probabilities do not depend on the number of other particles in the compartments. In this way, the $p_{ij}(t_\circ, t)$ serve to express the Markovian process. Indeed, the Markov process can be expressed in terms of the $m \times m$ transfer-intensity matrix $\mathbf{H}(t)$ with (i, j)th element $h_{ij}(t)$ given by

$$h_{ij}(t) = \lim_{\Delta t \to 0} \frac{p_{ij}(t, t + \Delta t)}{\Delta t} \qquad \text{and} \qquad h_{ii}(t) = h_{i0}(t) + \sum_{\substack{j=1 \\ j \neq i}}^{m} h_{ij}(t) . \tag{9.1}$$

[1] The terms "drug molecule" and "particle" will be used in this chapter interchangeably.

The so-defined elements $h_{ij}(t)$ of the transfer-intensity matrix are called the *hazard rates*, and define the conditional probability

$$\Pr\left[\text{transfer to } j \text{ by } t+\Delta t \mid \text{present in } i \text{ at } t\right] \triangleq h_{ij}(t)\,\Delta t + o(\Delta t)$$

that "a given particle resident in i at time t leaves by $t+\Delta t$ to go in j," where Δt is small and $o(\Delta t)$ denotes all possible higher-order terms of Δt.

- The state probability $p_{ij}(t)$ is the special case of the transfer probability where $t_{\circ}$ is the starting time, i.e., $t_{\circ} \equiv 0$. The state probability gives the probability that "a given particle starting in i at time 0 is resident in compartment j at time t." All these probabilities may be considered as the (i,j)th elements of the $m \times m$ state probabilities matrix $\mathbf{P}(t)$, with $p_{ij}(0) = 0$ when $i \neq j$. Also, to allow all possible movements, of particles starting from any initial position, the initial conditions $p_{ii}(0)$ are set to 1, i.e., $\mathbf{P}(0) = \mathbf{I}$.

In the above expressions, indices i and j may vary between 1 and m with $i \neq j$. Moreover, j may be set to 0, denoting the exterior space of the compartmental configuration.

To obtain equations for the state probabilities, write the equation for the state probability at $t+\Delta t$ as the sum of joint probabilities for all the mutually exclusive events that enumerate all the possible ways in which "a particle starting in i at 0 could pass through the various compartments at time t" AND "end up in j at $t+\Delta t$." These joint probabilities can be expressed as the product of a marginal by a conditional probability. The state probability $p_{ij}(t)$ that "a given particle starting in i at time 0 is resident in compartment s at time t" plays the role of the marginal and the transfer probability $h_{sj}(t)\,\Delta t$ that "a given particle resident in compartment s at time t will next transfer to compartment j, i.e., at time $t+\Delta t$" plays the role of the conditional probability

$$p_{ij}(t+\Delta t) = p_{ij}(t)\left[1 - h_{jj}(t)\,\Delta t\right] + \sum_{\substack{s=1\\ s\neq j}}^{m} p_{is}(t)\,h_{sj}(t)\,\Delta t + o(\Delta t)\,.$$

The first term in the right-hand side expresses the joint probability that "a particle starting in i at 0 is present in j at time t" AND "remains in j at $t+\Delta t$," and the second term expresses the sum of joint probabilities that "a particle starting in i at 0 is present in each compartment s, except j, at time t" AND "ends up in j at $t+\Delta t$."

Rearranging, taking the limit $\Delta t \to 0$ in the above difference equations, and neglecting the higher-order terms of Δt, one obtains m^2 differential equations, namely the *probabilistic transfer model*

$$\dot{p}_{ij}(t) = -h_{jj}(t)\,p_{ij}(t) + \sum_{\substack{s=1\\ s\neq j}}^{m} h_{sj}(t)\,p_{is}(t)\,.$$

These equations are linear differential equations with time-varying coefficients since the hazard rates are time-dependent and may be presented in matrix form as

$$\dot{\mathbf{P}}(t) = \mathbf{P}(t)\,\mathbf{H}(t)$$

with initial conditions $\mathbf{P}(0) = \mathbf{I}$. These models are referred to as *generalized compartmental models* and can be studied using the time-dependent Markov theory [345, 346] but are not of present interest.

In what follows, we will rather restrict ourselves mainly to the standard Markov process in the probabilistic transfer model with time-independent hazard rates. This is equivalent to assuming that the transfer probabilities do not depend on either the time the particle has been in the compartment or the previous history of the process, and

$$\mathbf{H}(t) \equiv \mathbf{H}. \tag{9.2}$$

These equations lead to the matrix solution

$$\mathbf{P}(t) = \exp(\mathbf{H}t). \tag{9.3}$$

In most pharmacokinetic applications, the system is open and the eigenvalues of $\mathbf{H}$ are real and negative. This implies that the solution has the form of a sum of negative exponentials.

9.1.2 The Basic Steps

To illustrate the successive steps in this procedure, we present the case of a simple two-compartment model, Figure 9.1. There will be four differential equations, one for each combination of the i and j previously introduced indices. For example, to obtain the differential equation for $j = 1$, one has to advocate the necessary events for a particle starting in i to pass through the two compartments at time t and end up in 1 at $(t + \Delta t)$:

- "the particle is present in 1 at time t," associated with the state probability $p_{i1}(t)$ AND "it remains in the compartment during the interval from t to $(t + \Delta t)$," associated with the transfer probability $[1 - (h_{10} + h_{12})\,\Delta t]$,

OR

- "the particle is present in 2 at time t," associated with the state probability $p_{i2}(t)$ AND "it goes to 1 during the interval from t to $(t + \Delta t)$," associated with the transfer probability $h_{21}\Delta t$.

Therefore, the probability of the desired joint event may be written as

$$p_{i1}(t + \Delta t) = p_{i1}(t)\,[1 - (h_{10} + h_{12})\,\Delta t] + p_{i2}(t)\,[h_{21}\Delta t].$$

To obtain the differential equation for $j = 2$, one has to advocate the necessary events for a particle starting in i to pass through the two compartments at time t and end up in 2 at $(t + \Delta t)$:

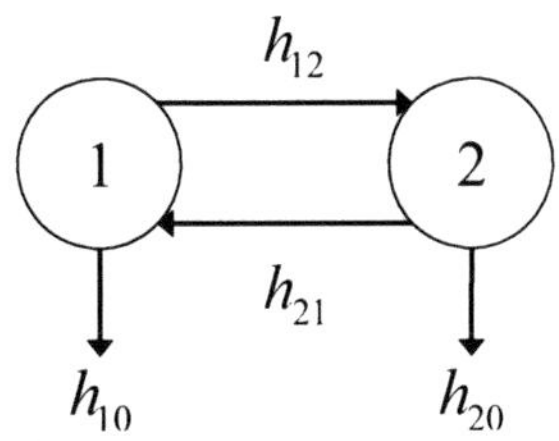

Figure 9.1: Two-compartment configuration.

- "the particle is present in 2 at time t," associated with the state probability $p_{i2}(t)$ AND "it remains in the compartment during the interval from t to $(t + \Delta t)$," associated with the transfer probability $[1 - (h_{20} + h_{21})\,\Delta t]$,

OR

- "the particle is present in 1 at time t," associated with the state probability $p_{i1}(t)$ AND "it goes to 2 during the interval from t to $(t + \Delta t)$," associated with the transfer probability $h_{12}\Delta t$.

Therefore, the probability of the desired joint event may be written as

$$p_{i2}(t + \Delta t) = p_{i2}(t)\,[1 - (h_{20} + h_{21})\,\Delta t] + p_{i1}(t)\,[h_{12}\Delta t]\,.$$

Rearranging and taking the limit $\Delta t \to 0$ for the above difference equations, one has

$$\begin{aligned} \dot{p}_{i1}(t) &= -(h_{10} + h_{12})\,p_{i1}(t) + h_{21}p_{i2}(t)\,, \\ \dot{p}_{i2}(t) &= h_{12}p_{i1}(t) - (h_{20} + h_{21})\,p_{i2}(t)\,, \end{aligned} \tag{9.4}$$

with initial conditions $p_{ii}(0) = 1$ and $p_{ij}(0) = 0$ for $i \neq j$, where $i = 1, 2$. The above differential equations have as solution

$$\begin{bmatrix} p_{11}(t) & p_{12}(t) \\ p_{21}(t) & p_{22}(t) \end{bmatrix} = \exp \begin{bmatrix} -(h_{10} + h_{12})\,t & h_{12}t \\ h_{21}t & -(h_{20} + h_{21})\,t \end{bmatrix}. \tag{9.5}$$

Markov processes have **H** matrices with real negative eigenvalues, which lead to models that are linear combinations of decaying exponentials, which are analogous to the deterministic models. In the presence of distinct multiple eigenvalues, the probability profiles are mixtures of exponentials multiplied by integer powers of time. The integer powers are the numbers of distinct eigenvalues [335]. Nevertheless, in practice, functions that include parts with noninteger powers of time have been needed to provide a satisfactory fit to data [298, 341]. In the face of these "impossible" experimental results, alternative working hypotheses should be created, e.g., retention-time distribution models.

9.2 Retention-Time Distribution Models

A stochastic model may also be defined on the basis of its retention-time distributions. In some ways, this conceptualization of the inherent chance mechanism is more satisfactory since it relies on a continuous-time probability distribution rather than on a conditional transfer probability in discretized units of size Δt.

One first needs the basic notions associated with a continuous probability distribution. Consider the age or the retention time of a molecule in the compartment as a random variable, A. Let:

- $f(a) = \mathrm{d}F(a)/\mathrm{d}a$ be the density function of ages A of the molecules in the compartment,
- $F(a) = \Pr[A < a]$ be the distribution function of A, i.e., the probability that "the molecule will leave the compartment prior to attaining age a," and
- $\mathcal{S}(a) = \Pr[A \geq a] = 1 - F(a)$ be the *survival function*, i.e., the probability that "the molecule survives in the compartment to age a."

From the above relations, the *hazard function* $h(a)$ is defined as

$$h(a) \triangleq f(a)/\mathcal{S}(a). \tag{9.6}$$

Also from this definition, the simple relationship

$$\frac{\mathrm{d}\log\mathcal{S}(a)}{\mathrm{d}a} = -h(a) \tag{9.7}$$

links the survival and the hazard functions.

9.2.1 Probabilistic vs. Retention-Time Models

We look now for the evaluation of the state probability $p(t)$ that "the molecule is in the compartment at time t" in the case of a one-compartment model. To this end, consider the partition $0 = a_1 < a_2 < \cdots < a_{n-1} < a_n = t$ and the $n-1$ mutually exclusive events that "the molecule leaves the compartment between its age instants a_{i-1} and a_i." The state probability $p(t)$ equals the probability of the complement of the above $n-1$ mutually exclusive events,

$$\begin{aligned} p(t) &= 1 - \sum_{i=2}^{n} \Pr[\text{leaves by } a_{i-1} \text{ to } a_i] = 1 - \sum_{i=2}^{n} [F(a_i) - F(a_{i-1})] \\ &= 1 - \sum_{i=2}^{n} [\mathcal{S}(a_{i-1}) - \mathcal{S}(a_i)] = 1 - [\mathcal{S}(a_1) - \mathcal{S}(a_n)] = \mathcal{S}(t). \end{aligned}$$

Therefore, the survival function $\mathcal{S}(a)$ plays the same role as the state probability $p(t)$. But the former independent variable a is defined as the endogenous or within-compartment measure of time after the particle introduction to the compartment, whereas the independent variable t denotes some exterior, exogenous time measure in a system. Only for the one-compartment model do a and t have the same meaning.

The link between the probabilistic transfer model and retention-time distribution model may be explicitly demonstrated by deriving the conditional probability implied in the one-compartment probabilistic transfer model. We look for the probability, $\mathcal{S}(a + \Delta a)$, that "a particle survives to age $(a + \Delta a)$." Clearly, the necessary events are that "the particle survives to age a," associated with the state probability $\mathcal{S}(a)$ AND that "it remains in the compartment during the interval from a to $(a + \Delta a)$," associated with the conditional probability $[1 - h\Delta a]$, where h is the probabilistic hazard rate. Therefore, the probability of the desired joint event may be written as

$$\mathcal{S}(a + \Delta a) = \mathcal{S}(a)[1 - h\Delta a].$$

Then, we can write

$$h\Delta a = 1 - \frac{\mathcal{S}(a + \Delta a)}{\mathcal{S}(a)} = 1 - \frac{1 - F(a + \Delta a)}{\mathcal{S}(a)} = \frac{\Delta F(a)}{\mathcal{S}(a)} \approx \frac{f(a)\Delta a}{\mathcal{S}(a)}.$$

Then, the probabilistic hazard rate h is the particular hazard function value $h(a)$ evaluated at a specified age a. For the retention-time distribution models, $h(a)\Delta a$ gives the conditional probability "that a molecule that has remained in the compartment for age a leaves by $a + \Delta a$." In other words, the probabilistic hazard rate is the instantaneous speed of transfer.

Noteworthy is that only for the exponential distribution is the hazard rate $h(a) = f(a)/\mathcal{S}(a) = \kappa$ not a function of the age a, i.e., the molecule "has no memory" and this is the main characteristic of Markovian processes. In other words, the assumption of an exponential retention time is equivalent to the assumption of an age-independent hazard rate. One practical restriction of this model is that the transfer mechanism must not discriminate on the basis of the accrued age of a molecule in the compartment. In summary, it is clear that the formulations in the probabilistic transfer model and in the retention-time distribution model are equivalent. In the probabilistic transfer model we assume an age-independent hazard rate and derive the exponential distribution, whereas in the retention-time distribution model we assume an exponential distribution and derive an age-independent hazard rate.

For multicompartment models, in addition to the retention-time distributions within each compartment, we require the specification of the *transition probabilities* ω_{ij} of transfer among compartments. These ω_{ij}'s, assumed age-invariant, give the probabilities of transfer from a donor compartment i to each possible recipient compartment j. From (9.1), it follows that $\omega_{ij} = h_{ij}/h_{ii}$ is the probability that a particle in i will transfer to j on the next departure.

9.2.2 Markov vs. Semi-Markov Models

Consider now a multicompartment structure aiming not only to describe the observed data but also to provide a rough mechanistic description of how the data were generated. This mechanistic system of compartments is envisaged with the drug flowing between the compartments. The stochastic elements describing these flows are the ω_{ij} transition probabilities as previously defined. In addition, with each compartment in this mechanistic structure, one can associate a retention-time distribution $f_i(a)$. The so-obtained multicompartment model is referred to as the *semi-Markov* formulation. The semi-Markov model has two properties, namely that:

- the transition probabilities ω_{ij} are time-invariant; this implies that the sequence of compartment visitations for a particle may be described by a Markov chain and
- the retention-time distributions are arbitrary.

The semi-Markov formulation in the compartmental context was originally proposed by Purdue [344] and Mehata and Selvan [347]. The present approach attempts to characterize fully the mechanistic flow pattern between compartments and to use nonmechanistic models with the smallest number of parameters to describe the within-compartment processes. The experimenter might first divide the system into compartments based on known theory. The retention-time distributions within each compartment are specified either through expert knowledge from hazard rates or by fitting alternative models to data. The ω_{ij} transition probabilities are then determined. One advantage of using these nonmechanistic retention times is the incorporation of inhomogeneous compartments and consequential particle age discrimination with a minimum number of additional parameters.

The Markov model is a special case of the semi-Markov model in which all the retention-time variables are exponentially distributed, $A_i \sim \text{Exp}(\kappa_i)$, and κ_i is the parameter of the exponential. In this case, the semi-Markov model parameters are $\kappa_i = h_{ii}$ and $\omega_{ij} = h_{ij}/\kappa_i$ for $j \neq i$ and $i, j = 1, \ldots, m$. This results from the assumption of the Markov model given in (9.1), which implies that the conditional transfer probability from i to j in a time increment Δt is time-invariant, or in other words is independent of the "age" of the particle in the compartment. Particles with such a constant flow rate, or hazard rate, are said to "lack memory" of their past retention time in the compartment.

Figure 9.2 shows a two-compartment Markov model with parameters h_{10}, h_{12}, and h_{21} and the semi-Markov model with exponentially distributed retention times with parameters κ_1, κ_2, and ω. The conversion relationships are

$$h_{10} = \kappa_1 (1 - \omega) \qquad \text{and} \qquad h_{12} = \kappa_1 \omega \qquad \text{and} \qquad h_{21} = \kappa_2$$

from semi-Markov to Markov, and

$$\kappa_1 = h_{10} + h_{12} \qquad \text{and} \qquad \kappa_2 = h_{21} \qquad \text{and} \qquad \omega = \frac{h_{12}}{h_{10} + h_{12}}$$

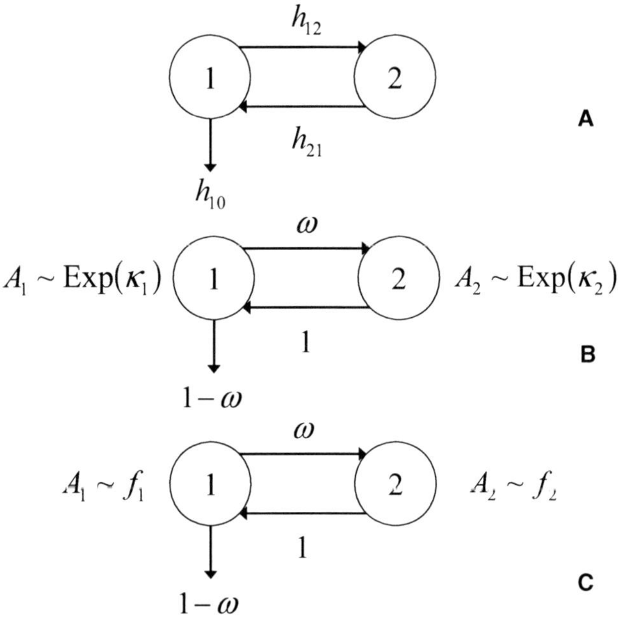

Figure 9.2: The two-compartment Markov model (**A**) vs. the semi-Markov model with exponential retention times (**B**) vs. the general semi-Markov model (**C**).

from Markov to semi-Markov with exponential retention-time distribution. We note that the compartmental structure of the semi-Markov model is simply determined by means of ω. The figure shows also the general semi-Markov model having the same structure determined by ω but allowing several distribution models f_1 and f_2 for the retention times A_1 and A_2, respectively in the compartments.

The causes of nonexponential retention times, and hence age-varying hazard rates, may be numerous. Two general reasons for such retention times in pharmacokinetic applications are noninstant mixing and compartmental heterogeneity. Noninstant mixing, for example, is likely to occur in compartmental models with oral dosing. Inhomogeneous compartments, on the other hand, are a natural consequence of the lumping inherent in dividing the body into two or three compartments, for example into a central and a peripheral compartment. Conversely but less likely, if all the drug particles in a compartment were homogeneous and also well stirred, then the transfer processes that determine how the drug particles leave the compartment could not discriminate on the basis of the accumulated age of a particle in the compartment. Hence such homogeneous, well-stirred compartments could be modeled using exponential transit times.

Table 9.1: Density, survival, and hazard functions.

	$f(a)$	$\mathcal{S}(a)$	$h(a)$
$\text{Exp}(\kappa)$	$\kappa \exp(-\kappa a)$	$\exp(-\kappa a)$	κ
$\text{Erl}(\lambda, \nu)$	$\frac{\lambda^{\nu} a^{\nu-1}}{(\nu-1)!} \exp(-\lambda a)$	$\exp(-\lambda a) \sum_{i=0}^{\nu-1} \frac{(\lambda a)^i}{i!}$	$\frac{\frac{\lambda^{\nu} a^{\nu-1}}{(\nu-1)!}}{\sum_{i=0}^{\nu-1} \frac{(\lambda a)^i}{i!}}$
$\text{Ray}(\lambda)$	$\lambda^2 a \exp\left[-\frac{1}{2}(\lambda a)^2\right]$	$\exp\left[-\frac{1}{2}(\lambda a)^2\right]$	$\lambda^2 a$
$\text{Wei}(\lambda, \mu)$	$\mu \lambda^{\mu} a^{\mu-1} \exp\left[-(\lambda a)^{\mu}\right]$	$\exp[-(\lambda a)^{\mu}]$	$\mu \lambda^{\mu} a^{\mu-1}$

9.2.3 Irreversible Models

One-Compartment Model

The one-compartment model is the typical simple irreversible model. For the one-compartment model and only when initial conditions are given, the exterior time t and the molecule ages a are the same. The state probability $p(t)$ that a molecule is in the compartment is $\mathcal{S}(t)$:

$$p(t) \equiv \mathcal{S}(t). \tag{9.8}$$

One has simply to assume a particular probability distribution for A with the survival function available in a closed form, namely the exponential, Erlang, Rayleigh, and Weibull. Table 9.1 summarizes the probability density functions, survival functions, and hazard rates for the above-mentioned distributions. In these expressions, λ is the scale parameter and μ and ν are shape parameters with $\kappa, \lambda, \mu > 0$ and $\nu = 1, 2, \ldots$.

- Exponential distribution. The survival function is a single exponential $p(t) = \exp(-\kappa t)$. A deterministic one-compartment model produces the same profile, so one can say that this model is the single-exponential distribution of residence times. However, following instantaneous administration of drugs, the time–concentration observed profiles sometimes present two decreasing phases on the semilogarithmic plot. This may be described using the one-compartment model and assuming a mixed distribution consisting of two exponential survival functions $p(t) = \gamma \exp(-\kappa_1 t) + (1-\gamma)\exp(-\kappa_2 t)$, where γ $(\gamma < 1)$ represents the relative contribution of the first exponential term. These biphasic profiles are usually attributed to the two-compartment models. However, there is no rigorous conjunction between the two-exponential and two-compartment models since more complex compartmental models may also give biphasic-like profiles with certain combinations of the microconstants. In the same way, one can use the single compartment model and conceive mixed survival functions containing three or more exponential forms leading to three- or more-phasic profiles. It follows therefore that one cannot discriminate on the basis of observed data alone between the situation in which the survival function in the single compartment is the sum of two exponentials and the situation

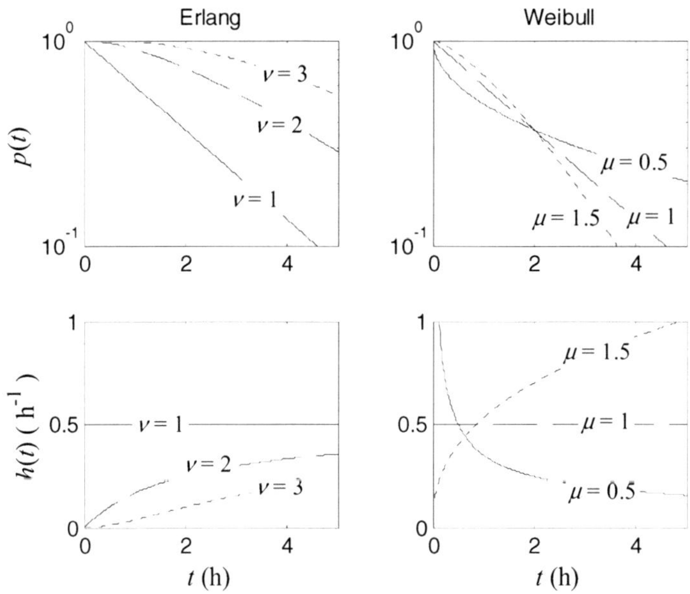

Figure 9.3: State probabilities and hazard functions with $\lambda = 0.5\,\mathrm{h}^{-1}$, and $\nu = 1, 2, 3$ and $\mu = 0.5, 1, 1.5$ for Erlang and Weibull distributions, respectively.

in which a single exponential survival function is associated with each of the two compartments present in the configuration.

- Erlang distribution. We assume that $A \sim \mathrm{Erl}(\lambda, \nu)$. The state probability is

$$p(t) = \exp(-\lambda t) \sum_{i=0}^{\nu-1} \frac{(\lambda t)^i}{i!}.$$

- Weibull and Rayleigh distributions. From Table 9.1, we have

$$p(t) = \exp\left[-(\lambda t)^{\mu}\right] \tag{9.9}$$

for the Weibull distribution and as a special case with $\mu = 2$, the Rayleigh distribution.

Figure 9.3 depicts state probability curves for the Erlang and the Weibull distributions. The hazard rates as functions of time are also illustrated. For $\nu = 1$ and $\mu = 1$, we obtain the behavior corresponding to an exponential retention-time distribution and to the one-compartment deterministic profile.

For $\nu > 1$, in case of an Erlang distribution, the rate function at age 0 is $h(0) = 0$, after which the rate increases and the kinetic profile has a log-concave form. This provides an initial dampening of the retention probability of newly introduced particles. Then, the rate is asymptotic to λ as the age increases. This implies that the age discrimination within the compartment diminishes, either rapidly or slowly depending on ν, as the retention time increases.

The Weibull distribution allows noninteger shape parameter values, and the kinetic profile is similar to that obtained by the Erlang distribution for $\mu > 1$. When $0 < \mu < 1$, the kinetic profile presents a log-convex form and the hazard rate decreases monotonically. This may be the consequence of some saturated clearance mechanisms that have limited capacity to eliminate the molecules from the compartment. Whatever the value of μ, all profiles have common ordinates, $p\,(1/\lambda) = \exp\,(-1)$.

These qualitative features are typical of data from inhomogeneous compartments and/or compartments with noninstant initial mixing. Reciprocally, many compartments that are not well stirred have these properties. For these reasons, the Erlang and Weibull retention-time distributions have been very useful in practice to fit to data. In a theoretical context, Weiss classifies the retention-time distributions according to the log-convexity or concavity of the corresponding time–concentration profiles. Moreover, that author attempts to explain these profiles by assuming time-varying mechanisms as time-varying volume of distribution or clearance capacities [348–350]. As in Section 7.5 for the empirical models, these investigations again reveal the strong link between time dependence and process heterogeneity.

Multicompartment Models

Consider the irreversible two-compartment model with survival, distribution, and density functions $\mathcal{S}_1\,(a)$, $F_1\,(a)$, $f_1\,(a)$ and $\mathcal{S}_2\,(a)$, $F_2\,(a)$, $f_2\,(a)$ for "ages" a of molecules in compartments 1 and 2, respectively. We will assume that at the starting time, the molecules are present only in the first compartment. The state probability $p_1\,(t)$ that "a molecule is in compartment 1 at time t" is $\mathcal{S}_1\,(a)$ with $t = a$; the external time t is the same with the age of the molecule in the compartment 1, i.e., $p_1\,(t) = \mathcal{S}_1\,(t)$. The state probability $p_2\,(t)$ that "a molecule survives in compartment 2 after time t" depends on the length of the time interval a between entry and the 1 to 2 transition, and the interval $t - a$ between this event and departure from the system. To evaluate this probability, consider the partition $0 = a_1 < a_2 < \cdots < a_{n-1} < a_n = t$ and the $n - 1$ mutually exclusive events that "the molecule leaves the compartment 1 between the time instants a_{i-1} and a_i." By applying the *total probability theorem* (cf. Appendix D), $p_2\,(t)$ is expressed as

$$\sum_{i=2}^{n} \Pr\left[\text{survive in 2 to } t \mid \text{leave 1 by } a_{i-1} \text{ to } a_i\right] \Pr\left[\text{leave 1 by } a_{i-1} \text{ to } a_i\right].$$

If $\max\,(a_{i-1} - a_i) \to 0$:

- $\Pr\left[\text{survive in 2 to } t \mid \text{leave 1 by } a_{i-1} \text{ to } a_i\right] = \mathcal{S}_2\,(t - a)$ and

- $\Pr[\text{leave 1 by } a_{i-1} \text{ to } a_i] = F_1(a_i) - F_1(a_{i-1}) = \mathrm{d}F_1(a)$

with $a \in [a_{i-1}, a_i]$. It follows that $p_2(t)$ is the Stieltjes integral of $\mathcal{S}_2$ with respect to F_1 (cf. Appendix E):

$$p_2(t) = \int_0^t \mathcal{S}_2(t-a)\,\mathrm{d}F_1(a) = \int_0^t \mathcal{S}_2(t-a)\,f_1(a)\,\mathrm{d}a = f_1 * \mathcal{S}_2(t), \quad (9.10)$$

that is the convolution of the density function f_1 in the input site with the survival function $\mathcal{S}_2$ in the sampling site. Similarly, the probability that "the molecule will leave the compartment 2 prior to time t" is

$$\int_0^t F_2(t-a)\,\mathrm{d}F_1(a) = f_1 * F_2(t),$$

that is, the convolution of the density function f_1 in the input site with the distribution function F_2 in the sampling site.

This result can be generalized in the case of a catenary irreversible m-compartment model [347]; the state probability in the compartment i ($i = 2, m$) at t is given by

$$p_i(t) = f_1 * \cdots * f_{i-1} * \mathcal{S}_i(t).$$

An elegant form of the previous expressions is obtained in the frequency domain. The convolution becomes the product of the Laplace transform of the survival and the density functions:

$$\widetilde{p}_i(s) = \widetilde{f}_1(s) \cdots \widetilde{f}_{i-1}(s)\,\widetilde{\mathcal{S}}_i(s), \quad (9.11)$$

where $\widetilde{f}(s)$ is the Laplace transform of $f(t)$, $\widetilde{f}(s) = L\{f(t)\}$ (cf. Appendix F).

For analyzing general irreversible compartmental configurations, Agrafiotis [351] developed a semi-Markov technique on the basis of conditional distributions on the retention time of the particles in the compartments before transferring into the next compartment. This approach uses the so-called *forces of separation*, and it is quite different from the one introduced at the beginning of this section, where the distribution of the retention time in each compartment is independent of the compartment that the particle is transferring to.

9.2.4 Reversible Models

Consider the reversible two-compartment model that is explained by way of the semi-Markov formulation as illustrated in Figure 9.2 C. We will assume that at the starting time all molecules are present in compartment 1. A single molecule that is present at the initial time in compartment 1 stays there for a length of time that has a single-passage density function $f_1(a)$. Then, it has the possibility to definitively leave the system with probability $1 - \omega$ or reach the compartment 2 with probability ω. The retention time in this compartment is governed by the single-passage density function $f_2(a)$. At the end of its stay in compartment 2, the molecule reenters compartment 1. Our goal is to evaluate the probability $p_1(t)$ that "a molecule survives in 1 after time t." This event is the compilation of the following mutually exclusive events:

- "survive in 1 without visit in 2" with probability $\mathcal{S}_1(t)$,
- "survive in 1 with 1 visit in 2" with probability $\omega \mathcal{S}_1 * f_1 * f_2(t)$,
- "survive in 1 with 2 visits in 2" with probability $\omega^2 \mathcal{S}_1 * f_1 * f_2 * f_1 * f_2(t)$,

etc. for an infinite number of visits. The probability of a composite event is then

$$p_1(t) = \mathcal{S}_1(t) * \sum_{i=0}^{\infty} \omega^i [f_1 * f_2(t)]^{*i},$$

where f^{*m} denotes the m-fold convolution of f with itself. This last expression has the structure of the renewal density for positive random variables, which is studied in probability theory [346]. Taking the Laplace transform of this expression, the probability has a simpler form in the frequency domain:

$$\widetilde{p}_1(s) = \widetilde{\mathcal{S}}_1(s) \sum_{i=0}^{\infty} \omega^i \widetilde{f}_1^i(s) \widetilde{f}_2^i(s).$$

Because of the boundness of density functions $0 < \int_0^t f(a)\mathrm{d}a < 1$, the infinite sum has a closed-form expression:

$$\widetilde{p}_1(s) = \frac{\widetilde{\mathcal{S}}_1(s)}{1 - \omega \widetilde{f}_1(s) \widetilde{f}_2(s)}. \tag{9.12}$$

The probability $p_2(t)$ that "a molecule survives in 2 after time t" will be given by using the inverse Laplace transform of

$$\widetilde{p}_2(s) = \frac{\omega \widetilde{f}_1(s) \widetilde{\mathcal{S}}_2(s)}{1 - \omega \widetilde{f}_1(s) \widetilde{f}_2(s)}.$$

If at the starting time the molecules are present only in compartment 2, the Laplace transform of the state probabilities in the compartments are

$$\widetilde{p}_1(s) = \frac{\widetilde{f}_2(s) \widetilde{\mathcal{S}}_1(s)}{1 - \omega \widetilde{f}_1(s) \widetilde{f}_2(s)} \quad \text{and} \quad \widetilde{p}_2(s) = \frac{\widetilde{\mathcal{S}}_2(s)}{1 - \omega \widetilde{f}_1(s) \widetilde{f}_2(s)}.$$

To deal with more complex compartmental configurations, the block diagrams and transfer functions are now introduced. Block diagrams are extensively used in the automatic control field [352] to represent the functionality of a process. These are diagrams involving a set of elements each of them representing a given function. When the process is described by a mathematical model, each element of the block diagram represents a mathematical operation such as scaling, integration, addition, multiplication. Here the block diagrams are used to represent a compartmental configuration by specifying the pathways and the retention sites that a molecule can encounter when it is administered in the system. Associated with the block diagrams, the concept of a transfer function is of fundamental importance in the analysis of control systems and feedback problems in general. After specifying in the block diagram a site of external action, i.e., the input or administration site, and a site of observation, i.e., the output or sampling site, the transfer function is defined as the ratio of the Laplace transform of the output of a system to its input.

For instance, in the simple one-compartment model associated with a gamma retention-time distribution $A \sim \text{Gam}(\lambda, \mu)$,

$$f(a) = \frac{\lambda}{\Gamma(\mu)} (\lambda a)^{\mu-1} \exp(-\lambda a),$$

the transfer function is

$$\widetilde{f}(s) = \frac{\lambda^{\mu}}{(s+\lambda)^{\mu}}.$$

The usefulness of the transfer functions lies in the fact that:

- The problem of obtaining the transfer function of a complex system, composed of two or more simple elements, consists in combining the transfer functions of its elements following some elementary rules [352].
- Once the transfer function is known for a particular complex system, then the response of the system to any known input is readily found by multiplying the transfer function by the Laplace transform of the input.

The same problem arises when the pharmacokinetic system is decomposed into subsystems that can be characterized by transfer functions, but where no closed-form solution emerges for the complete system. This holds for the evaluation of the time–amount curve after oral administration when the models of the input and disposition process are known. In these cases, numerical techniques may be of substantial help in performing inverse Laplace transforms. These methods fall into two classes, (1°) approximations by Fourier series expansion and (2°) numerical integration in the complex plane. The validation of these methods and the performances of the available software has been tested directly with pharmacokinetic models [353]. These evaluations showed that simulations as well as parameter estimations from functions defined only in the Laplace domain can be associated with the same degree of reliability as in the conventional case, in which the models are directly given as functions of time. These techniques are commonly used in pharmacokinetics for recirculation drug modeling [354, 355] or physiological modeling [356, 357]. With few exceptions [358], all these approaches are deterministic, and consequently, they use exponential retention-time distributions.

These findings can be summarized in the following procedure to obtain equations for a semi-Markov stochastic model:

1. Represent the underlying mechanistic model with the desired physiological structure through a set of phenomenological compartments with their interconnections.

2. Obtain the equivalent semi-Markov representation by specifying the transition probabilities ω_{ij} and the single-passage retention-time distributions $f_i(a)$ for each compartment. Obtain the block diagram representation using the transition probabilities as gain factors and the Laplace transforms of single-passage density functions as transfer functions.

3. Solve the system of algebraic equations in the frequency domain to obtain the transfer function between the input and sampling sites. The Laplace transform of the probability that "a molecule survives in the sampling time after time t" is this transfer function where substitution of the multiplying $\widetilde{f}_i(s)$ factor in the sampling site by the corresponding $\widetilde{S}_i(s)$ was made.

4. Evaluate in the time domain the time–amount course by applying traditional inverse Laplace transforms [359] or numerical inversion techniques [353, 360].

Conceiving models based on block diagrams may be quite complex, involving feedback loops and time delays. A paper [361] shows in detail how such a model can be constructed for a pharmacokinetic system. On the other hand, retention-time reversible models can be very powerful and flexible for simulation and data fitting.

Example 6 *Simulate a Complex System*

The procedure is presented in a complex system involving three compartments. Figure 9.4 illustrates the original model (upper panel) and the semi-Markov model (lower panel), and Figure 9.5 shows the block diagram representation. If we denote by $r(s)$ the input function and by $y_2(s)$ and $y_3(s)$ the output functions at the sampling sites 2 and 3, respectively, we can write

$$y_2(s) = \left[\omega_{12}\widetilde{f}_1(s)\, r(s) + \omega_{32} y_3(s)\right] \widetilde{f}_2(s)$$

and

$$y_3(s) = \left[\omega_{13}\widetilde{f}_1(s)\, r(s) + \omega_{23} y_2(s)\right] \widetilde{f}_3(s).$$

Solving with respect to $y_2(s)$ and $y_3(s)$, we obtain the transfer functions between the administration and sampling sites:

$$\frac{y_2(s)}{r(s)} = \frac{\widetilde{f}_1(s)\left[\omega_{12} + \omega_{13}\omega_{32}\widetilde{f}_3(s)\right]\widetilde{f}_2(s)}{1 - \omega_{23}\omega_{32}\widetilde{f}_2(s)\,\widetilde{f}_3(s)} \tag{9.13}$$

and

$$\frac{y_3(s)}{r(s)} = \frac{\widetilde{f}_1(s)\left[\omega_{13} + \omega_{12}\omega_{23}\widetilde{f}_2(s)\right]\widetilde{f}_3(s)}{1 - \omega_{23}\omega_{32}\widetilde{f}_2(s)\,\widetilde{f}_3(s)}. \tag{9.14}$$

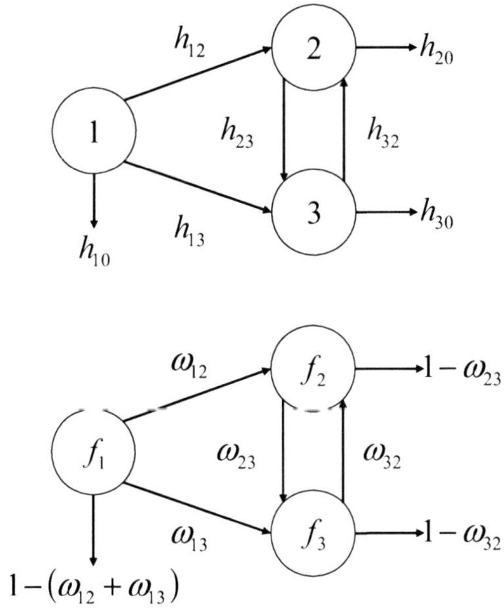

Figure 9.4: Complex 3-compartment configuration. The administration site is in compartment 1 and the sampling sites are compartments 2 and 3.

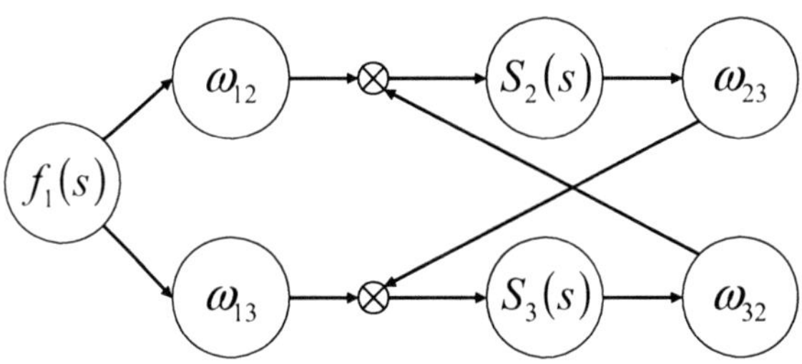

Figure 9.5: Block diagram representation of the complex system shown in Figure 9.4.

The Laplace transform of the probabilities $p_2(t)$ and $p_3(t)$ that "a molecule survives in 2 and 3, respectively, after time t" are obtained by substituting in (9.13), $y_2(s)/r(s)$ and $\widetilde{f}_2(s)$ by $\widetilde{p}_2(s)$ and $\widetilde{\mathcal{S}}_2(s)$, respectively,

$$\widetilde{p}_2(s) = \frac{\widetilde{f}_1(s)\left[\omega_{12} + \omega_{13}\omega_{32}\widetilde{f}_3(s)\right]\widetilde{\mathcal{S}}_2(s)}{1 - \omega_{23}\omega_{32}\widetilde{f}_2(s)\widetilde{f}_3(s)},$$

and in (9.14), $y_3(s)/r(s)$ and $\widetilde{f}_3(s)$ by $\widetilde{p}_3(s)$ and $\widetilde{\mathcal{S}}_3(s)$, respectively,

$$\widetilde{p}_3(s) = \frac{\widetilde{f}_1(s)\left[\omega_{13} + \omega_{12}\omega_{23}\widetilde{f}_2(s)\right]\widetilde{\mathcal{S}}_3(s)}{1 - \omega_{23}\omega_{32}\widetilde{f}_2(s)\widetilde{f}_3(s)}.$$

■

First, Purdue [362] reviewed the use of semi-Markov theory, from which in principle, the requisite $p_{ij}(t)$ regression function may be determined for arbitrary (nonexponential) retention-time distributions. Although the semi-Markov formulation is elegant, the mechanism determining the sequential location of the particles in the compartmental structure is highly complex, and it may be difficult to write down explicit expressions when one is dealing with a general multicompartment system. The solutions are given in general terms involving an infinite sum of convolutions, and the complexity generally rules out an analytical solution for the $p_{ij}(t)$ function.

9.2.5 Time-Varying Hazard Rates

The initial idea is to use the differential equations of a probabilistic transfer model with hazard rates varying with the age of the molecules, i.e., to enlarge the limiting hypothesis (9.2). The objective is to find nonexponential families of survival distributions that are mathematically tractable and yet sufficiently flexible to fit the observed data. In the simplest case, the differential equation (9.7) links hazard rates and survival distributions. Nevertheless, this relation was at the origin of an erroneous use of the hazard function. In fact, substituting in this relation the age a by the exogenous time t, we obtain

$$\dot{\mathcal{S}}(t) = -h(t)\,\mathcal{S}(t)\,,$$

which looks like the deterministic one-compartment model (8.4) with time-varying fractional flow rate $k(t)$, where the amount of the substance $q(t)$ and $k(t)$ are associated with the $\mathcal{S}(t)$ and $h(t)$, respectively. This correspondence is valid only in exceptional cases, and particularly for multicompartment configurations, the use of a hazard function $h(a)$ as a time-varying fractional flow rate $k(t)$ must be handled with extreme care.

One-Compartment Model

Since the exterior time t and the age of the molecules a are the same for the one-compartment model, we can use the previous equation to write

$$\dot{p}(t) = -h(t)\, p(t), \qquad p(0) = 1. \tag{9.15}$$

The solution is given by

$$p(t) = \exp\left[-\int_0^t h(a)\, \mathrm{d}a\right].$$

The closed-form solutions are more difficult to obtain than those previously obtained by means of the survival functions. Numerical integration or quadrature can be used to solve the differential equation or the integral. For instance:

- If the hazard rate is Weibull:

  $$\dot{p}(t) = -\mu\lambda^{\mu} t^{\mu-1} p(t), \qquad p(0) = 1.$$

  This form is very similar to the model often used when the molecules move across fractal media, e.g., the dissolution rate using a time-dependent coefficient given by (5.12) to describe phenomena that take place under dimensional constraints or understirred conditions [16]. The previous differential equation has the solution given by (9.9).

- If the hazard rate is $h(t) = \left(\frac{\alpha}{t} + \beta\right)$:

  $$p(t) = \gamma t^{-\alpha} \exp(-\beta t),$$

  where γ is a normalizing constant. This model involving terms like $t^{-\alpha}$ or $t^{-\alpha} \exp(-\beta t)$ contains only two parameters and seems to be applicable to fit some of the data much better for many drugs [244].

In a pioneer work, Marcus established the link between some usual time-varying forms of $h(t)$ and $f(a)$ in a single compartment [300]. For instance in $h(t) = \left(\frac{\alpha}{t} + \beta\right)$, $\alpha = 1$ leads to $A \sim$Gam(λ, β) and $1 < \alpha < 2$ defines the standard extreme stable-law density with exponent α. In the case of $a = 1.5$, the obtained distribution is known as the retention-time distribution of a Wiener process with drift.

The use of age-dependent hazard rates provides great increase in modeling flexibility, and such models are currently investigated with increasing interest for the following two reasons, among others:

- Many processes have rates that are inherently age-dependent, e.g., various digestion and enzyme-kinetic processes.

- Some complex standard models with many compartments may be simplified by using approximate age-dependent models with fewer parameters, and thus often with superior subsequent statistical analysis. One such application is the description of mixing in passage models.

From a practical point of view, starting from observed data, we are looking for the retention-time distribution $f(a)$ of molecule ages. Using the data and (9.15), recursive techniques may be applied to reveal an approximative time profile of $h(t)$ (cf. Appendix C). On a second level, this profile can be identified using retention-time distributions from Table 9.1.

Multicompartment Models

This formulation is the time-varying alternative to the probabilistic transfer models assuming constant hazard rates as defined by (9.1), and it can be accommodated by generalizing the Markov processes. These models with age-varying hazard rates are expressed by a set of linear differential equations with time-varying coefficients. One may call them *generalized compartmental models* since they satisfy the equations of a deterministic model with k_{ij} being a function of age a. Nevertheless, reference must always be made to the stochastic origin of these equations and confusion avoided between the exogenous time t and ages a of the molecules in the compartments.

Let us examine now the conditions for which a probabilistic transfer model is equivalent to a retention-time model, both using the same hazard functions. More precisely, for the irreversible multicompartment structures, the study can be reduced to the analysis of an irreversible two-compartment model, where the compartment n°1 embodies all compartments before the compartment n°2. One has to compare two situations:

- The probabilistic transfer model whose differential form is

$$\dot{p}_2(t) = h_1(t)\, p_1(t) - h_2(t)\, p_2(t) = f_1(t) - h_2(t)\, p_2(t),$$

  where no distinction is made between the time t and ages a.

- The retention-time model expressed by the convolution (9.10). The derivation of this convolution product leads to (cf. Appendix E)

$$\dot{p}_2(t) = f_1(t) + f_1 * \dot{\mathcal{S}}_2(t),$$

  where by definition $\dot{\mathcal{S}}_2(t) = -h_2(t)\, \mathcal{S}_2(t)$.

By merging the last three equations, one has

$$f_1 * [h_2(t)\, \mathcal{S}_2(t)] = h_2(t)\, f_1 * \mathcal{S}_2(t),$$

which is the condition of equivalence, i.e., $h_2(t)$ must commute to ensure the equivalence between the probabilistic transfer and the retention-time models, both using the same hazard functions. Among the usual hazard functions the exponential distribution has this property. For the reversible multicompartment structures, such a condition further reduces the set of possible distribution functions. For exponential distributions alone we may have such equivalence, but the model degenerates into a Markovian one.

9.2.6 Pseudocompartment Techniques

This section proposes the use of a semi-Markov model with Erlang- and phase-type retention-time distributions as a generic model for the kinetics of systems with inhomogeneous, poorly stirred compartments. These distributions are justified heuristically on the basis of their shape characteristics. The overall objective is to find nonexponential retention-time distributions that adequately describe the flow within a compartment (or pool). These distributions are then combined into a more mechanistic (or physiologically based) model that describes the pattern of drug distribution between compartments. The new semi-Markov model provides a generalized compartmental analysis that can be applied to compartments that are not well stirred.

The Erlang-Type Retention-Time Distributions

The Erlang distributions used as retention-time distributions $f_i(a)$ have interesting mathematical properties considerably simplifying the modeling. For the Erlang distribution, it is well known that if ν independent random variables Z_i are distributed according to the exponential distribution

$$Z_i \sim \mathrm{Exp}(\lambda), \qquad i = 1, \ldots, \nu,$$

then their summation follows an Erlang distribution:

$$Z = \sum_{i=1}^{\nu} Z_i \sim \mathrm{Erl}(\lambda, \nu).$$

The application of this statement in the present context enables one to represent the process responsible for the retention of molecules by a chain of ν catenary compartments, each of them associated with an exponential retention-time distribution with parameter λ. This compartment chain is well known as the *pseudocompartment* chain, but no physical or mechanistic meaning may be associated with this chain. It simply represents a formal way to take into account the Erlang retention-time distribution.

The Phase-Type Retention-Time Distributions

A more general yet tractable approach to semi-Markov models is the phase-type distribution developed by Neuts [363], who showed that any nondegenerate distribution $f(a)$ of a retention time A with nonnegative support can be approximated, arbitrarily closely, by a distribution of phase type. Consequently, all semi-Markov models in the recent literature are special phase-type distribution models. However, the phase-type representation is not unique, and in any case it will be convenient to consider some restricted class of phase-type distributions.

The phase-type distribution has an interpretation in terms of the compartmental model. Indeed, if the phenomenological compartment in the model, which is associated with a nonexponential retention-time distribution, is considered as consisting of a number of pseudocompartments (phases) with movement of particles between these pseudocompartments or out of them, then the retention time of a particle within the entire phenomenological compartment will have a phase-type distribution. The pseudocompartments do not have a *mechanistic* interpretation but rather are a mathematical artifice to generate the desired retention-time distribution.

Using the Markovian formulation, the expanded set of pseudocompartments leads to the solution

$$\mathbf{P}^* (t) = \exp (\mathbf{H}^* t) ,$$

where $\mathbf{H}^*$ and $\mathbf{P}^* (t)$ are the transfer-intensity and the state probability matrices, respectively, both associated with the pseudocompartment structure. It has been shown that the solutions for the state probability $p(t)$ of the phenomenological compartment with the assumed retention-time distribution may be obtained by finding appropriate linear combinations of the $p^*_{ij} (t)$. Mathematically, one has

$$p(t) = \underline{b}_1^T \mathbf{P}^* (t) \underline{b}_2,$$

where $\underline{b}_1^T = \begin{bmatrix} 1 & 0 & \dots & 0 \end{bmatrix}$ and $\underline{b}_2^T = \begin{bmatrix} 1 & 1 & \dots & 1 \end{bmatrix}$ are m-dimensional vectors of indicator variables, i.e., 0's or 1's. The elements of $\underline{b}_1$ indicate the origin of particles in the pseudocompartment structure and $\underline{b}_2$ indicates that all the pseudocompartments contribute to build the phenomenological compartment.

Perhaps the most commonly used example of a phase-type distribution is the $\mathrm{Erl}(\lambda, \nu)$ distribution, defined by the catenary system consisting of ν pseudocompartments. According to the phase-type concept for generating distributions, one can find phase-type distributions that exhibit rich kinetic behaviors using concatenation of Erlang distributions associated with several λ's: this case is reported as the *generalized Erlang distribution*. Further kinetic flexibility can be achieved by using feedback pathways and partition of hazard rates in the pseudocompartmental structures.

To describe heterogeneity within a compartment, Figure 9.6 illustrates three pseudocompartment configurations, each of them involving four pseudocompartments in all.

- The first one (A) is a catenary system with pseudocompartments associated with a λ_1 hazard rate. The transfer-intensity matrix $\mathbf{H}^*$ is

$$\mathbf{H}^* = \begin{bmatrix} -\lambda_1 & \lambda_1 & 0 & 0 \\ 0 & -\lambda_1 & \lambda_1 & 0 \\ 0 & 0 & -\lambda_1 & \lambda_1 \\ 0 & 0 & 0 & -\lambda_1 \end{bmatrix},$$

  and the phase-type distribution generated by this structure is $\mathrm{Erl}(\lambda_1, 4)$.

- Like the previous system, the second (B) is also a catenary one, but two pseudocompartments are associated with the λ_1 hazard rate, and two oth-

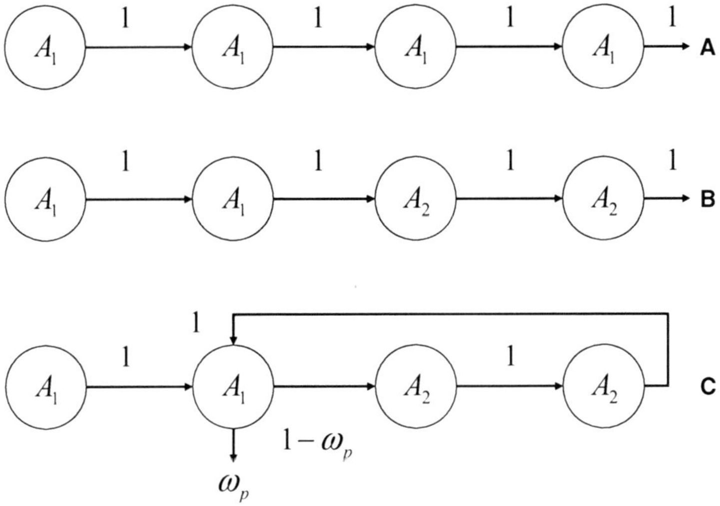

Figure 9.6: Pseudocompartment configurations generating Erlang (**A**), generalized Erlang (**B**), and phase-type (**C**) distributions for retention times in phenomenological compartments. Retention times are distributed according to A_1 ~Exp(λ_1) and A_2 ~Exp(λ_2).

ers with the λ_2 hazard rate. The transfer-intensity matrix is

$$\mathbf{H}^* = \begin{bmatrix} -\lambda_1 & \lambda_1 & 0 & 0 \\ 0 & -\lambda_1 & \lambda_1 & 0 \\ 0 & 0 & -\lambda_2 & \lambda_2 \\ 0 & 0 & 0 & -\lambda_2 \end{bmatrix},$$

and the generalized Erlang density function is more dispersed with the actual parameter values than the previous one.

- The third configuration (C) is unusual because the phenomenological compartment output takes place from the second pseudocompartment and the output of the last pseudocompartment is fed back to the second pseudocompartment. The transfer-intensity matrix is

$$\mathbf{H}^* = \begin{bmatrix} -\lambda_1 & \lambda_1 & 0 & 0 \\ 0 & -\lambda_1 & (1-\omega_p)\,\lambda_1 & 0 \\ 0 & 0 & -\lambda_2 & \lambda_2 \\ 0 & \lambda_2 & 0 & -\lambda_2 \end{bmatrix},$$

and the resulting density is "long-tailed."

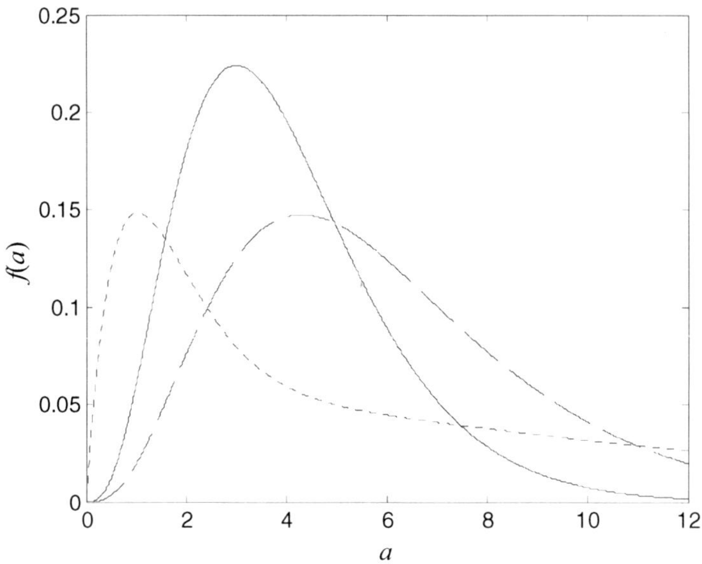

Figure 9.7: Retention-time densities generated by pseudocompartment configurations: Erlang (*solid line*), generalized Erlang (*dashed line*), and phase-type densities (*dotted line*).

For the three pseudocompartment configurations presented above, Figure 9.7 depicts the obtained density functions with parameters set to $\lambda_1 = 1$, $\lambda_2 = 0.25\,\mathrm{h}^{-1}$, and $\omega_p = 0.3$.

The phase-type distributions are designed to serve as retention-time distributions in semi-Markov models. To obtain the equations of the model for a phenomenological compartmental configuration, one has to follow the following procedure:

1. Represent the underlying mechanistic model with the desired physiological structure through a set of phenomenological compartments with their interconnections.

2. Express the retention-time distribution for each phenomenological compartment by using phase-type distributions. However, the phase-type distributions for these sites are determined empirically. There is no assurance of finding the "best" phase-type distribution. This step leads to the expanded model involving pseudocompartments generating the desired phase-type distribution.

3. For the resulting model with phase-type distributions, find the expanded

transfer-intensity and the state probability matrices of the equivalent Markov model $\mathbf{H}^*$ and $\mathbf{P}^*(t)$, respectively.

4. Simulate the kinetic behavior by combining the $\mathbf{P}^*(t)$ probability functions for the pseudocompartments to obtain the state probabilities $\mathbf{P}(t)$ of a particle belonging to the phenomenological compartments at time t. That is defined by means of appropriate matrices $\mathbf{B}_1$ and $\mathbf{B}_2$ with indicator variables, i.e., 0's or 1's:

$$\mathbf{P}(t) = \mathbf{B}_1 \mathbf{P}^*(t) \mathbf{B}_2.$$

The elements of $\mathbf{B}_1$ indicate the origin of particles in the pseudocompartment structure and establish the correspondence between the numbering of the original compartments and the sequence of the pseudocompartments. The elements of $\mathbf{B}_2$ indicate the summing of pseudocompartments to yield the phenomenological compartment.

Structured Models

Although the structured models are at the origin of the pseudocompartment concept, these models are less well known [364, 365]. The structured models are compartmental systems, but with a structure that describes the dynamics in a physically reasonable way. Imposing some structure on the compartmental model is certainly a way of dealing with possible ill-conditioning of more general models. The resulting model has only a few parameters, and is capable of fitting well some observed data. The proposed structure is more holistic, in the sense that the compartments themselves may not have an obvious physical interpretation, but the system as a whole does. Although the number of compartments is increasing, the number of estimated parameters does not because the model is structured, unlike traditional compartmental modeling. Models of this type include some well-known systems [341], and they have been used as examples in other work [311]. This structured compartmental model has some similarities to the dispersion model [268], but it does have certain advantages.

Faddy [364] consider the compartmental configuration, shown schematically in Figure 9.8, where compartments are numbered 1, 2, ..., m are pseudocompartments, with the starting compartment numbered i_s ($2 \le i_s \le m$). The material is transferred between compartments over time according to a Markov process, where the positive parameters h_+, h_-, and h_0 are the hazard rates. Thus a molecule administered to the system would be able to clear the system only via the series of compartments $i_s + 1, \ldots, m$, corresponding to $\mathrm{Erl}(h_0, m - i_s + 1)$ distributed retention times. As previously noted, a large value of m would be exemplified by a "hump" in any observed retention data, corresponding to a delay in clearance of the drug. Pseudocompartments $1, 2, \ldots, i_s$ correspond to the states of a random walk with reflecting barrier at 1, which describes the retention of the drug by movement of elements between nearest neighbor sites within a heterogeneous peripheral medium. For large i_s, this random walk can be thought of as approximating a diffusion. Such a model thus describes drug kinetics in terms of two components:

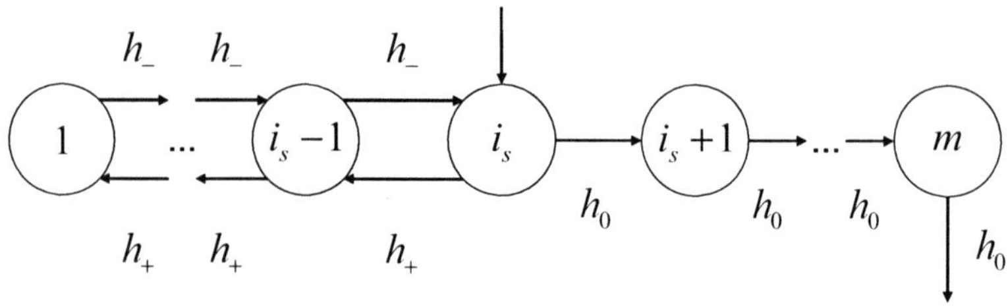

Figure 9.8: Structured Markovian model. Diffusion is expressed by means of h_+, h_-, and compartments 1 to i_s. Erlang-type elimination is represented by means of h_0 and compartments i_s to m. The drug is given in compartment i_s and cleared from compartment m.

- diffusion within the heterogeneous peripheral medium and
- Erlang distributed retention times describing the elimination from the system.

Retention data that after a possible delay in concentration show a sharp decline followed by a long tail would be modeled by $i_s \gg 2$ and $h_0 \gg h_- > h_+$. The condition $h_- > h_+$ ensures that the drift of the random walk (or diffusion) is away from the reflecting barrier. Figure 9.9 illustrates the probability profiles in the distribution and elimination compartments when $m = 20$, $i_s = 15$, $h_+ = 0.1$, $h_- = 0.2$, and $h_0 = 1$.

In summary, any stochastic semi-Markov model may be represented as an expanded Markov model. This simply involves subdividing each compartment into a number of pseudocompartments, leading to a matrix that essentially defines a new expanded compartmental system, but with many more compartments [364]. After passing through a sequence of pseudocompartments, a particle would transfer according to the ω_{ij} transition probabilities. Thus, multicompartment modeling may be done using the definitions and the methodology developed for the probabilistic transfer models. Therefore, the formulation of the probabilistic transfer model is immediate and hence the questions associated with the nature of the eigenvalues and the complexity of the analytical solutions may be attempted using suitable numerical procedures and computer software. The assumption of the Erlang retention-time distributions has several consequences:

- for irreversible models, the eigenvalues of the **H** matrix may be multiple real, leading to a "time-power" solution,
- for reversible models, the eigenvalues may be real or complex multiple values, with negative parts leading to damped oscillations.

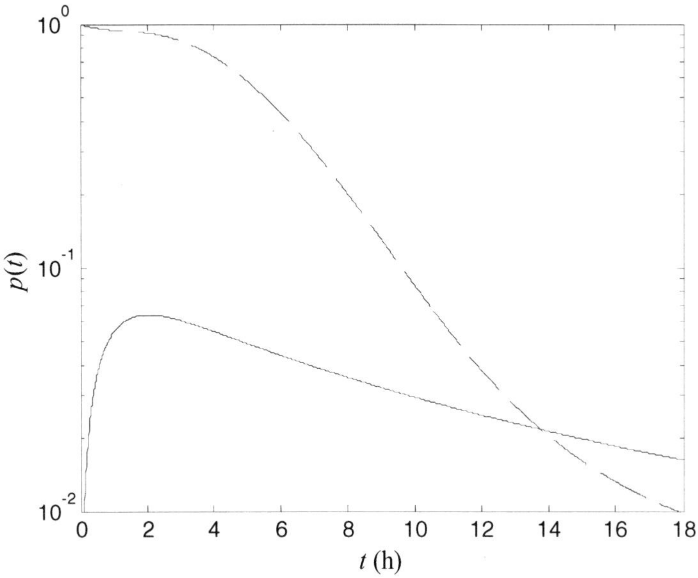

Figure 9.9: The total probabilities in the distribution (*solid line*) and elimination (*dashed line*) compartments of the Faddy structured model.

Erlang- and phase-type distributions provide a versatile class of distributions, and are shown to fit naturally into a Markovian compartmental system, where particles move between a series of compartments, so that phase-type compartmental retention-time distributions can be incorporated simply by increasing the size of the system. This class of distributions is sufficiently rich to allow for a wide range of behaviors, and at the same time offers computational convenience for data analysis. Such distributions have been used extensively in theoretical studies (e.g., [366]), because of their range of behavior, as well as in experimental work (e.g., [367]). Especially for compartmental models, the phase-type distributions were used by Faddy [364] and Matis [301, 306] as examples of "long-tailed" distributions with high coefficients of variation.

9.2.7 A Typical Two-Compartment Model

The mechanistic model is the traditional reversible two-compartment model. For this model, Karalis et al. [368] hypothesized a well-stirred compartment, the central compartment, and a heterogeneous, peripheral compartment. In general, one would assume that the sampling site is a well-stirred medium ensuring sampling feasibility technology where the particles mix quickly and homogeneously with blood, e.g., the central compartment. But such an assumption is not valid

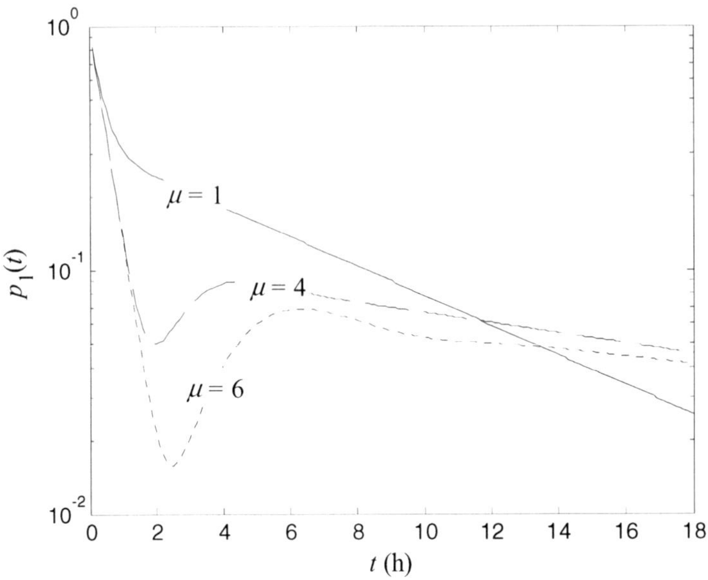

Figure 9.10: Simulation of time–$p_1(t)$ profiles for $\mu = 1, 4, 6$.

for the peripheral compartment that represents soft tissues, muscles, or bone or other organs, Figure 9.2 A. We assume that all molecules are present in compartment 1 at time 0. In the following, we express the heterogeneity in the peripheral compartment in several manners.

Semi-Markov Formulation

We propose to use as single-passage retention-time distributions the $A_1 \sim \text{Exp}(\kappa)$ for the central compartment and the $A_2 \sim \text{Gam}(\lambda, \mu)$ distribution for the peripheral compartment and we assume that all molecules are present in compartment 1 at initial time. According to (9.12),

$$\widetilde{p}_1(s) = \frac{1}{s + \kappa - \omega_e \kappa \left(\frac{\lambda}{s+\lambda} \right)^{\mu}}. \tag{9.16}$$

Using the numerical inverse Laplace transform and $\kappa = 2\,\text{h}^{-1}$, $\lambda = 1\,\text{h}^{-1}$, $\omega_e = 0.8$, and $\mu = 1, 4, 6$, Figure 9.10 illustrates the $p_1(t)$ time profiles.

Instead of the gamma single-passage distribution for the peripheral compartment, Wise [298] proposed the mixed random walk in series distribution,

$$f(t) \propto t^{-w} \exp\left[-\phi \left(\frac{t}{\mu} + \frac{\mu}{t} \right) \right],$$

to justify the gamma-type function $\gamma t^{-\alpha} \exp(-\beta t)$ often used as an empirical model to fit several series of data. These gamma profiles can also be interpreted in terms of a recirculation process, where the single-passage retention time is the generalized inverse Gaussian distribution [246].

Erlang-Type Distribution

We propose the retention-time distributions $A_1 \sim \text{Exp}(\kappa)$ and $A_2 \sim \text{Erl}(\lambda, \nu)$ for the first and second compartments, respectively. The peripheral compartment 2 is then constituted by the ν pseudocompartments that are required to express $\text{Erl}(\lambda, \nu)$. It follows that

$$\kappa = h_{10} + h_{12} \qquad \text{and} \qquad \lambda = h_{21} \qquad \text{and} \qquad \omega_e = \frac{h_{12}}{h_{10} + h_{12}}.$$

The system now becomes an $m = \nu + 1$ compartment model and the probabilistic transfer differential equations are

$$\dot{p}_1 = -\kappa p_1 + \lambda p_m, \qquad \dot{p}_2 = \omega_e \kappa p_1 - \lambda p_2, \qquad \dot{p}_j = \lambda (p_{j-1} - p_j), \qquad j = 3, m.$$

In the above equations, p_i represents the probability that a molecule starting in compartment 1 is in compartment i at time t. By using $\tau = \lambda t$ and $\mu = \kappa / \lambda$, one obtains the dimensionless system of differential equations

$$\dot{p}_1 = -\mu p_1 + p_m, \qquad \dot{p}_2 = \omega_e \mu p_1 - p_2, \qquad \dot{p}_j = p_{j-1} - p_j, \qquad j = 3, m.$$

This model is a special case of the model studied by Matis and Wehrly [369] in which $A_1 \sim \text{Erl}(\lambda_1, \nu_1)$ and $A_2 \sim \text{Erl}(\lambda_2, \nu_2)$ retention-time distributions are associated with the first and second compartments, respectively. The analysis of the characteristic polynomial of this model implies that there are at least two complex eigenvalues, except for the case $\nu = 2$ with parameters satisfying the condition

$$\omega_e < \frac{4(\mu - 1)^3}{27\mu}.$$

The practical significance is that for the above two-compartment models with large ν or large ω_e, the p_i do not have the simple commonly used sum of exponential forms but damped oscillatory ones. According to the μ and ω_e values, one obtains a broad spectrum of models able to fit unusual data profiles. Figure 9.11 illustrates the p_1 profiles for $\nu = 2, 4, 6$ associated with $\kappa = 2\,\text{h}^{-1}$, $\lambda = 1\,\text{h}^{-1}$, and $\omega_e = 0.8$. These simulations are identical to those obtained with the transfer functions in Figure 9.10. Therefore, Erlang distributions are useful for a class of problems in which there is initial dampening of the conditional transfer probability due to such phenomena as noninstant mixing.

Phase-Type Distribution

We propose the use of the phase-type distributions previously developed as retention-time distributions associated with the peripheral compartment. The

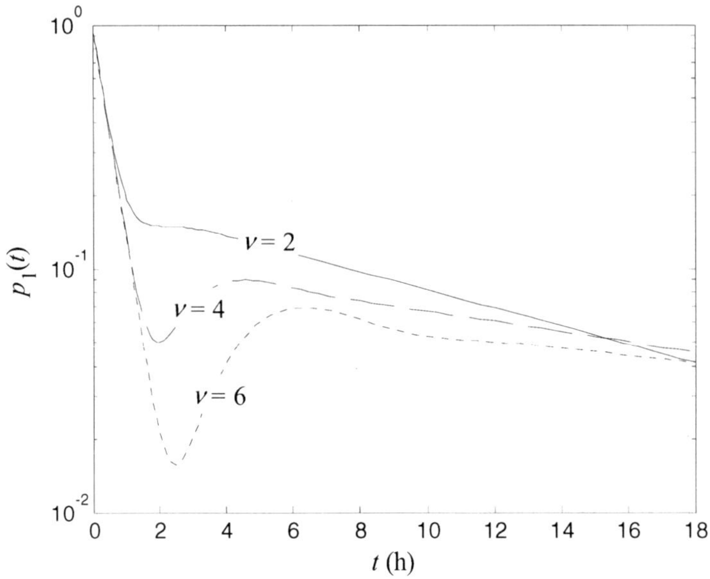

Figure 9.11: Simulation of time–$p_1(t)$ profiles for $\nu = 2, 4, 6$.

numeric values of the parameters are $\kappa = 2\,\mathrm{h}^{-1}$, $\lambda_1 = 1\,\mathrm{h}^{-1}$, $\omega_e = 0.8$, $\lambda_2 = 0.25\,\mathrm{h}^{-1}$, and $\omega_p = 0.3$. For the three cases, the transfer-intensity matrices $\mathbf{H}^*$ of the equivalent Markov model are

$$\mathbf{H}^* = \begin{bmatrix} -\kappa & \omega_e\kappa & 0 & 0 & 0 \\ 0 & -\lambda_1 & \lambda_1 & 0 & 0 \\ 0 & 0 & -\lambda_1 & \lambda_1 & 0 \\ 0 & 0 & 0 & -\lambda_1 & \lambda_1 \\ \lambda_1 & 0 & 0 & 0 & -\lambda_1 \end{bmatrix},$$

$$\mathbf{H}^* = \begin{bmatrix} -\kappa & \omega_e\kappa & 0 & 0 & 0 \\ 0 & -\lambda_1 & \lambda_1 & 0 & 0 \\ 0 & 0 & -\lambda_1 & \lambda_1 & 0 \\ 0 & 0 & 0 & -\lambda_2 & \lambda_2 \\ \lambda_2 & 0 & 0 & 0 & -\lambda_2 \end{bmatrix},$$

$$\mathbf{H}^* = \begin{bmatrix} -\kappa & \omega_e\kappa & 0 & 0 & 0 \\ 0 & -\lambda_1 & \lambda_1 & 0 & 0 \\ \omega_p\lambda_1 & 0 & -\lambda_1 & (1-\omega_p)\,\lambda_1 & 0 \\ 0 & 0 & 0 & -\lambda_2 & \lambda_2 \\ 0 & 0 & \lambda_2 & 0 & -\lambda_2 \end{bmatrix},$$

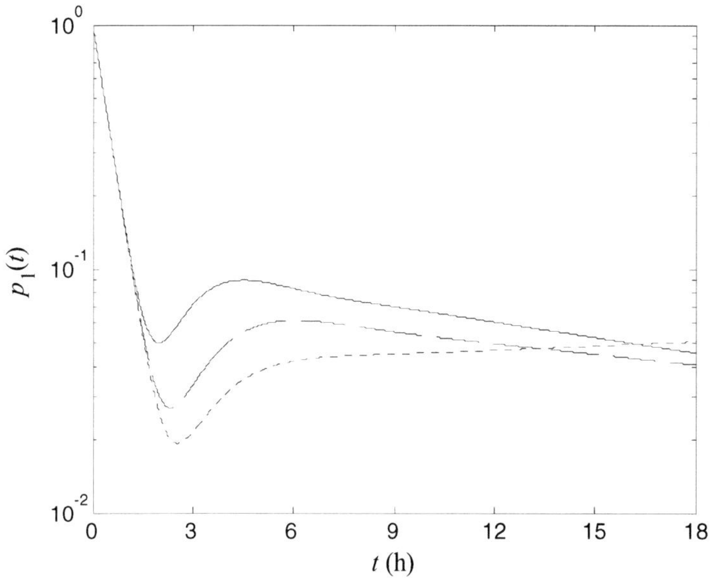

Figure 9.12: Simulation of time–$p_1(t)$ profiles using pseudocompartments to generate Erlang (*solid line*), generalized Erlang (*dashed line*), and phase-type densities (*dotted line*); cf. Figure 9.7.

respectively. Also, the $\mathbf{B}_1$ and $\mathbf{B}_2$ matrices with the indicator variables are

$$\mathbf{B}_1 = \begin{bmatrix} 1 & 0 & 0 & 0 & 0 \\ 0 & 1 & 0 & 0 & 0 \end{bmatrix} \quad \text{and} \quad \mathbf{B}_2^T = \begin{bmatrix} 1 & 0 & 0 & 0 & 0 \\ 0 & 1 & 1 & 1 & 1 \end{bmatrix}.$$

Figure 9.12 illustrates the $p_1(t)$ state probability of having a particle in the sampled compartment at time t. These $p_1(t)$ use the retention-time distributions presented in Figure 9.7. The profile of $p_1(t)$ that corresponds to $\mathrm{Erl}(\lambda_1, 4)$ is the same as that drawn in Figure 9.11.

9.3 Time–Concentration Profiles

The probabilistic transfer and retention-time models are models evaluating the transition or retention probabilities that are associated with a single particle. This is why these models are called the *particle models*. In order to account for all the particles in the process and administered amounts, one needs to make further statistical and practical considerations.

9.3.1 Routes of Administration

Let us consider some drug administration practicalities. Up to now, the administered amounts were considered as initial units introduced simultaneously into several compartments at the beginning of the experiment. These amounts were considered as initial conditions to the differential equations describing the studied processes. Nevertheless, this concept seems to have limited applications in pharmacokinetics. In this section, we develop the probabilistic transfer and retention-time models associated with an extravascular or intravascular route of administration.

In both cases, an extra compartment is introduced: the absorption or the infusion balloon compartment for the extravascular and intravascular route, respectively. To model these disposition processes, we again apply probabilistic analysis for these compartments looking for the probability $p(t+\Delta t)$ that a particle is present at time $(t+\Delta t)$ in that compartment. Clearly, the necessary events are "that the particle is present at time t," associated with the state probability $p(t)$ AND "that it remains in the compartment during the interval from t to $(t+\Delta t)$," associated with the conditional probability $[1-h\Delta t]$. Therefore, the probability of the desired joint event may be written as

$$p(t+\Delta t) = p(t)[1-h\Delta t]. \tag{9.17}$$

For the extravascular and intravascular routes, $p(t)$ will be referred to as $p_{ev}(t)$ and $p_{iv}(t)$, and h will be referred to as h_{ev} and h_{iv}, respectively.

The two routes of administration can be formulated as follows:

- In the extravascular case, the compartment is the absorption compartment and the hazard rate h_{ev} represents the absorption rate constant. If we assume that h_{ev} is not dependent on time, rearranging (9.17), taking the limit $\Delta t \to 0$, and solving the so obtained differential equation with initial condition $p_{ev}(0) = 1$, we obtain

$$p_{ev}(t) = \exp(-h_{ev}t).$$

The retention-time distribution follows

$$f_{ev}(t) = h_{ev}p_{ev}(t) = h_{ev}\exp(-h_{ev}t).$$

The Laplace transform of the extravascular retention-time distribution is

$$\widetilde{f}_{ev}(s) = \frac{h_{ev}}{s+h_{ev}}.$$

- In the intravascular case with a constant rate infusion between the starting time T_S and the ending time T_E, the state probability $p_{iv}(t)$ is given by

$$p_{iv}(t) = \frac{T_E - T_S - t}{T_E - T_S}[u(t-T_S) - u(t-T_E)].$$

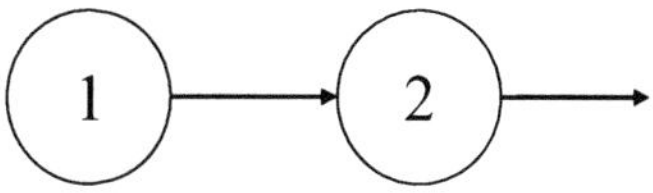

Figure 9.13: Two-compartment irreversible system.

Solving (9.17) for h_{iv}, we obtain a time-varying hazard rate h_{iv},

$$h_{iv}(t) = \frac{1}{T_E - T_S - t}\left[u(t - T_S) - u(t - T_E)\right],$$

and the retention-time distribution

$$f_{iv}(t) = h_{iv}(t)\,p_{iv}(t) = \frac{1}{T_E - T_S}\left[u(t - T_S) - u(t - T_E)\right]. \qquad (9.18)$$

In the above relationships, $u(t)$ is the step Heaviside function. The Laplace transform of the intravascular retention-time distribution is

$$\widetilde{f}_{iv}(s) = \frac{1}{s(T_E - T_S)}\left[\exp(-T_S s) - \exp(-T_E s)\right]. \qquad (9.19)$$

For both cases, the retention-time distribution functions $f_{ev}(t)$ and $f_{iv}(t)$ are similar to the input functions $v_{ev}(t)$ and $v_{iv}(t)$, respectively, defined for the deterministic models. The only difference is that in the stochastic consideration, the drug amounts are not included is these input functions.

In conclusion, in order to account for the usual routes of administration, one has to expand the system by artifactual compartments and associated retention-time distributions corresponding to the extravascular or intravascular routes. So, the expanded system may now be considered without environmental links and the administration protocol is simply expressed by the initial conditions in the input compartments. In this case, at least one of the m compartments must be considered as the input compartment.

9.3.2 Some Typical Drug Administration Schemes

In the following, we present how to apply the above relationships for the compartmental model shown in Figure 9.13.

Extravascular Case

The most frequent situation is the heterogeneous absorption materialized by retention-time distributions $A_1 \sim \text{Erl}(\lambda, \nu)$ and $A_2 \sim \text{Exp}(\kappa)$ for compartments 1 and 2, respectively. In this configuration, compartment 1 represents the heterogeneous absorption compartment and compartment 2 represents the distribution compartment that is the sampled compartment. The state probability

$p(t)$ that "a molecule initially introduced in compartment 1 is in compartment 2 at time t" is evaluated using (9.10) with

$$f_1(a) = \frac{\lambda^\nu a^{\nu-1}}{(\nu-1)!}\exp(-\lambda a) \qquad \text{and} \qquad \mathcal{S}_2(a) = \exp(-\kappa a).$$

The convolution integral can be evaluated using the Laplace transform. In fact,

$$L\{f_1(a)\} = \frac{\lambda^\nu}{(s+\lambda)^\nu} \qquad \text{and} \qquad L\{\mathcal{S}_2(a)\} = \frac{1}{s+\kappa}.$$

In these expressions, L and s denote the Laplace operator and Laplace variable, respectively. The solution is given by

$$p(t) = L^{-1}\left\{\frac{\lambda^\nu}{(s+\kappa)(s+\lambda)^\nu}\right\}. \tag{9.20}$$

The inverse Laplace calculus of (9.20) leads to

$$p(t) = \gamma^\nu \exp(-\kappa t) - \exp(-\lambda t)\sum_{i=1}^{\nu}\gamma^i\frac{(\lambda t)^{\nu-i}}{(\nu-i)!}$$

with $\gamma = \lambda/(\lambda-\kappa)$.

Intravascular Case

Here the drug is administered by a constant rate infusion over T hours. This model may be conceived in two different ways:

- Probabilistic transfer model. The model is a special case of the two-compartment model presented in Figure 9.1, where compartment 1 is associated with the infusion balloon and compartment 2 is associated with the central compartment. The links between compartments are specified as $h_{12} = h_{iv}(t)$, $h_{21} = 0$, $h_{10} = 0$, and $h_{20} = h$. The state probabilities associated with compartment 1 are $p_{11}(t) = p_{iv}(t)$ and $p_{21}(t) = 0$. The probabilistic transfer equation for the central compartment 2 is obtained directly from (9.4):

  $$\dot{p}_{12}(t) = h_{iv}(t)\,p_{iv}(t) - hp_{12}(t).$$

  Given (9.18), the solution of the differential equation is

  $$p_{12}(t) = \frac{1}{Th}\{\exp[-h(t-T)\,u(t-T)] - \exp(-ht)\}. \tag{9.21}$$

  This equation gives the probability that "a molecule set in the infusion balloon 1 at time 0 is present in the central compartment 2 at time t."

- Retention-time model. The model is an irreversible two-com-partment model whose solution is given by (9.11):

$$\widetilde{p}_2(s) = \widetilde{f}_1(s)\,\widetilde{\mathcal{S}}_2(s),$$

where $\widetilde{f}_1(s) \equiv \widetilde{f}_{iv}(s)$ and $\widetilde{\mathcal{S}}_2(s) = (s+h)^{-1}$. The solution is given by

$$p_2(t) = \frac{1}{Th}\left\{\exp\left[-h(t-T)\,u(t-T)\right] - \exp(-ht)\right\}.$$

Example 7 *Infusion for the Typical Two-Compartment Model*

This example concerns the typical two-compartment model previously presented under the semi-Markov formulation (cf. Section 9.2.7). By assuming that molecules are initially present in the central compartment, (9.16) is the Laplace transform of the survival function in that compartment. If now the drug molecules are administered by a constant rate infusion between T_S and T_E, the Laplace transform of the survival function in the central compartment becomes

$$\widetilde{p}_1^*(s) = \widetilde{f}_{iv}(s)\,\widetilde{p}_1(s) = \frac{\exp(-T_S s) - \exp(-T_E s)}{s(T_E - T_S)} \frac{1}{s + \kappa - \omega_e \kappa \left(\frac{\lambda}{s+\lambda}\right)^{\mu}}.$$

This expression is obtained by reporting (9.16) and (9.19) into (9.10). Using the numerical inverse Laplace transform and $\kappa = 2\,\mathrm{h}^{-1}$, $\lambda = 1\,\mathrm{h}^{-1}$, $\omega_e = 0.8$, and $\mu = 1, 4, 6$, Figure 9.14 illustrates the $p_1^*(t)$ time profiles for a 6- h constant-rate infusion. This figure takes into account the infusion duration, whereas Figure 9.10 considers that all molecules are in compartment 1 at initial times. ■

9.3.3 Time-Amount Functions

After specifying the route of drug administration, we now turn to some statistical considerations in order to express the behavior of all particles administered in the system.

Number of Particles

Let $\underline{n}_0$ be an m-dimensional deterministic vector representing the number of particles contained in the drug amount $\underline{q}_0$ initially given in each compartment. Also, let $\underline{N}_i(t)$ be an m-dimensional random vector that takes on zero and positive integer values. $\underline{N}_i(t)$ represents, at time t, the random distribution among the m compartments of the number of molecules starting in i. Since all of the molecules are independent by assumption, $\underline{N}_i(t)$ follows a multinomial distribution:

$$\underline{N}_i(t) \sim \text{multinomial}\left[n_{0i}, \underline{p}_i(t)\right], \tag{9.22}$$

where $\underline{p}_i(t)$ is the vector of state probabilities for the molecules starting in i. The expectation vector, variances, and covariances of $\underline{N}_i(t)$ have simple well-known

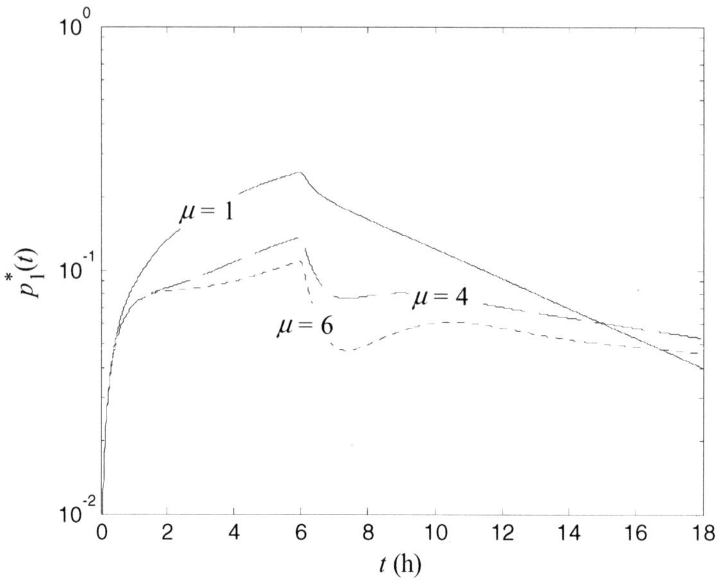

Figure 9.14: Simulation of time–$p_1^*(t)$ profiles for $\mu = 1, 4, 6$ obtained with a 6- h infusion.

forms:

$$\begin{aligned} E\left[\underline{N}_i(t)\right] &= n_{0i}\underline{p}_i(t), \\ Var\left[N_{ij}(t)\right] &= n_{0i}p_{ij}(t)\left[1 - p_{ij}(t)\right], \\ Cov\left[N_{ij}(t)\, N_{ik}(t)\right] &= -n_{0i}p_{ij}(t)\, p_{ik}(t). \end{aligned}$$

Particles starting in each compartment i contribute to obtaining the number of particles in each compartment:

$$\underline{N}(t) = \sum_{i=1}^{m} \underline{N}_i(t),$$

where $\underline{N}(t)$ is a random vector having expectation, variance, and covariance

$$\begin{aligned} E\left[\underline{N}(t)\right] &= \sum_{i=1}^{m} n_{0i}\underline{p}_i(t), \qquad (9.23) \\ Var\left[N_j(t)\right] &= \sum_{i=1}^{m} n_{0i}p_{ij}(t)\left[1 - p_{ij}(t)\right], \\ Cov\left[N_j(t)\, N_k(t)\right] &= -\sum_{i=1}^{m} n_{0i}p_{ij}(t)\, p_{ik}(t), \end{aligned}$$

respectively.

Repeated Dosage

When drugs are given in repeated dosage, we have to compile the repeated schemes. We assume linearity in mixing multinomial distributions, i.e., if

$$\underline{N}_{ik}(t) \sim \text{multinomial}\left[n_{0ik}, \underline{p}_i(t)\right]$$

for $k = 1, \ldots, m_r$, then

$$\underline{N}_i(t) = \sum_{k=1}^{m_r} \underline{N}_{ik}(t) \sim \text{multinomial}\left[\sum_{k=1}^{m_r} n_{0ik}, \underline{p}_i(t)\right]$$

with expectation vector, variances, and covariances

$$\begin{aligned} E\left[\underline{N}_i(t)\right] &= \underline{p}_i(t) \sum_{k=1}^{m_r} n_{0ik}, \\ Var\left[N_{ij}(t)\right] &= p_{ij}(t)\left[1 - p_{ij}(t)\right] \sum_{k=1}^{m_r} n_{0ik}, \\ Cov\left[N_{ij}(t) N_{ik}(t)\right] &= -p_{ij}(t) p_{ik}(t) \sum_{k=1}^{m_r} n_{0ik}, \end{aligned}$$

respectively. Moreover, if the m_r administrations are delayed by t_k°, one has to substitute in the previous expressions $\underline{p}_i(t)$ by $\underline{p}_i(t - t_k^\circ) u(t - t_k^\circ)$.

Example 8 *Repeated Infusions for the One-Compartment Model*

For the one-compartment model of (9.21), assume that n_0 particles of drug was initially in compartment 2 and then two constant-rate infusions delayed by t° were given in compartment 1. Let n_1 and n_2 be the infused amounts and T_1 and T_2 the infusion times. According to the previous relations, the expectation of the time–amount curve will be

$$\begin{aligned} E\left[N_2(t)\right] &= n_0 \exp(-ht) + \frac{n_1}{T_1 h}\left\{\exp\left[-h(t - T_1) u(t - T_1)\right] - \exp(-ht)\right\} \\ &\quad + \frac{n_2}{T_2 h}\left\{\exp\left[-h(t' - T_2) u(t' - T_2)\right] - \exp\left[-ht' u(t')\right]\right\} \end{aligned}$$

with $t' = t - t^\circ$. ■

Drug Amounts

Given the gram-molecular weight of the drug and using Avogadro's number, one converts the number of particles n_{0i} and $\underline{N}(t)$ to the equivalent amounts q_{0i} and

$\underline{Q}(t)$, respectively. Thus, the expectation vector, variances, and covariances of the drug amount $\underline{Q}(t)$ in the compartments at time t are

$$\begin{aligned} E\left[\underline{Q}(t)\right] &= \sum_{i=1}^{m} q_{0i}\underline{p}_i(t), \\ Var\left[Q_j(t)\right] &= \sum_{i=1}^{m} q_{0i}p_{ij}(t)\left[1-p_{ij}(t)\right], \\ Cov\left[Q_j(t)Q_k(t)\right] &= -\sum_{i=1}^{m} q_{0i}p_{ij}(t)p_{ik}(t), \end{aligned}$$

respectively. In matrix notation, $E\left[\underline{Q}^T(t)\right]$ may also be written as $\underline{q}_0^T\mathbf{P}(t)$. Taking into account (9.3), the expectation of the drug amount becomes

$$E\left[\underline{Q}^T(t)\right] = \underline{q}_0^T \exp(\mathbf{H}t),$$

a similar form to that of deterministic models (8.5).

In conclusion, the solutions $E\left[\underline{Q}^T(t)\right]$ for the expected values for such stochastic models are the same as the solutions $\underline{q}^T(t)$ for the corresponding deterministic models, and the *transfer-intensity* matrix $\mathbf{H}$ is analogous to the *fractional flow rates* matrix $\mathbf{K}$ of the deterministic model. If the hazard rates are constant in time, we have the stochastic analogues of linear deterministic systems with constant coefficients. If the hazard rates depend on time, we have the stochastic analogues of linear deterministic systems with time-dependent coefficients.

So, it is possible to associate some probabilistic interpretations in the deterministic model. From the probabilistic viewpoint $k_{ij}\Delta t$ is the conditional probability that a molecule will be transferred from i to j in the interval t to $t+\Delta t$. Thus $k_{ii}\Delta t$ is the conditional probability that a molecule leaves i in that interval.

If the hazard rate of any single particle out of a compartment depends on the state of the system, the equations of the probabilistic transfer model are still linear, but we have nonlinear rate laws for the transfer processes involved and such systems are the stochastic analogues of nonlinear compartmental systems. For such systems, the solutions for the deterministic model are not the same as the solutions for the mean values of the stochastic model.

Example 9 *Two-Compartment Reversible Model*

For the model presented in Section 9.2.4 and in the presence of q_{01} and q_{02} amounts of molecules at the starting time in compartments 1 and 2, respectively, the expectation of the time–amount curve in the two compartments will be the inverse Laplace transform of

$$E\left[\widetilde{Q}_1(s)\right] = \frac{\left[q_{01}+q_{02}\widetilde{f}_2(s)\right]\widetilde{S}_1(s)}{1-\omega\widetilde{f}_1(s)\widetilde{f}_2(s)}$$

and

$$E\left[\widetilde{Q}_2(s)\right] = \frac{\left[q_{01}\omega\widetilde{f}_1(s) + q_{02}\right]\widetilde{S}_2(s)}{1 - \omega\widetilde{f}_1(s)\widetilde{f}_2(s)}.$$

■

9.3.4 Process Uncertainty or Stochastic Error

So, we find that the mean behavior of the stochastic model is described by the deterministic model we have already developed. The fundamental difference between the stochastic and the deterministic model arises from the chance mechanism in the stochastic model that generates so-called *process uncertainty*, or *stochastic error*.

Spatial Error

The stochastic error is expressed in (9.23) by the variance $Var[N_j(t)]$ and covariance $Cov[N_j(t)N_k(t)]$ that did not exist in the deterministic model. This error could also be named *spatial stochastic error*, since it describes the process uncertainty among compartments for the same t and it depends on the number of drug particles initially administered in the system. For the sake of simplicity, assume $n_{0i} = n_0$ for each compartment i. From the previous relations, the coefficient of variation $CV_j(t)$ associated with a time curve $N_j(t)$ in compartment j is

$$CV_j(t) = \frac{\sqrt{Var[N_j(t)]}}{E[N_j(t)]} = \sqrt{\frac{1}{n_0}\frac{\sum_{i=1}^{m}[1 - p_{ij}(t)]}{\sum_{i=1}^{m} p_{ij}(t)}}.$$

CV varies as $1/\sqrt{n_0}$ and it is not a small number for dosages involving few particles or drugs administered at very low doses; otherwise, $CV \ll 1$, as is typical in pharmacokinetics [370, 371]. From a mechanistic point of view, if the number of molecules present is not large, the concentration as a function of time will show the random fluctuations we expect from chance occurrences. However, if the number is very large, these fluctuations will be negligible, and for purposes of estimation, the stochastic error may be omitted in comparison with the measurement error.

Serial Error

An important generalization concerns the multinomial distribution of observations at different times. To deal with this, we analyze in the Markovian context the prediction of the statistical behavior of particles at time $t + t^\circ$ based on the observations at t, i.e., the state about the conditional random variable $[\underline{N}_i(t + t^\circ) \mid n_i(t)]$. As previously, in common use is the multinomial distribution

$$[\underline{N}_i(t + t^\circ) \mid n_i(t)] \sim \text{multinomial}\left[n_i(t), \underline{p}_i(t, t + t^\circ)\right]$$

using the transfer probability $\underline{p}_i(t, t + t^\circ)$ with elements $p_{ij}(t, t + t^\circ)$. For the standard Markov process, the above expression is reduced to

$$[\underline{N}_i(t + t^\circ) \mid n_i(t)] \sim \text{multinomial}\left[n_i(t), \underline{p}_i(t^\circ)\right], \tag{9.24}$$

where $\underline{p}_i(t^\circ)$ is the state probability with elements $p_{ij}(t^\circ)$. The conditional expectation of $E[\underline{N}_i(t + t^\circ) \mid n_i(t)]$ is $n_i(t)\,\underline{p}_i(t^\circ)$, and whatever the particles' origin,

$$\begin{aligned} E[\underline{N}(t + t^\circ) \mid \underline{n}(t)] &= \sum_{i=1}^{m} n_i(t)\,\underline{p}_i(t^\circ), \qquad (9.25) \\ Var[N_j(t + t^\circ) \mid \underline{n}(t)] &= \sum_{i=1}^{m} n_i(t)\,p_{ij}(t^\circ)\left[1 - p_{ij}(t^\circ)\right], \\ Cov[N_j(t + t^\circ)\,N_k(t + t^\circ) \mid \underline{n}(t)] &= -\sum_{i=1}^{m} n_i(t)\,p_{ij}(t^\circ)\,p_{ik}(t^\circ). \end{aligned}$$

The expressions (9.24) and (9.25) correspond to (9.22) and (9.23), respectively. The latter expressions can be obtained from the former ones by substituting t by 0 and t° by t. Since $\underline{N}(t + t^\circ)$ is conditioned to the random $\underline{n}(t)$, the total expectation theorem leads unconditionally to (cf. Appendix D)

$$E[\underline{N}(t + t^\circ)] = \sum_{i=1}^{m} E[N_i(t)]\,\underline{p}_i(t^\circ),$$

and the total variance theorem leads to

$$\begin{aligned} Var[N_j(t + t^\circ)] &= \sum_{i=1}^{m} E[N_i(t)]\,p_{ij}(t^\circ)\left[1 - p_{ij}(t^\circ)\right] \\ &\quad + \sum_{i=1}^{m} Var[N_i(t)]\,p_{ij}(t^\circ), \\ Cov[N_j(t + t^\circ)\,N_k(t + t^\circ)] &= -\sum_{i=1}^{m} E[N_i(t)]\,p_{ij}(t^\circ)\,p_{ik}(t^\circ) \\ &\quad - \sum_{i=1}^{m} Var[N_i(t)]\,p_{ij}(t^\circ)\,p_{ik}(t^\circ). \end{aligned}$$

The covariance structure following the chain binomial distribution [305, 346] introduces a serial covariance process error [372]. It is defined by

$$E[N_j(t)\,N_k(t + t^\circ) \mid \underline{n}(t)] = \sum_{i=1}^{m} N_j(t)\,N_i(t)\,p_{ik}(t^\circ),$$

and using the same unconditional approach,

$$E[N_j(t)\,N_k(t + t^\circ)] = E\left[N_j^2(t)\right] p_{jk}(t^\circ) + \sum_{\substack{s=1 \\ s \neq j}}^{m} E[N_j(t)\,N_i(t)]\,p_{ik}(t^\circ).$$

Hence for all $j = 1, \ldots, m$ and $k = 1, \ldots, m$,

$$Cov\left[N_j(t), N_k(t+t^\circ)\right] = p_{jk}(t^\circ) Var\left[N_j(t)\right] + \sum_{\substack{s=1 \\ s \neq j}}^{m} Cov\left[N_j(t), N_k(t)\right] p_{ik}(t^\circ),$$

which can be expressed in terms of the n_{0i} and $p_{ij}(t)$ $(i, j = 1, \ldots, m)$ using (9.23). This error could be named *temporal stochastic error*, since it describes the error correlation between two time instants for a couple of compartments. These results agree with the equations of Kodell and Matis [373] in the two-compartment case that they discussed. The above derivations apply equally to the time-dependent Markov process if we replace $p_{ij}(t^\circ)$ by $p_{ij}(t, t+t^\circ)$. The additional difficulties in the time-dependent case come in the computation of $p_{ij}(t_\circ, t)$. It is important to note that in the general case, the stochastic errors have slight serial correlation and hence are not independent. In the pharmacokinetic context where the number of molecules is large, the serial error may be neglected in comparison with the measurement error.

In principle, the general objective is to solve for the distribution of the random vector $\underline{N}(t)$, which might then be compared with the deterministic solution. However, the first and the second moments are sufficient for many applications using least squares procedures, since the mean value function gives the regression model and the second moments provide information useful in weighting the data and in identifying the model. Hence, one focus only on these moments, and, for simplicity, one considers only the expectations and variances. The covariance structure, where the $\underline{N}(t)$ are interrelated both temporally as well as serially, must be used together with the measurement error.

Finally, note also that we do not use the count of particles that have gone to the environment. This can be recovered from the original counts and the counts in the other compartments. Use of that count would introduce an exact linear dependence in the data.

9.3.5 Distribution of Particles and Process Uncertainty

To illustrate the process uncertainty, we present the case of the two-compartment model, Figure 9.1. Equations (9.5) associated with the transfer-intensity matrix **H** were used to simulate the random distribution of particles, which expresses the process uncertainty.

The Time Profile of the Distribution of Particles

After obtaining the state probabilities and setting the distributional assumptions, it is interesting to simulate the probabilistic behavior of the system, i.e., evaluate $\Pr\left[N_j(t) = n\right]$, $n = 0, \ldots, \infty$ and $j = 1, 2$. For a given n, $\Pr\left[N_j(t) = n\right]$ is the joint probability of the $n+1$ possible mutually exclusive events that "i particles originated in compartment 1" AND "$n-i$ particles originated in compartment 2 are present in compartment j at t" with $i = 0, \ldots, n$.

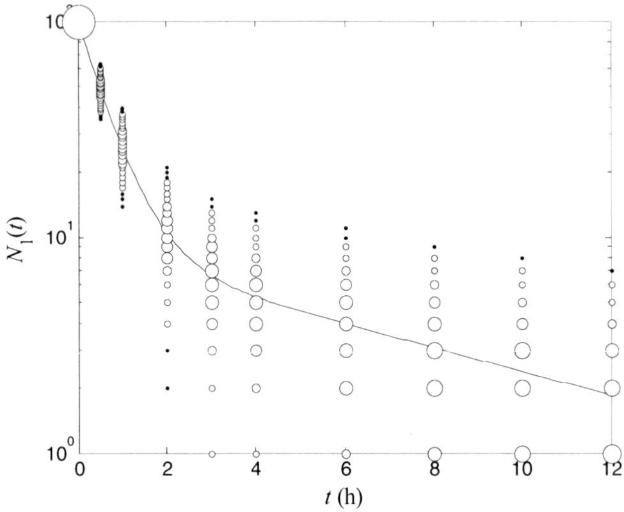

Figure 9.15: Probabilistic behavior of the particles observed in compartment 1. The *solid line* is the solution of the deterministic model. The area of a disk located at coordinates (t, n) is proportional to $\Pr[N_1(t) = n]$.

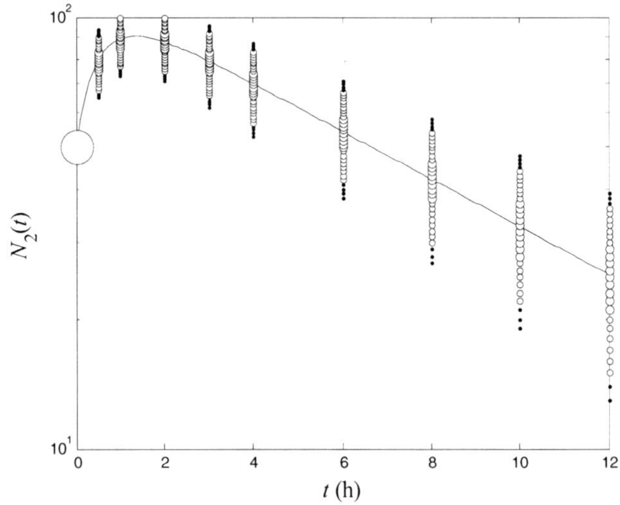

Figure 9.16: Probabilistic behavior of the particles observed in compartment 2. The *solid line* is the solution of the deterministic model. The area of a disk located at coordinates (t, n) is proportional to $\Pr[N_2(t) = n]$.

Because the particles behave independently

$$\Pr\left[N_j(t) = n\right] = \sum_{i=0}^{n} \Pr\left[i; n_{01}, p_{1j}(t)\right] \Pr\left[n - i; n_{02}, p_{2j}(t)\right],$$

where $\Pr[i; n, p]$ is the binomial distribution giving the probability of obtaining i tiles among n with prior probability p.

Using hazard rates $h_{10} = 0.5$, $h_{20} = 0.1$, $h_{12} = 1$, and $h_{21} = 0.1\,\mathrm{h}^{-1}$, and initial conditions $\underline{n}_0^T = [100\ 50]$, Figures 9.15 and 9.16 show the time profile of $\Pr[N_j(t) = n]$ (the $n = 0$ levels were not shown). In these figures for a given time t and a fixed level n, the disk area is proportional to the associated probability $\Pr[N_j(t) = n]$. Thus for each t, the sum of areas is equal to 1. It is noted that a fixed n has chances to occur at several t, and for a fixed t, the probability is widespread over a range of n values. This phenomenon is the process uncertainty or stochastic error.

The Process Uncertainty and the Serial Correlation

Assuming as initial conditions first $\underline{n}_0^T = [10\ 5]$ and then $10\underline{n}_0$, Figures 9.17 and 9.18 illustrate:

- the time-particle-count profiles for the two compartments $E\left[\underline{N}_j(t)\right]$, $j = 1, 2$,
- the confidence intervals computed as $E\left[\underline{N}_j(t)\right] \pm \sqrt{Var\left[N_j(t)\right]}$, $j = 1, 2$, and
- random data generated from the binomial distribution, $\mathrm{Bin}[n_{0i}, p_{ij}(t)]$, with prior probabilities computed from (9.5).

These profiles were normalized with respect to the initial condition in each compartment. The wider confidence intervals correspond to the initial conditions $\underline{n}_0$, and the narrower confidence intervals to $10\underline{n}_0$. Even without measurement error, fluctuations in the predicted amounts expressing the process uncertainty were observed: the lower the number of molecules initially present in the compartments, the higher the observed fluctuations.

For $t^{\circ} = 1, 2, 4$, Figure 9.19 illustrates the correlation coefficients

$$Cor\left[N_j(t), N_k(t + t^{\circ})\right]$$

computed from covariances

$$Cov\left[N_j(t), N_k(t + t^{\circ})\right]$$

between the same and different compartments. The autocorrelations

$$Cor\left[N_1(t), N_1(t + t^{\circ})\right] \quad \text{and} \quad Cor\left[N_2(t), N_2(t + t^{\circ})\right]$$

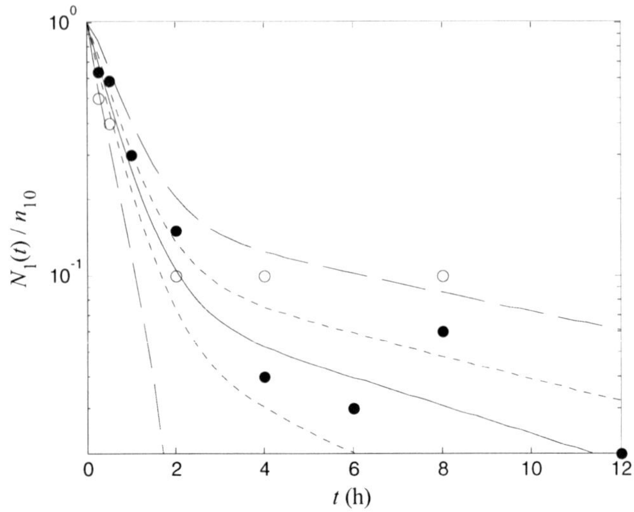

Figure 9.17: Normalized particle-count profiles in compartment 1. *Dashed line* and *open circles* for low initial conditions, and *dotted line* and *full circles* for high initial conditions.

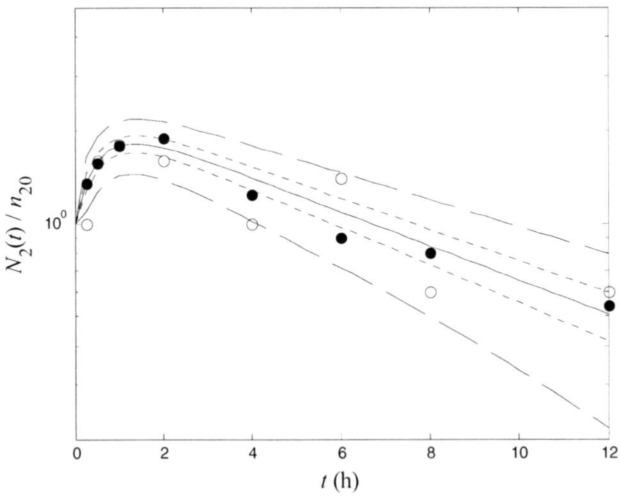

Figure 9.18: Normalized particle-count profiles in compartment 2. Symbols as in Figure 9.17.

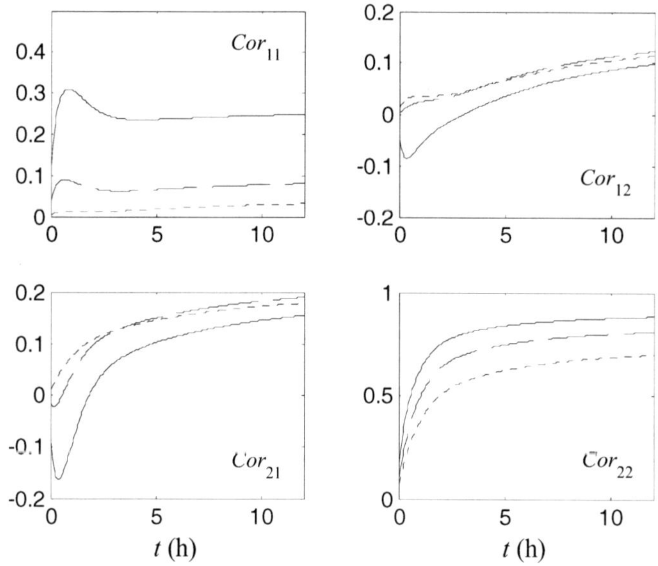

Figure 9.19: Autocorrelations and cross-compartment serial correlations with increased values of delay $t^{\circ} = 1, 2, 4$ (*solid*, *dashed*, and *dotted lines*, respectively).

vanish with increasing t° and they are always positive. $Cor\left[N_2(t), N_2(t+t^{\circ})\right]$ reaches high levels because particles stay longer in compartment 2 when trapped by the slow hazard rate h_{21}. The cross-correlations $Cor\left[N_j(t), N_k(t+t^{\circ})\right]$, $j \neq k$, are low in absolute value. Sensitivity analysis reveals that the inter-compartment hazard rates h_{12} and h_{21} highly influence autocorrelations, while cross-correlations are more influenced by h_{10} and h_{20}, the elimination rates of particles to the environment.

9.3.6 Time Profiles of the Model

According to definitions (8.3) and (9.6), the relationship between clearance, volume of distribution, and hazard rate is again recalled:

$$CL(t) = V(t)\, h(t).$$

This relationship is now considered as time-dependent because of $h(t)$, the age-dependent hazard rate in the retention-time models, or because of $V(t)$, the time-varying volume of distribution. For all the above models, the time–concentration curve $E[C(t)]$ in each observed compartment is obtained by dividing $E[Q(t)]$ by $V(t)$. For the simplest one-compartment model, two different

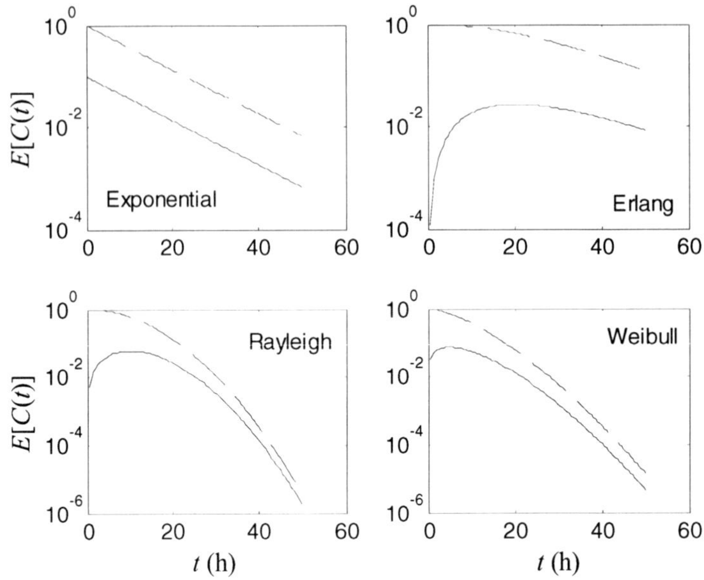

Figure 9.20: Time–concentration curves for the hypotheses of a constant V (*dashed line*) and a constant CL (*solid line*).

interpretations may arise:

- The volume of distribution is assumed constant. In this case,

$$E\left[C\left(t\right)\right]=\frac{E\left[Q\left(t\right)\right]}{V}=\frac{q_0\mathcal{S}\left(t\right)}{V}$$

and $E\left[C\left(t\right)\right]$ is directly proportional to the survival function $\mathcal{S}\left(t\right)$. Also, the clearance $CL\left(t\right)$ becomes an age-dependent parameter proportional to the hazard rate $h\left(t\right)$.

- The clearance is assumed constant. In this case,

$$E\left[C\left(t\right)\right]=\frac{E\left[Q\left(t\right)\right]}{V\left(t\right)}=q_0\mathcal{S}\left(t\right)\frac{h\left(t\right)}{CL}=\frac{q_0}{CL}f\left(t\right)$$

and $E\left[C\left(t\right)\right]$ is directly proportional to the density function $f\left(t\right)$. Also, the volume $V\left(t\right)$ becomes an age-dependent parameter inversely proportional to the hazard rate $h\left(t\right)$.

In other words, the expectation of the amount behaves always as the survival function $\mathcal{S}\left(t\right)$ but the expectation of the concentration behaves either as the density function $f(t)$ if CL is assumed constant, or as $\mathcal{S}(t)$ if V is assumed constant. Consequently, given a set of observed data, we may have indication that the process has a constant V if the best fitting is obtained by using the survival function. Conversely, if the best fitting is obtained by using the density function, the process is rather driven by a constant CL. Figure 9.20 simulates one-compartment retention-time models with initial conditions and compares the time–concentration curves obtained under the hypothesis of a constant V or a constant CL. It is noticeable that:

- the exponential retention-time distribution did not discriminate between the two hypotheses, and
- for the other distributions, the constant CL hypothesis yields a maximum in the time–concentration curve.

After bolus administration and keeping the CL constant, Weiss [245] obtained the simple time–concentration profile

$$c(t) = \gamma t^{-(1-\mu)} \exp(-\lambda t)$$

by assuming the Gam(λ, μ) retention-time distribution for the particle ages in the single compartment. Under the same conditions, Piotrovskii [374] assumed the Wei(λ, μ) retention-time distribution, but both models constrained the shape parameter $(0 < \mu < 1)$ in order to ensure monotonically decreasing kinetic profiles. Nevertheless, there is no indication of biological or numerical nature excluding cases with $\mu > 1$ that lead to profiles similar to those shown in Figure 9.20 and, in several cases [298], experimental data lead to negative powers of time less than -1 that contrast with the positivity of μ. Therefore, some assumptions become questionable, e.g., the simple compartmental structure, or the time–constancy of CL, or the choice for the retention-time distribution.

9.4 Random Hazard-Rate Models

In the models of the previous section, the stochastic nature of the system was due to the random movement of homogeneous individual particles. They are probabilistic transfer models or retention-time models expressing that the molecules are retained, or trapped by cells or otherwise fixed components of the process. In this way, these models express the structural heterogeneity that may originate from the time courses of particles through media that are inhomogeneous or from the retention of particles by organs in the body that are characterized by heterogeneous or fractal structures, e.g., the liver and lung.

We consider now a class of models that introduce particle heterogeneity through random rate coefficients. In this conceptualization, the particles are assumed different due to variability in such characteristics as age, size, molecular conformation, or chemical composition. The hazard rates h are now considered to be random variables that vary influenced by extraneous sources of fluctuation as though stochastic processes were added on to the hazard rates. This approach corresponds to a physiologically realistic mechanism by which the hazard rates fluctuate in an apparently random manner because of influences from other parts of the real system affecting them but that are not included in the model. The random variable h is associated with a specific probability density function $f(h)$.

Hazard rates are heterogeneous particle models expressing a functional heterogeneity. As such they contrast with probabilistic transfer and retention-time models, which assume homogeneous particles and express a structural heterogeneity. As pointed out in Chapter 7, these heterogeneities may be described by simple empirical models with time-varying parameters. Using stochastic modeling, these heterogeneities may also be expressed in a different manner. In fact, the combination of the resulting stochasticities will provide a rich collection of models. Matis and Wehrly [304] call P1 stochasticity the variability induced by structural heterogeneity, and P2 stochasticity the variability induced by functional heterogeneity.

We now consider models that combine the sources of stochastic variability identified previously [375]. The experimental context reproducing the randomness of h can be conceived as follows:

- Assume that m_0 independent *units* were introduced initially into the system with a transfer mechanism whose hazard rate h applies to *all units* in the experiment. The random movement of individual *units* in the heterogeneous process will result in a state probability $p(t,h)$ depending on the specific h of all units in that experiment. Using the binomial distribution, the *conditional* expectation and variance are

$$\begin{aligned} E[N(t) \mid h] &= m_0 p(t,h), \\ Var[N(t) \mid h] &= m_0 p(t,h)[1 - p(t,h)]. \end{aligned}$$

- Let also n_0 be replicates of the above experiment where the hazard rate varies from experiment to experiment with probability density function $f(h)$. From the previous relations, the *unconditional* expectation and variance are (cf. Appendix D)

$$\begin{aligned} E[N(t)] &= E_h E[N(t) \mid h] = m_0 p(t), \\ Var[N(t)] &= E_h Var[N(t) \mid h] + Var_h E[N(t) \mid h] \\ &= m_0 p_S(t) + m_0^2 p_F(t), \end{aligned}$$

with

$$\begin{aligned} p(t) &= \int_h p(t,h) f(h) \mathrm{d}h, \qquad (9.26) \\ p_S(t) &= \int_h p(t,h)[1 - p(t,h)] f(h) \mathrm{d}h = p(t) - \int_h p^2(t,h) f(h) \mathrm{d}h, \\ p_F(t) &= \int_h [p(t,h) - p(t)]^2 f(h) \mathrm{d}h = \int_h p^2(t,h) f(h) \mathrm{d}h - [p(t)]^2. \end{aligned}$$

Table 9.2: Density and moment generating functions.

	$f(h)$	$\mathcal{M}(-t)$
$\mathrm{Exp}(\kappa)$	$\kappa \exp(-\kappa h)$	$(1+t/\kappa)^{-1}$
$\mathrm{Erl}(\lambda,\nu)$	$\frac{\lambda^{\nu}h^{\nu-1}}{(\nu-1)!}\exp(-\lambda h)$	$(1+t/\lambda)^{-\nu}$
$\mathrm{Gam}(\lambda,\mu)$	$\lambda(\lambda h)^{\mu-1}\exp(-\lambda h)/\Gamma(\mu)$	$(1+t/\lambda)^{-\mu}$
$\mathrm{Rec}(\alpha,\beta)$	$1/\beta,\ 0 \leq (h-\alpha)/\beta \leq 1$	$\exp(-\alpha t)\left[1-\exp(-\beta t)\right]/(\beta t)$

The variance expression is composed of two terms; $m_0 p_S(t)$ generalizes the variance of a standard binomial distribution and is attributable to the stochastic transfer mechanism (structural heterogeneity) and $m_0^2 p_F(t)$ reflects the random nature of h (functional heterogeneity).

The random hazard rate model is easily obtained from the above by considering a *single unit*, $m_0 = 1$, and n_0 particles initially administered into the system. The first two moments are obtained by summing n_0 independent and identically distributed experiments:

$$\begin{aligned} E[N(t)] &= n_0 p(t), \\ Var[N(t)] &= n_0[p_S(t)+p_F(t)] = n_0 p(t)[1-p(t)]. \end{aligned} \tag{9.27}$$

These relations are analogous to (9.23); the only difference is that in (9.27), $p(t)$ mixes the conditional $p(t,h)$ with the distribution $f(h)$.

9.4.1 Probabilistic Models with Random Hazard Rates

The solution of the probabilistic transfer equations leads to the exponential model (9.3). The presence of negative exponentials in the model may simplify somewhat the choice of distribution associated with the random hazard rate. In fact, the elements $p(t,h)$ of the state probability matrix $\exp(\mathbf{H}t)$ in (9.3) are exponentials, and integrating (9.27) over the random variable h, we obtain

$$\int_h \exp(-ht) f(h)\,\mathrm{d}h = \mathcal{M}(-t),$$

where $\mathcal{M}(-t)$ is the *moment generating function* of h. Hence, the parameters of the assumed "mixing" distribution $f(h)$ for the population of heterogeneous particles may be estimated directly by fitting $\mathcal{M}(-t)$ to data.

For the one-compartment model with n_0 initial conditions, the distribution of the random hazard rate h can be simply mixed with the state probability $p(t,h) = \exp(-ht)$, and relations (9.27) become

$$\begin{aligned} E[N(t)] &= n_0 \mathcal{M}(-t), \\ Var[N(t)] &= n_0\left[\mathcal{M}(-t) - \mathcal{M}^2(-t)\right]. \end{aligned}$$

By using the moment generating functions of Table 9.2, one directly obtains the following cases (the variance expressions were omitted for simplicity):

- Discrete distribution:

$$E\left[N\left(t\right)\right] = n_0 \sum_{i=1}^{m} p_i \exp(-\kappa_i t)$$

associated with the distribution function $\Pr\left[h = \kappa_i\right] = p_i$, $i = 1, \ldots, m$. In principle, one could fit this model with multiple rates to data and estimate the κ_i parameters. However, in practice the estimation can be difficult even for $m = 3$, and becomes particularly hazardous for any real application with $m > 3$ [375].

- Rectangular distribution:

$$E\left[N\left(t\right)\right] = n_0 \exp(-\alpha t) \frac{1 - \exp(-\beta t)}{\beta t}.$$

This model is an analogue to the previous model with multiple rates in that the m specified fractions are replaced by a continuous rectangular rate distribution [375, 376].

- Gamma distribution:

$$E\left[N\left(t\right)\right] = n_0 \left(1 + t/\lambda\right)^{-\mu}. \tag{9.28}$$

This is the most widely applied distribution for h. When the shape parameter is an integer, one obtains the Erlang distribution. Hence, the one-compartment stochastic model leads to power-law profiles involving λ and μ parameters.

In the following, we show how to apply probabilistic transfer models with random hazard rates associated with the administration and elimination processes in a single-compartment configuration.

Hazard Rate for the Absorption Process

We report the one-compartment probabilistic transfer model receiving the drug particles by an absorption process. In this model, the elimination rate h was fixed and the absorption constant h_{ev} was random. For the stochastic context, the difference $h_{ev} - h = w$ is assumed to follow the gamma distribution, i.e., $W \sim$Gam(λ, μ) with density $f\left(w; \lambda, \mu\right)$ and $E\left[W\right] = \mu/\lambda$.

The state probability for a particle with given h_{ev} to be in the central compartment at time t is

$$\begin{aligned} p\left(t, w\right) &= \frac{h_{ev}}{h_{ev} - h} \left[\exp\left(-ht\right) - \exp\left(-h_{ev} t\right)\right] \\ &= \frac{h + w}{w} \exp\left(-ht\right) \left[1 - \exp\left(-wt\right)\right]. \end{aligned} \tag{9.29}$$

Irrespective of the individual h_{ev}, the state probability is the mixture

$$p(t) = \int_w p(t,w) f(w;\lambda,\mu)\,\mathrm{d}w.$$

But for the gamma density the following hold:

$$\begin{aligned} xf(x;\lambda,\mu) &= \frac{\mu}{\lambda} f(x;\lambda,\mu+1), \\ \frac{1}{x} f(x;\lambda,\mu) &= \frac{\lambda}{\mu-1} f(x;\lambda,\mu-1), \end{aligned} \tag{9.30}$$

allowing us to compute

$$\begin{aligned} &\int_h p(t,h) f(h)\,\mathrm{d}h \\ = \;& \exp(-ht)\left\{1 - \mathcal{M}(-t;\lambda,\mu) + \frac{h\lambda}{\mu-1}\left[1 - \mathcal{M}(-t;\lambda,\mu-1)\right]\right\} \\ &\int_h p^2(t,h) f(h)\,\mathrm{d}h \\ = \;& \exp(-2ht)\left\{1 - 2\mathcal{M}(-t;\lambda,\mu) + \mathcal{M}(-2t;\lambda,\mu)\right. \\ & + \frac{2h\lambda}{(\mu-1)}\left[1 - 2\mathcal{M}(-t;\lambda,\mu-1) + \mathcal{M}(-2t;\lambda,\mu-1)\right] \\ & \left. + \frac{h^2\lambda^2}{(\mu-1)(\mu-2)}\left[1 - 2\mathcal{M}(-t;\lambda,\mu-2) + \mathcal{M}(-2t;\lambda,\mu-2)\right]\right\} \end{aligned}$$

with $\mu > 2$ and $\mathcal{M}(-t;\lambda,\mu)$ given in Table 9.2. From (9.26) and (9.27), one obtains the expected profile $E[N(t)] = n_0 p(t)$ and the standard deviations $\sqrt{n_0 p_S(t)}$ and $\sqrt{n_0 p_F(t)}$ associated with the structural and functional heterogeneity, respectively. For $n_0 = 10$, $h = 0.1\,\mathrm{h}^{-1}$, $\lambda = 1.5\,\mathrm{h}^{-1}$, and $\mu = 2.5$, Figure 9.21 shows the expected profile, the confidence corridors computed from the previous standard deviations, and the profile $n_0 p(t, \mu/\lambda)$ obtained from (9.29) using the expected value of the random variable W. All these profiles were normalized with respect to the initial condition n_0. We note the larger variability associated with the functional heterogeneity compared to that associated with the structural one, and the difference between the expected profile of the model with a random rate coefficient and the profile of the model with a fixed coefficient evaluated at the mean rate.

Hazard Rate for the Elimination Process

We present the one-compartment case in which the drug amount n_0 is given over a period T by a constant-rate infusion. Assuming a random hazard rate h over the molecules, the state probability that "a molecule associated with a hazard rate h is in the compartment at time t" is

$$p(t,h) = \frac{1}{Th}\begin{cases} 1 - \exp(-ht), & t \leq T, \\ \exp[-h(t-T)] - \exp(-ht), & T < t. \end{cases} \tag{9.31}$$

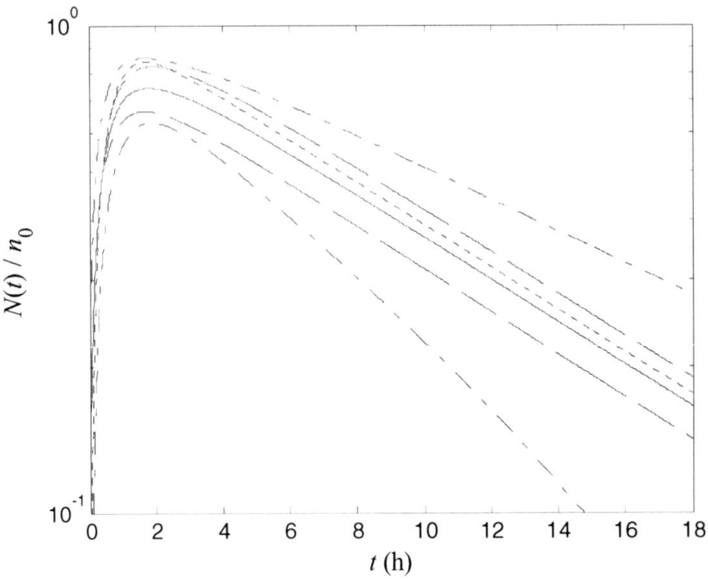

Figure 9.21: For the absorption model, expected profile (*solid line*), confidence corridors (*mixed* and *dashed lines* for functional and structural heterogeneity, respectively) and profile with the mean coefficient value (*dotted line*).

Let H be gamma distributed, i.e., $H \sim \text{Gam}(\lambda, \mu)$. By using properties (9.30) with $\mu > 1$ for the gamma distribution, the resulting probability that "a molecule regardless of its hazard rate is in the compartment at time t" will be

$$p(t) = \frac{\lambda}{T(\mu - 1)} \begin{cases} 1 - (1 + t/\lambda)^{-(\mu-1)}, & t \leq T, \\ [1 + (t - T)/\lambda]^{-(\mu-1)} - (1 + t/\lambda)^{-(\mu-1)}, & T < t. \end{cases}$$

From (9.26) and (9.27), we obtain the expected profile and standard deviation. For $n_0 = 10$, $\lambda = 1.5\,\text{h}^{-1}$, and $\mu = 2.5$, Figure 9.22 shows the expected profile, the confidence corridors computed from the standard deviation, and the profile $n_0 p(t, \mu/\lambda)$ obtained from (9.31) using the expected value of the random variable H. All these profiles were normalized with respect to the initial condition n_0. As for the absorption process, we note the difference between the expected profile of the model with a random rate coefficient and the profile of the model with a fixed coefficient evaluated at the mean rate.

For $\mu = 1$, the gamma distribution is reduced to an exponential one, and following the same procedure,

$$p(t) = \frac{\lambda}{T} \begin{cases} \ln(1 + t/\lambda), & t \leq T, \\ \ln(1 + t/\lambda) - \ln[1 + (t - T)/\lambda], & T < t. \end{cases}$$

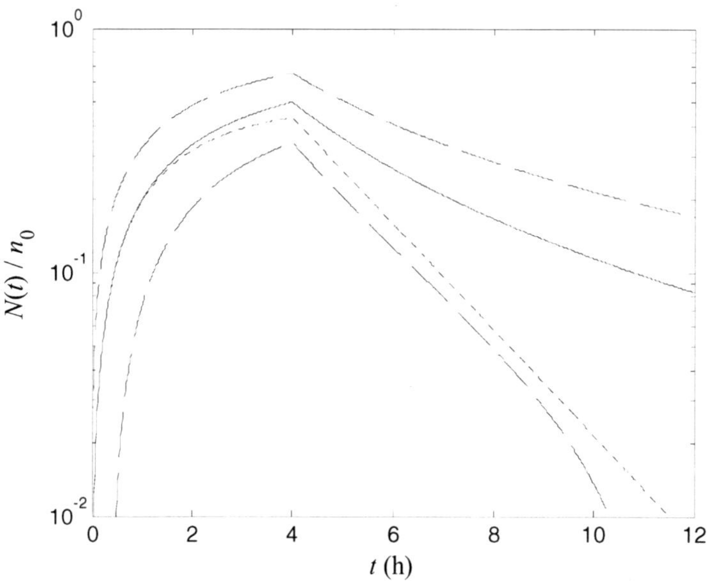

Figure 9.22: For the elimination model, expected profile (*solid line*), confidence corridors (*dashed lines*), and profile with the mean coefficient value (*dotted line*).

In this case, following long-term infusion, no asymptotic behavior can be reached as t goes to infinity, i.e., no steady state exists.

Note that when the drug is given by a short infusion, i.e., $T \to 0$, the above expressions for $p(t)$ are reduced to (9.28).

Each molecule has its own hazard rate, and if we assume a constant volume of distribution V, each molecule will have its own clearance defined as $CL = Vh$. Then CL becomes a random variable, and there follows the distribution of h with expectation $E[CL] = VE[h] = V\mu/\lambda$. Regardless of the molecule's clearance, the systemic clearance may be obtained on the basis of the expected profile $E[N(t)]$ using either the plateau evaluation during a long-term infusion or the total area under the curve. Both evaluations give $CL = V(\mu - 1)/\lambda$. Note that for $\mu = 1$, the systemic clearance cannot be defined albeit individual molecular clearances exist. The discrepancy between $E[CL]$ and CL is due to the randomness of the model parameter h.

The discrepancy mentioned above in the parameter space is at the origin of the often reported discrepancy in the output space. When a rate coefficient is a random variable, the *expected amount* of a model with a *random rate* coefficient will always exceed the *amount* of a model with a fixed coefficient evaluated at the *mean rate*. It is a widespread conjecture in modeling that for systems with linear kinetics, the deterministic solution is identical to the mean value from any stochastic formulation. This conjecture, however, clearly does not hold when the rate coefficient is a random variable. In fact, the function $\exp(-kt)$ is convex, and using *Jensen's inequality* [377] we can prove that for any t,

$$E\left[\exp\left(-kt\right)\right] \geq \exp\left(-E\left[k\right]t\right),$$

which permits us to conclude that the kinetic profile of a homogeneous substance is always faster than that of a heterogeneous compound for which the mean rate is the same as the rate of the homogeneous one. Therefore, the mean of the stochastic model exceeds the deterministic model evaluated at the mean rates, $E\left[N\left(t\right)\right] > n\left(t, E\left[CL\right]\right)$, and this is why $CL < E\left[CL\right]$.

For models using the pseudocompartment techniques to express the retention-time distribution, the same procedure as for the probabilistic transfer models can be applied to incorporate the randomness of the distribution parameter. Also, for simple situations, several assumed probability density functions of h that are rich in form yet parsimonious in parameters have been suggested by Matis [304,376,378]. Although these models are lengthy, they have few parameters and may be fitted to data using standard nonlinear least squares computer programs. Clearly, these models represent the union of many mechanisms that have been observed in experimental studies to be of interest in retention-time modeling. These models have considerable appeal analytically because the parameters are identifiable, the regression functions are not necessarily monotonic, and most of the previous models are special cases of this mixture model.

9.4.2 Retention-Time Models with Random Hazard Rates

Like the previous ones, these models are two-level models. Now, the retention-time model substitutes the probabilistic transfer model in the first level, and in the second level, parameters of this model are assumed to be random and they are associated with a given distribution. Consider, for instance, the one-compartment model with Erlang retention times where the parameter λ is a random variable expressing the heterogeneity of the molecules. Nevertheless, even for the simplest one-compartment case, the model may reach extreme complexity. In these cases, analytical solutions do not exist and numerical procedures have to be used to evaluate the state probability profiles.

This approach is presented for the two-compartment model of Section 9.2.7. At the second level in (9.16), we assume that λ is a gamma-distributed random variable, $\Lambda \sim \text{Gam}(\lambda_2, \mu_2)$. The Laplace transform of the state probability is

$$p_1\left(t\right) = \frac{\lambda_2}{\Gamma\left(\mu_2\right)} \int_\lambda \left(\lambda_2 \lambda\right)^{\mu_2 - 1} \exp\left(-\lambda_2 \lambda\right) L^{-1} \left\{ \frac{1}{s + \kappa - \omega_e \kappa \left(\frac{\lambda}{s+\lambda}\right)^{\mu}} \right\} d\lambda.$$

This expression was computed by the numerical inverse Laplace transform embedded in the numerical quadrature. As previously, we used $\mu = 1, 4, 6$, $E\left[\Lambda\right] = \mu_2/\lambda_2 = 1$, and $\lambda_2 = 4\,\text{h}^{-1}$. Figure 9.23 illustrates the influence of the μ parameter on the shape of the state probability profile: the larger μ, the most

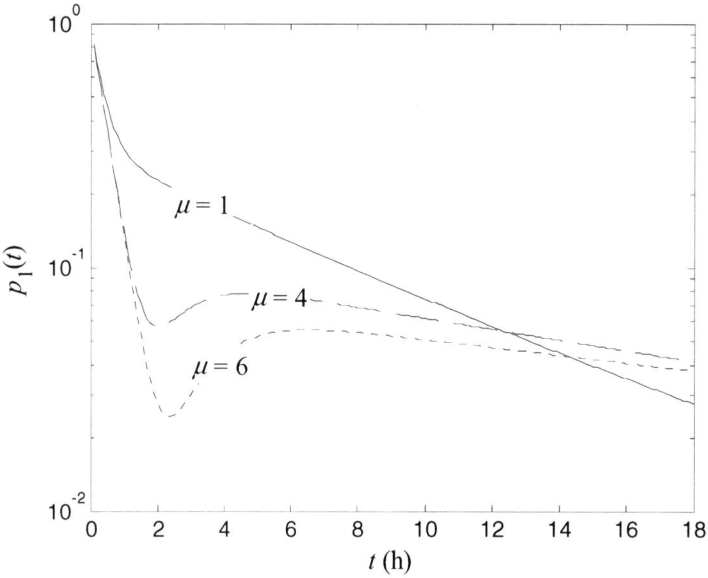

Figure 9.23: Simulation of time–$p_1(t)$ profiles by assuming $\lambda \sim$Gam(λ_2, μ_2) and for $\mu = 1, 4, 6$.

pronounced the rebound form of the profile. For comparison, cf. Figure 9.10, which was obtained with fixed $\lambda = 1\,\text{h}^{-1}$.

Actually, the inverse problem should be solved, i.e., given the data $n(t)$ containing errors, obtain a plausible candidate $f(h)$ associated with a known function $p(t, h)$. This function, termed *kernel*, is assumed to be a retention-time distribution other than an exponential one; otherwise, the problem has a tractable solution by means of the moment generating functions as presented earlier. This part aims to supply some indications on how to select the density of h. For a given probability density function $f(h)$, one has to mix the kernel with $f(h)$:

$$n(t) = n_0 \int_h p(t, h) f(h) \, \mathrm{d}h. \tag{9.32}$$

Equation (9.32) is a linear Fredholm integral equation of the first kind. It is also known as an *unfolding* or *deconvolution* equation. One can preanalyze the data and try to solve this first-kind integral equation. Besides the complexity of this equation, there is a paucity of numerical methods for determining the unknown function $f(h)$ [208, 379] with special emphasis on methods based on the principle of maximum entropy [207, 380]. The so-obtained density function may be approximated by several models, gamma, Weibull, Erlang, etc., or by phase-type distributions.

Although attractive at first, there are some problems associated with the random rate models. Here, we really have a two-level stochastic model in that the parameters of the basic model contain a stochastic process. So, the first important problem is how to partition the contribution of the basic probabilistic or retention-time model, and the contribution of the random rate distribution model. This cannot be decided on the basis of empirical time–concentration data alone. The case is that of fitting a sum of exponentials model to the time–concentration data and then to assume that the number of compartments in the system is at least as large as the number of exponential terms required to achieve an acceptable fitting. This practice is inappropriate and may be very misleading when a random rate coefficient is present. Indeed, for a biphasic distribution of time–concentration data, the biexponential model is used with the common hypothesis that the underlying mechanism is a deterministic two-compartment model. But it is apparent that a one-compartment model with h having two possible outcomes has a biexponential function for its mean value. It follows therefore that one cannot imply that a multiexponential fitting of the observed mean value is sufficient evidence of a multicompartment system [375]. A second problem is related to the choice of stochastic processes to be added to a transfer coefficient. Since no transfer coefficient may ever be negative, distributions such as the normal are excluded, but log-normal, gamma, or Weibull would be acceptable.

9.5 The Kolmogorov or Master Equations

Given a compartmental structure, the probabilistic transfer, the retention-time, as well as the random hazard rate models were first conceived to express the probability of a particle transferring between compartments. In a next step and for the multinomial distribution, the model was extended to the whole of particles administered in the system and expectations, variances, and covariances were obtained. In the last step, $\Pr[N_j(t) = n]$, $n = 0, \ldots, \infty$ and $j = 1, \ldots, m$, was obtained, i.e., the probability of having a given number of particles n in a given compartment j at time t. In a reverse way and for a given compartmental structure, one could model $\Pr[N_j(t) = n]$ and subsequently obtain the statistical distribution of particle transfer among compartments. The probabilistic transfer formulation is rather focused on compartmental modeling and describes mainly diffusion processes. Additionally, probabilistic transfer models can also be proposed for processes involving chemical, metabolic, and enzymatic reactions as well as for release, transport, and absorption phenomena. In this section, enlarged modeling concepts will be used to take into account all these processes without exclusively referring to the special case of compartmental modeling.

In a general context, suppose a given volume V contains a spatially homogeneous mixture of N_i particles from m different populations of initial size n_{0i} $(i = 1, \ldots, m)$. Suppose further that these m populations can interact through $m_\circ$ specified reaction or diffusion channels R_l $(l = 1, \ldots, m_\circ)$. These processes are assumed to be characterized by the *probability of an elementary event per unit time* that depends only on the physical properties of the diffusing or reacting particles and on the real system environment such as temperature, pressure, etc. Then, we may assert the existence of $m_\circ$ constants h_l that are the hazard rates as they were defined in (9.1) for the probabilistic transfer models. Now they are reformulated and the elementary transfer event is designated by a single index l, instead the double index ij denoting the start and end compartments.

The h_l are used to express the conditional probability of "changes in the population sizes for the R_l reaction from t to $t + \Delta t$ given the system in $\underline{n}(t)$ at t." These probabilities are described by means of the *intensity functions* $I_{\varphi_{l,1},\ldots,\varphi_{l,m}}(N_1, \ldots, N_m)$, whereby

$$\begin{aligned} & \Pr\left[N_1 \text{ changes by } \varphi_{l,1}, \ldots, N_m \text{ changes by } \varphi_{l,m}\right] \\ = \; & I_{\varphi_{l,1},\ldots,\varphi_{l,m}}(N_1, \ldots, N_m)\, \Delta t + o(\Delta t) \end{aligned}$$

with $\varphi_{l,i}$ denoting the changes in population i by the R_l reaction. Analysis of the mechanistic behavior of a population of reacting particles inside V leads to the intensity functions [381] of the form

$$I_{\varphi_{l,1},\ldots,\varphi_{l,m}}(N_1, \ldots, N_m) = h_l N_1^{\psi_{l,1}} \cdots N_m^{\psi_{l,m}} \tag{9.33}$$

involving the hazard rates h_l and the number of particles $\psi_{l,i}$ from the population i implied in the R_l reaction. The intensity functions are ψ_l-order elementary processes in the previous equation, with

$$\psi_l = \sum_{i=1}^{m} \psi_{l,i}.$$

The definition of the h_l hazard rates and the model of (9.33) are the only required hypotheses to formulate the stochastic movement or reaction of particles in a spatial homogeneous mixture of m-particle populations interacting through $m_\circ$ reactions.

To calculate the stochastic time evolution of the system, the key element is the *grand probability function*

$$p_{n_1,\ldots,n_m}(t) = \Pr\left[N_1(t) = n_1, \ldots, N_m(t) = n_m\right],$$

i.e., the joint probability that "there will be in the system n_1 particles of the 1st population, ..., and n_m particles of the mth population at time t." The abundance of particles at t can be viewed as a random vector $\underline{N}(t) = [N_1(t), \ldots, N_m(t)]^T$ and the objective is to solve for $p_{\underline{n}}(t)$, for any $t > 0$. One standard approach for solving for the grand probability function is to use equations known as the *Kolmogorov differential equations* or known also as the *master equation* in chemical engineering. This equation may be obtained by using the addition and multiplication laws of probability theory to write $p_{\underline{n}}(t + \Delta t)$ as the sum of the probabilities of the $1 + m_\circ$ different ways in which the system can arrive at the state $\underline{n}(t)$ at time $t + \Delta t$:

$$p_{\underline{n}}(t + \Delta t) = p_{\underline{n}}(t) \left[1 - \Delta t \sum_{l=1}^{m_\circ} a_l \right] + \Delta t \sum_{l=1}^{m_\circ} b_l. \tag{9.34}$$

Here we have defined the quantities a_l and b_l by

$$\begin{aligned} a_l &\equiv I_{\varphi_{l,1},\ldots,\varphi_{l,m}}(N_1, \ldots, N_m), \\ b_l &\equiv p_{n_1-\varphi_{l,1},\ldots,n_m-\varphi_{l,m}}(t)\, I_{\varphi_{l,1},\ldots,\varphi_{l,m}}\left(n_1 - \varphi_{l,1}, \ldots, n_m - \varphi_{l,m}\right). \end{aligned}$$

Thus:

- The quantity $a_l \Delta t$ is the probability that "an R_l reaction occurs in Δt, given the system in $\underline{n}(t)$" and the first term in (9.34) is the probability that "the system will be in the state $\underline{n}(t)$ at time t, and then remains in that state in $(t, t + \Delta t)$."
- The quantity $b_l \Delta t$ gives the probability that "the system has one R_l reaction *removed* from the state $\underline{n}(t)$ at time t, and then undergoes an R_l reaction in $(t, t + \Delta t)$." Thus, b_l will be the product of $p_{\underline{n}}(t)$ evaluated at the appropriate once-removed state at t, TIMES the lth intensity function evaluated in that once-removed state.

Subtracting $p_{\underline{n}}(t)$ in (9.34), dividing by Δt, and taking the limit as $\Delta t \to 0$, one has

$$\dot{p}_{\underline{n}}(t) = -p_{\underline{n}}(t) \sum_{l=1}^{m_\circ} a_l + \sum_{l=1}^{m_\circ} b_l \tag{9.35}$$

for $n_i > 0$ and the appropriate boundary conditions for each $n_i = 0$. The initial conditions are

$$p_{\underline{n}}(0) = \prod_{i=1}^{m} \delta(n_i - n_{0i}),$$

where $\delta(n - n_0)$ is the *Dirac delta function.* These equations yield the desired probability distribution for $\underline{N}(t)$. This is an infinite system of linear differential equations in the state probabilities expressed by the Kolmogorov equations. Although the system is infinite, the probabilities associated with states much larger than $\sum_i n_{0i}$ become minute.

An important property of the stochastic version of compartmental models with linear rate laws is that the mean of the stochastic version follows the same time course as the solution of the corresponding deterministic model. That is not true for stochastic models with nonlinear rate laws, e.g., when the probability of transfer of a particle depends on the state of the system. However, under fairly general conditions the mean of the stochastic version approaches the solution of the deterministic model as the number of particles increases. It is important to emphasize for the nonlinear case that whereas the deterministic formulation leads to a *finite* set of *nonlinear* differential equations, the master equation generates an *infinite* set of *linear* differential equations although the rate laws are nonlinear.

Besides the hypothesis of spatially homogeneous processes in this stochastic formulation, the particle model introduces a structural heterogeneity in the media through the scarcity of particles when their number is low. In fact, the number of differential equations in the stochastic formulation for the state probability keeps track of all of the particles in the system, and therefore it accounts for the particle scarcity. The presence of several differential equations in the stochastic formulation is at the origin of the uncertainty, or stochastic error, in the process. The deterministic version of the model is unable to deal with the stochastic error, but as stated in Section 9.3.4, that is reduced to zero when the number of particles is very large. Only in this last case can the set of Kolmogorov differential equations be adequately approximated by the deterministic formulation, involving a set of differential equations of fixed size for the states of the process.

9.5.1 Master Equation and Diffusion

As an example application, we will develop the master equation for a fragment of a two-way catenary compartment model around three compartments spaced by Δz, as illustrated in Figure 9.24. By assuming only one particle in movement, the master equation gives

$$\begin{aligned} p_{\ldots 010\ldots}(t+\Delta t) &= p_{\ldots 010\ldots}(t)\left[1-\Delta t\left(h_{-}+h_{+}\right)\right] \\ &\quad + p_{\ldots 100\ldots}(t)\, h_{+}\Delta t + p_{\ldots 001\ldots}(t)\, h_{-}\Delta t. \end{aligned}$$

The subscripts $\ldots 010\ldots$, $\ldots 100\ldots$, and $\ldots 001\ldots$ indicate that the particle is located in the z, $z-\Delta z$, or $z+\Delta z$ compartment, respectively. For writing convenience, we denote these probabilities by $p(z,t)$, $p(z-\Delta z,t)$, and $p(z+\Delta z,t)$. Assuming equal probabilities that the particle jumps to the nearest site to its left or right, i.e., $h_{-}\Delta t = h_{+}\Delta t = 0.5$, the previous equation becomes

$$p(z,t+\Delta t) - p(z,t) = \frac{1}{2}\left[p(z-\Delta z,t) - 2p(z,t) + p(z+\Delta z,t)\right].$$

Similarly to (2.7), we define

$$\mathcal{D} \triangleq \frac{1}{2}\frac{(\Delta z)^2}{\Delta t}.$$

Dividing these last two equations term by term, we obtain

$$\frac{p(z,t+\Delta t) - p(z,t)}{\Delta t} = \mathcal{D}\frac{p(z-\Delta z,t) - 2p(z,t) + p(z+\Delta z,t)}{(\Delta z)^2}.$$

Taking the limits $\Delta t \to 0$ and $\Delta z \to 0$ and keeping $(\Delta z)^2/\Delta t = 2\mathcal{D}$ constant in the limiting case, the previous equation gives the diffusion equation (2.18) in one dimension.

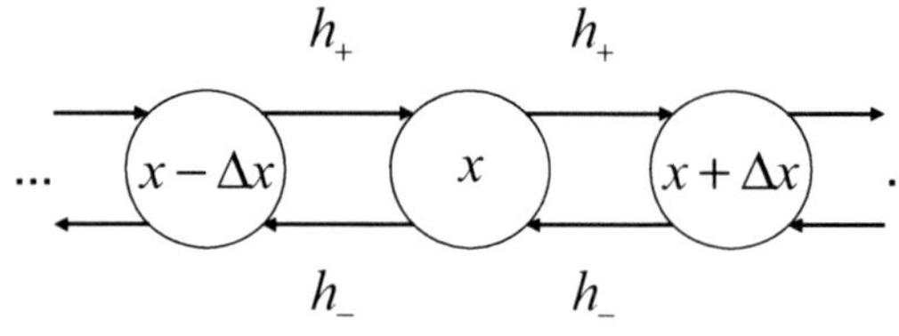

Figure 9.24: Two-way catenary compartment model.

Therefore, the solution of the master equation can be thought of as a Markovian random walk in the space of reacting or diffusing species. It measures the probability $p(z,t)$ of finding the walker in a particular position z at any given time t. Furthermore, by taking into account the number of particles in the compartments, probabilities can be converted to concentrations to obtain the second Fick's law, (2.16). If we consider an asymmetric walk where $h_- \neq h_+$, we obtain the diffusion equation with the drift velocity of the walker [382]. Moreover, if the transfer probabilities h_- and h_+ depend on the number of walkers present at a given time, the master equation corresponds to a nonlinear situation leading to anomalous diffusion, as presented in Section 2.2 for fractals and disordered media.

A good review of the master equation approach to chemical kinetics has been given by McQuarrie [383]. Jacquez [335] presents the master equation for the general m-compartment, the catenary, and the mammillary models. That author further develops the equation for the one- and two-compartment models to obtain the expectation and variance of the number of particles in the model. Many others consider the m-compartment case [342, 345, 384], and Matis [385] gives a complete methodological rule to solve the Kolmogorov equations.

In any particular case, the master equation is fairly easy to write; however, solving it is quite another matter. The number of problems for which the master equation can be solved analytically is even less than the number of problems for which the deterministic corresponding equations can be solved analytically. In addition, unlike the reaction equations (linear, nonlinear, etc.), the master equation does not readily lend itself to numerical solution, owing to the number and nature of its independent variables. In fact, the master equation is a generic form that when expanded, leads to the set of Kolmogorov differential equations whose number is equal to the product of population size for all the reactants. In short, although the master equation is both exact and elegant, it is usually not very useful for making practical numerical calculations.

We can, however, analyze these problems within the framework of the stochastic formulation by looking for an exact solution, or by using the probability generating functions, or the stochastic simulation algorithm.

9.5.2 Exact Solution in Matrix Form

For simple cases with populations of small sizes, one can express the Kolmogorov equations in a matrix form. The elements of the grand probability function $p_{n_1,\ldots,n_m}(t)$ can be considered in a vector form:

$$\begin{array}{rlllll} \underline{p}(t)^T = & [p_{0,\ldots,0}(t), & p_{1,\ldots,0}(t), & \ldots, & p_{n_{01},\ldots,0}(t), & \ldots, \\ & \ldots, & \ldots, & \ldots, & \ldots, & \ldots, \\ & p_{0,\ldots,n_{0m}}(t), & p_{1,\ldots,n_{0m}}(t), & \ldots, & p_{n_{01},\ldots,n_{0m}}(t), & \ldots]. \end{array}$$

And the Kolmogorov equations may be written as

$$\dot{\underline{p}}(t) = \underline{p}(t)\,\mathbf{R},$$

where $\mathbf{R}$ is a constant-coefficient matrix. For models with finite $\mathbf{R}$, e.g., when only initial conditions are present in an open system, one can proceed to find numerical solutions for the probability distributions by the direct solution of the above differential equation. In the general context, the dimension of $\mathbf{R}$ is infinite and the previous equation rules out a direct exact solution. One option for such problems is to truncate the set of differential equations for some large upper bound of population size, and then proceed to find directly a close approximate solution for the size distributions. Therefore, one can truncate the distribution of $\underline{N}(t)$ at some large upper value and solve the resulting finite system. This useful option will be illustrated in Section 9.5.5.

9.5.3 Cumulant Generating Functions

A very useful tool for finding analytically the distribution of $\underline{N}(t)$ is to obtain and solve partial differential equations for the associated cumulant generating functions. The *moment generating function*, denoted by $\mathcal{M}(\underline{\theta}, t)$, is defined for a multivariate integer-valued variable $\underline{N}(t)$ as

$$\mathcal{M}(\underline{\theta}, t) = \sum_{(n_1,\ldots,n_m \geq 0)} p_{n_1,\ldots,n_m}(t) \prod_{i=1}^{m} \exp(\theta_i n_i), \tag{9.36}$$

where $\underline{\theta}$ is a dummy variable. The *cumulant generating function*, denoted by $\mathcal{K}(\underline{\theta}, t)$, is defined as

$$\mathcal{K}(\underline{\theta}, t) = \log \mathcal{M}(\underline{\theta}, t) \tag{9.37}$$

with power series expansion

$$\mathcal{K}(\underline{\theta}, t) = \sum_{(s_1,\ldots,s_m \geq 0)} \kappa_{s_1,\ldots,s_m}(t) \prod_{i=1}^{m} \frac{\theta_i^{s_i}}{s_i!}. \tag{9.38}$$

This equation formally defines the joint $(s_1, \ldots, s_m)$th cumulants, $\kappa_{s_1,\ldots,s_m}(t)$ as the coefficients in the series expansion of $\mathcal{K}(\underline{\theta}, t)$. The multiple summations in (9.36) and (9.38) on n_i and s_i, respectively, require that at least one n_i or s_i

be different from 0. Hence, the approach to deriving differential equations for cumulants is simple in practice.

Statistical characteristics of the random vector $\underline{N}(t)$ can be directly obtained from cumulants $\kappa_{s_1,\ldots,s_m}(t)$ with all $s_i = 0$ except:

- $s_j = 1$ to calculate expectation $E[N_j(t)]$, or
- $s_j = 2$ to calculate variance $Var[N_j(t)]$, or
- $s_j = 1$ and $s_k = 1$ to calculate covariance $Cov[N_j(t)N_k(t)]$, etc.

There are also a number of advantages to using cumulant generating functions instead of probability or moment generating functions. For instance, in the univariate case:

- The cumulant functions provide a basis for parameter estimation using weighted least squares. The expected value function $\kappa_1(t)$ could serve as the regression function, the variance function $\kappa_2(t)$ supplies the weights, and $\kappa_3(t)$ provides a simple indicator of possible departure from an assumed symmetric distribution.
- The cumulant structure provides a convenient characterization for some common distributions:
  1. for the Poisson distribution, all cumulants are equal, i.e., $\kappa_i = c$ for all i, and
  2. for the Gaussian distribution, all cumulants above order two are zero, i.e., $\kappa_i = 0$ for $i > 2$.
- The low-order cumulants may be utilized to give saddle-point approximations of the underlying distribution [385, 386].

Partial differential equations may be written directly using an infinitesimal generator technique, called the *random-variable technique*, given in Bailey [387]. For intensity functions of the form (9.33), we define the operator notation

$$I_{\varphi_{l,1},\ldots,\varphi_{l,m}}\left(\frac{\partial}{\partial\theta_1},\ldots,\frac{\partial}{\partial\theta_m}\right)\mathcal{M}(\underline{\theta},t) = h_l\frac{\partial^{\psi_l}\mathcal{M}(\underline{\theta},t)}{\partial\theta_i^{\psi_{l,1}}\ldots\partial\theta_i^{\psi_{l,m}}}.$$

Using this notation, the moment generating function is given in [387] (p. 73):

$$\frac{\partial\mathcal{M}(\underline{\theta},t)}{\partial t} = \sum_{l=1}^{m_\circ}\left[\exp\left(\sum_{i=1}^{m}\theta_i\varphi_{l,i}\right)-1\right]I_{\varphi_{l,1},\ldots,\varphi_{l,m}}\left(\frac{\partial}{\partial\theta_1},\ldots,\frac{\partial}{\partial\theta_m}\right)\mathcal{M}(\underline{\theta},t). \tag{9.39}$$

The boundary condition for this partial differential equation is obtained from (9.36). Multiplying both sides of this relationship by $\mathcal{M}^{-1}$ and using the definition of the cumulant generating function, the partial differential equation of the cumulant generating function is derived. To use the operator equation is a very useful approach. Therefore, this approach is applied easily to density-dependent models, for which the intensity functions involve higher powers of $\underline{N}$ leading to nonlinear partial differential equations. The approach also extends to multiple populations. In cases in which the analytical solutions are not available, one might consider using series expansions and after equating coefficients of powers of $\underline{\theta}$, solve for the cumulant of desired order.

For linear systems, the differential equation for the jth cumulant function is linear and it involves terms up to the jth cumulant. The same procedure will be followed subsequently with other models to obtain analogous differential equations, which will be solved numerically if analytical solutions are not tractable. Historically, numerical methods were used to construct solutions to the master equations, but these solutions have pitfalls that include the need to approximate higher-order moments as a product of lower moments, and convergence issues [383]. What was needed was a general method that would solve this sort of problem, and that came with the stochastic simulation algorithm.

9.5.4 Stochastic Simulation Algorithm

A computational method was developed by Gillespie in the 1970s [381, 388] from premises that take explicit account of the fact that the time evolution of a spatially homogeneous process is a discrete, stochastic process instead of a continuous, deterministic process. This computational method, which is referred to as the *stochastic simulation algorithm*, offers an alternative to the Kolmogorov differential equations that is free of the difficulties mentioned above. The simulation algorithm is based on the *reaction probability density function* defined below.

Let us now consider how we might go about simulating the stochastic time evolution of a dynamic system. If we are given that the system is in the state $\underline{n}(t)$ at time t, then essentially all we need in order to "move the system forward in time" are the answers to two questions: "when will the next random event occur," and "what kind of event will it be?" Because of the randomness of the events, we may expect that these two questions will be answered in only some probabilistic sense.

Prompted by these considerations, Gillespie [388] introduced the reaction probability density function $p(\varkappa, l)$, which is a joint probability distribution on the space of the continuous variable $\varkappa$ $(0 \leq \varkappa < \infty)$ and the discrete variable l $(l = 1, \ldots, m_\circ)$. This function is used as $p(\varkappa, l)\,\Delta\varkappa$ to define the probability that "given the state $\underline{n}(t)$ at time t, the next event will occur in the infinitesimal time interval $(t + \varkappa, t + \varkappa + \Delta\varkappa)$, AND will be an R_l event." Our first step toward finding a legitimate method for assigning numerical values to $\varkappa$ and l is to derive, from the elementary conditional probability $h_l \Delta t$, an analytical expression for $p(\varkappa, l)$. To this end, we now calculate the probability $p(\varkappa, l)\,\Delta\varkappa$ as the product $p_0(\varkappa)$, the probability at time t that "no event will occur in the time interval $(t, t + \varkappa)$" TIMES $a_l \Delta\varkappa$, the subsequent probability that "an R_l

event will occur in the next differential time interval $(t + \varkappa, t + \varkappa + \Delta\varkappa)$":

$$p(\varkappa, l)\,\Delta\varkappa = p_0(\varkappa)\,a_l \Delta\varkappa. \tag{9.40}$$

The probability of more than one reaction occurring in $(t + \varkappa, t + \varkappa + \Delta\varkappa)$ is $o(\Delta\varkappa)$.

In order to appreciate $p_0(\varkappa)$, the probability that "no event occurs in $(t, t + \varkappa)$," imagine the interval $(t, t + \varkappa)$ to be divided into L subintervals of equal length $\varepsilon = \varkappa/L$. The probability that none of the events $R_1, \ldots, R_{m_\circ}$ occurs in the first ε subinterval $(t, t + \varepsilon)$ is

$$\prod_{\nu=1}^{m_\circ} [1 - a_\nu \varepsilon + o(\varepsilon)] = 1 - \sum_{\nu=1}^{m_\circ} a_\nu \varepsilon + o(\varepsilon) = 1 - a_0 \varepsilon + o(\varepsilon)$$

if we put $a_0 \equiv \sum_{\nu=1}^{m_\circ} a_\nu$. This is also the subsequent probability that no event occurs in $(t + \varepsilon, t + 2\varepsilon)$, and then in $(t + 2\varepsilon, t + 3\varepsilon)$, and so on. Since there are L such ε subintervals between t and $t + \varkappa$, then $p_0(\varkappa)$ can be written as

$$p_0(\varkappa) = [1 - a_0 \varepsilon + o(\varepsilon)]^L = \left[1 - \frac{a_0 \varkappa}{L} + o\left(L^{-1}\right)\right]^L.$$

This is true for any $L > 1$, and in particular, for infinitely large L. By using the limit formula for the exponential function,

$$\lim_{\nu \to \infty} \left(1 - \frac{x}{\nu}\right)^\nu = \exp(-x),$$

the probability $p_0(\varkappa)$ becomes

$$\lim_{L \to \infty} p_0(\varkappa) = \exp(-a_0 \varkappa).$$

Inserting the previous expression in (9.40), we arrive at the following exact expression for the reaction probability density function:

$$p(\varkappa, l) = a_l \exp(-a_0 \varkappa). \tag{9.41}$$

Thus, we observe that $p(\varkappa, l)$ depends, through the quantity in the exponential, on the parameters for all events (not just R_l) and on the current sizes of populations for all particles (not just the R_l reactants).

Even though it may be impossible to solve a complicated dynamic system exactly, Gillespie's method can be used to numerically simulate the time evolution of the system [381]. In this method, implied events are thought of as occurring with certain probabilities, and the events that occur change the probabilities of subsequent events. This stochastic simulation algorithm has been shown to be physically and mathematically well grounded from a kinetic point of view, and rigorously equivalent to the spatial homogeneous master equation, yet surprisingly simple and straightforward to implement on a computer [381,383,388]. In the limit of large numbers of reactant molecules, the supplied results are entirely equivalent to the solution of the traditional kinetic differential equations derived from the mass-balance law [381]. As presented here, the stochastic simulation algorithm is applicable only to spatially homogeneous systems. Work toward extending the algorithm to accommodate particle diffusion in spatially heterogeneous systems is currently in progress, and will be reported on.

For most macroscopic dynamic systems, the neglect of correlations and fluctuations is a legitimate approximation [383]. For these cases the deterministic and stochastic approaches are essentially equivalent, and one is free to use whichever approach turns out to be more convenient or efficient. If an analytical solution is required, then the deterministic approach will always be much easier than the stochastic approach. For systems that are driven to conditions of instability, correlations and fluctuations will give rise to transitions between nonequilibrium steady states and the usual deterministic approach is incapable of accurately describing the time behavior. On the other hand, the stochastic simulation algorithm is directly applicable to these studies.

Implementation

This algorithm can easily be implemented in an efficient modularized form to accommodate quite large reaction sets of considerable complexity [388]. For an easy implementation, the joint distribution can be broken into two disjoint probabilities using Bayes's rule $p(\varkappa, l) = p(\varkappa)p(l \mid \varkappa)$. But note that $p(\varkappa)$ may be considered as the marginal probability of $p(\varkappa, l)$, i.e.,

$$p(\varkappa) = \sum_{l=1}^{m_\circ} p(\varkappa, l),$$

and substituting this into (9.41) leads to values for its component parts:

$$p(\varkappa) = a_0 \exp(-a_0 \varkappa) \tag{9.42}$$

and

$$p(l \mid \varkappa) = \frac{a_l}{a_0}. \tag{9.43}$$

Given these fundamental probability density functions, the following algorithm can be used to carry out the reaction set simulation:

- Initialization:

  1. Set values for the h_l.
  2. Set the initial number n_{0i} of the m reactants.
  3. Set $t = 0$, and select a value for t_{sim}, the maximum simulation time.

- Loop:

  1. Compute the intensity functions a_l and $a_0 \equiv \sum_{\nu=1}^{m_\circ} a_\nu$.
  2. Generate two random numbers r_1 and r_2 from a uniform distribution on $[0, 1]$.

3. Compute the next time interval $\varkappa = \ln(1/r_1)/a_0$. Draw from the probability density function (9.42).
4. Select the reaction to be run by computing l such that

$$\sum_{\nu=1}^{l-1} a_\nu < r_2 a_0 \leq \sum_{\nu=1}^{l} a_\nu .$$

Draw from the probability density function (9.43).

5. Adjust $t = t + \varkappa$ and update the n_l values according to the R_l reaction that just occurred.
6. If $t > t_{\mathrm{sim}}$, then terminate. Otherwise, go to 1.

By carrying out the above procedure from time 0 to time t_{sim}, we evidently obtain only one possible realization of the stochastic process. In order to get a statistically complete picture of the temporal evolution of the system, we must actually carry out several independent realizations or "runs." These runs must use the same initial conditions of the problem but different starting numbers for the uniform random number generator in order for the algorithm to result in different but statistically equivalent chains. If we make K runs in all, and record the population sizes $n_i(k,t)$ in run k at time t ($i = 1, \ldots, m$ and $k = 1, \ldots, K$), then we may assert that the average number of particles at time t is

$$\overline{n}_i(t) \approx \frac{1}{K} \sum_{k=1}^{K} n_i(k,t),$$

and the fluctuations that may reasonably be expected to occur about this average are

$$s_i(t) \approx \left\{ \frac{1}{K} \sum_{k=1}^{K} \left[n_i(k,t) - \overline{n}_i(t) \right]^2 \right\}^{1/2} .$$

The approximately equal signs in the previous relations become equality signs in the limit $K \to \infty$. However, the fact that we obviously cannot pass to this limit of infinitely many runs is not a practical source of difficulty. On the one hand, if $s_i(t) \ll \overline{n}_i(t)$, then the results $n_i(k,t)$ will not vary much with k; in that case the estimate of $\overline{n}_i(t)$ would be accurate even for $K = 1$. On the other hand, if $s_i(t) \gtrsim \overline{n}_i(t)$, a highly accurate estimate of $\overline{n}_i(t)$ is not necessary. Of more practical significance and utility in this case would be the approximate range over which the numbers $n_i(k,t)$ are scattered for several runs k. In practice, somewhere between 3 and 10 runs should provide a statistically adequate picture of the state of the system at time t.

The computer storage space required by the simulation algorithm is quite small. This is an important consideration, since charges at most large computer facilities are based not only on how long a job runs but also on how much memory storage is used.

Because the speed of the stochastic simulation algorithm is linear with respect to the number of reactions, adding new reaction channels will not greatly increase the runtime of the simulation, i.e., doubling either the number of reactions or the number of reactant species, doubles (approximately) the total runtime of the algorithm. The speed of the algorithm depends more on the number of molecules. This is seen by noting that the computation of the next time interval in $\varkappa = \ln(1/r_1)/a_0$ depends on the reciprocal of a_0, a term representing the number of molecules in the simulation. If the reaction set contains at least one second-order reaction, then a_0 will contain at least one multiplication product of two species in the population. In this case the speed of the simulation will fall off like the reciprocal of the square of the population. Recent improvements to the algorithm are helping to keep the runtime in check [389, 390].

Extensions

The simulation algorithm might allow one to deal in an approximate way with spatial heterogeneities. The basic idea is to divide the volume V into a number of subvolumes V_μ $(\mu = 1, \ldots, M)$ in such a way that spatial homogeneity may be assumed within each subvolume. Each subvolume V_μ would then be characterized by its own (uniformly distributed) particle populations $N_{1\mu}(t), \ldots, N_{m\mu}(t)$ and also a set of hazard rates $h_{l\mu}$ appropriate to the physicochemical characteristics inside V_μ. For instance, in order to apply the simulation algorithm to a collection of cells, the original algorithm must be extended to accommodate the introduction of spatial dependencies of the concentration variables. Introducing the spatial context into the stochastic simulation algorithm using the subvolumes V_μ may be materialized by a rectangular array of square cells with only nearest-neighbor, cell–cell interactions. In this model of interacting cells, it is assumed that each cell is running its own internal program of biochemical reactions.

The fact that simulation of any given reaction generates its own "local" simulation time steps poses the problem of synchronization of the internal simulation times of cells. In the simplest case with no specific interaction affecting the order of the reaction, converting the algorithm from what is essentially a spatial-scanning method to a temporal-scanning method can solve this problem. This is accomplished by first making an initial spatial scan through all of the cells in the array, and inserting the cells into a priority queue that is ordered from shortest to longest local cell time. All succeeding iterations are then based on the temporal order of the cells in the priority queue. In other words, a cell is drawn from the queue, calculations are performed on the reaction set for that cell, and then the cell is placed back on the queue in its new temporally ordered position.

The use of a priority queue to order the cells was a unique innovation, and it solves the synchronizing problem inherent in a multicellular situation. Not only does this allow an easy mechanism for intercellular signaling, but this methodology can also readily accommodate local inhomogeneities in the molecular populations. Work has been done that extends the stochastic simulation algorithm to reaction-diffusion processes, and the modification to the method is straightforward. Diffusion is considered to be just another possible chemical event with an associated probability [391]. As with all the other chemical events, the diffusion is assumed to be intracellular and the basic idea behind this approach is incorporated into the simulation.

The justification for using the stochastic approach, as opposed to the simpler mathematical deterministic approach, was that the former presumably took account of *fluctuations* and *correlations*, whereas the latter did not. It was subsequently demonstrated by Oppenheim et al. [392] that the stochastic formulation reduces to the deterministic formulation in the thermodynamic limit (wherein the size of particle populations and the containing volume all approach infinity in such a way that the particle concentrations approach finite values). Experience indicates that for most systems, the constituent particle populations need to have sizes only in the hundreds or thousands in order for the deterministic approach to be adequate; thus, for most systems the differences between the deterministic and stochastic formulations are purely academic, and one is free to use whichever formulation turns out to be more convenient or efficient. However, near state instabilities in certain nonlinear systems, fluctuations, and correlations can produce dramatic effects, even for a huge number of particles [393]; for these systems the stochastic formulation would be the more appropriate choice.

Among the three presented approaches to solve the Kolmogorov or master equations, the partial differential equations for cumulant generating functions are most adequate for the estimation problem. The exact solution using the **R** matrix can never be applied in a real context because of the astronomic requirement of memory for storing and matrix processing operations. The stochastic simulation algorithm is an elegant tool for simulating and analyzing the system, but as a nonparametric approach it is not adequate for the estimation problem (cf. Appendix G).

9.5.5 Simulation of Linear and Nonlinear Models

The two-compartment model and the model of the enzymatic reaction (cf. Sections 9.1.2 and 8.5.1, respectively) will be presented as typical cases for linear and nonlinear models, respectively. For these simulations, the model parameters were set as follows:

- For the compartmental system: $h_{10} = 0.5$, $h_{20} = 0.1$, $h_{12} = 1$, $h_{21} = 0.1\,\mathrm{h}^{-1}$.
- For the enzyme reaction: $k_{+1} = 1$, $k_{-1} = 0.5$, $k_{+2} = 1\,\mathrm{h}^{-1}$.

Exact Solution

Initial conditions for the compartmental model and the enzymatic reaction were set to $\underline{n}_0^T = [10\ 5]$, and $s_0 = 10$, $e_0 = 5$, and $c_0 = 0$, respectively. These values are very low regarding the experimental reality, but they were deliberately chosen as such to facilitate the computation of the exact solution.

Two-Compartment Model First, we develop full probabilistic transfer modeling. Consider the number of particles in the first and second compartments being n_1 and n_2, respectively, at time $t + \Delta t$, where Δt is some small time interval. There are a number of mutually exclusive ways in which this event could have come about, starting from time t. Specifically, they are:

- to have size (n_1, n_2) at time t with no change from t to $t + \Delta t$,
- to have size $(n_1 + 1, n_2)$ at time t with only a single irreversible elimination by the h_{10} way in the next interval Δt,
- to have size $(n_1, n_2 + 1)$ at time t with only a single irreversible elimination by the h_{20} way in the next interval Δt,
- to have size $(n_1 + 1, n_2 - 1)$ at time t with only a single reversible particle transfer to compartment 2 by the h_{12} way in Δt,
- to have size $(n_1 - 1, n_2 + 1)$ at time t with only a single reversible particle transfer to compartment 1 by the h_{21} way in Δt, and
- other ways that involve two or more independent changes of unit size in the interval Δt.

Because this set of mutually exclusive "pathways" to the desired event at $t + \Delta t$ is exhaustive, the probability of size n_1, n_2 at $t + \Delta t$ may be written as the sum of the individual probabilities of these pathways. Symbolically, using the assumptions for possible changes, one has for suitably small Δt,

$$\begin{aligned}
p_{n_1,n_2}(t + \Delta t) &= p_{n_1,n_2}(t)\left[1 - h_{10}n_1\Delta t - h_{20}n_2\Delta t - h_{12}n_1\Delta t - h_{21}n_2\Delta t\right] \\
&\quad + p_{n_1+1,n_2}(t)\left[h_{10}(n_1 + 1)\Delta t\right] \\
&\quad + p_{n_1,n_2+1}(t)\left[h_{20}(n_2 + 1)\Delta t\right] \\
&\quad + p_{n_1+1,n_2-1}(t)\left[h_{12}(n_1 + 1)\Delta t\right] \\
&\quad + p_{n_1-1,n_2+1}(t)\left[h_{21}(n_2 + 1)\Delta t\right] + o(\Delta t),
\end{aligned}$$

where $o(\Delta t)$ denotes terms of higher order than Δt associated with multiple independent changes. Subtracting $p_{n_1,n_2}(t)$, dividing by Δt, and taking the limit as $\Delta t \to 0$, one has

$$\begin{aligned}
\dot{p}_{n_1,n_2}(t) &= h_{10}\left[(n_1 + 1)p_{n_1+1,n_2}(t) - n_1p_{n_1,n_2}(t)\right] \\
&\quad + h_{20}\left[(n_2 + 1)p_{n_1,n_2+1}(t) - n_2p_{n_1,n_2}(t)\right] \\
&\quad + h_{12}\left[(n_1 + 1)p_{n_1+1,n_2-1}(t) - n_1p_{n_1,n_2}(t)\right] \\
&\quad + h_{21}\left[(n_2 + 1)p_{n_1-1,n_2+1}(t) - n_2p_{n_1,n_2}(t)\right]
\end{aligned} \tag{9.44}$$

for $n_1, n_2 > 0$, with boundary conditions for either $n_1 = 0$ or $n_2 = 0$. Initial conditions are $p_{n_{01},n_{02}}(0) = 1$ and $p_{n_1,n_2}(0) = 0$ for $n_1 \neq n_{01}$ and $n_2 \neq n_{02}$. The solution of this set of differential equations yields the desired probability distribution for $\left[N_1(t), N_2(t)\right]^T$.

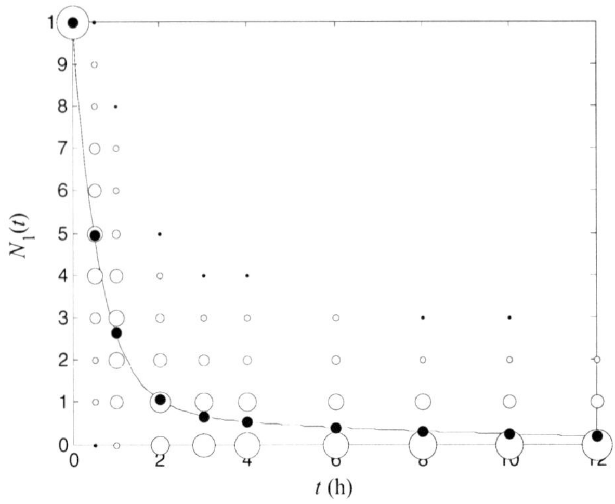

Figure 9.25: The exact solution of the Kolmogorov equations associating marginal probabilities with the number of particles in compartment 1. The *solid line* is the solution of the deterministic model. The areas of disks located at coordinates (t, n_1) are proportional to $p_{n_1}(t)$.

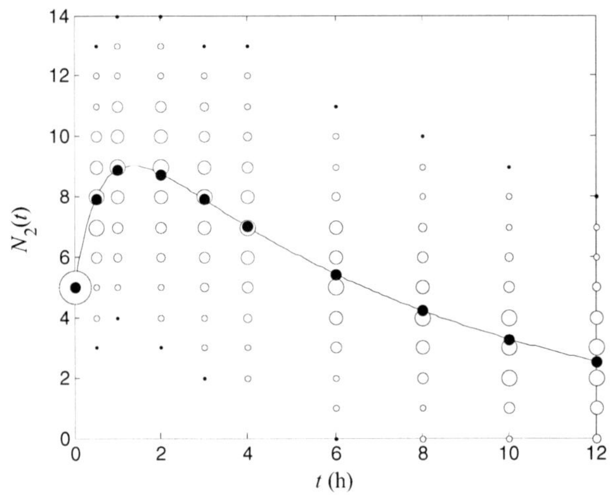

Figure 9.26: The exact solution of the Kolmogorov equations associating marginal probabilities with the number of particles in compartment 2. The *solid line* is the solution of the deterministic model. The areas of disks located at coordinates (t, n_2) are proportional to $p_{n_2}(t)$.

Second, considering now the exchange processes between compartment and environment as the set of first-order reactions,

$$\begin{array}{ll} N_1 \overset{h_{10}}{\rightarrow} \text{environment}, & I_{-1,0} = h_{10}N_1, \\ N_2 \overset{h_{20}}{\rightarrow} \text{environment}, & I_{0,-1} = h_{20}N_2, \\ N_1 \overset{h_{12}}{\rightarrow} N_2, & I_{-1,1} = h_{12}N_1, \\ N_2 \overset{h_{21}}{\rightarrow} N_1, & I_{1,-1} = h_{21}N_2, \end{array} \tag{9.45}$$

we can obtain the master equation (9.44) directly from the (9.35) formulation. For this model, there were two interacting populations ($m = 2$) and four ($m_\circ = 4$ in equation 9.45) intensity functions. Since only one particle from each population was implied in these intensity functions, all $\psi_{l,i}$ exponents were equal to one.

The possible states in each compartment are $n_{01} + n_{02} + 1$. Therefore $\mathbf{R}$ is a 256-dimensional matrix. The initial condition for the master equation is $p_{10,5}(0) = 1$. Figures 9.25 and 9.26 show the associated probabilities for each state as functions of time for the central and peripheral compartments, respectively. In these figures the disk area is proportional to the associated probability, the full markers are the expected values, and the solid lines the solution of the deterministic model. As already mentioned, we note that the expectation of the stochastic model follows the time profile of the deterministic system.

Enzymatic Reaction The usual stochastic approach begins by focusing attention on the probability function $p_{s,e,c}(t)$, which is defined to be the probability of finding s molecules of substrate S, e molecules of enzyme E, and c molecules of complex C at time t. From (8.7), the intensity functions are

$$\begin{aligned} I_{-1,-1,1} &= k_{+1}se, \\ I_{1,1,-1} &= k_{-1}c, \\ I_{0,1,-1} &= k_{+2}c. \end{aligned}$$

From the conservation law of enzyme sites, $e = e_0 - c$, the enzyme population size can be substituted by the previous relation involving the initial enzyme amount e_0 and the current complex population size c. The intensity functions become

$$\begin{aligned} I_{-1,1} &= k_{+1}s(e_0 - c), \\ I_{1,-1} &= k_{-1}c, \\ I_{0,-1} &= k_{+2}c, \end{aligned} \tag{9.46}$$

and applying the standard rules of probability theory and the (9.35) formulation, it is a straightforward matter to deduce the master equation:

$$\begin{aligned} \dot{p}_{s,c}(t) &= k_{+1}\left[(s+1)(e_0 - c + 1)p_{s+1,c-1}(t) - s(e_0 - c)p_{s,c}(t)\right] \\ &\quad + k_{-1}\left[(c+1)p_{s-1,c+1}(t) - cp_{s,c}(t)\right] \\ &\quad + k_{+2}\left[(c+1)p_{s,c+1}(t) - cp_{s,c}(t)\right]. \end{aligned} \tag{9.47}$$

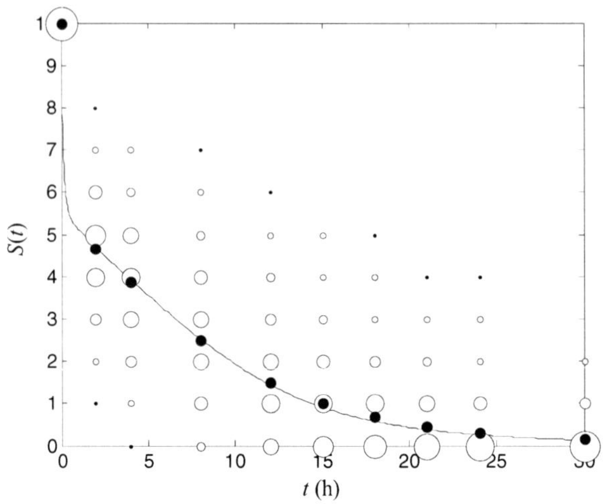

Figure 9.27: The exact solution for the substrate, $s(t)$, of the Kolmogorov equations associating marginal probabilities with the number of particles. The *solid line* is the solution of the deterministic model. The areas of disks located at coordinates (t, s) are proportional to $p_s(t)$.

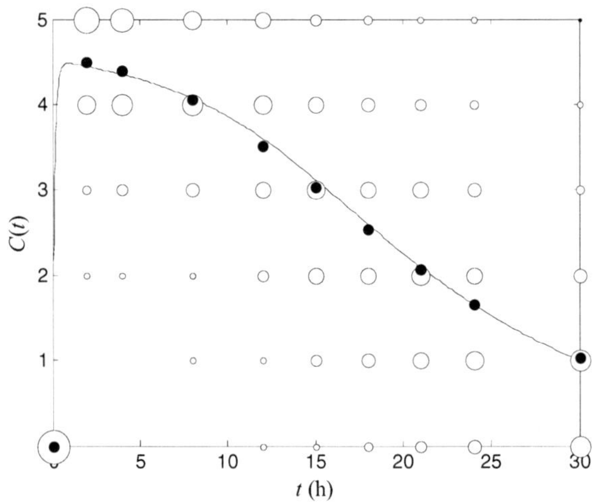

Figure 9.28: The exact solution for the complex, $c(t)$, of the Kolmogorov equations associating marginal probabilities with the number of particles. The *solid line* is the solution of the deterministic model. The areas of disks located at coordinates (t, c) are proportional to $p_c(t)$.

In principle, this time-evolution equation can be solved subject to the given initial condition $p_{s,c}(0) = \delta(s - s_0)\,\delta(c - c_0)$ to obtain $p_{s,c}(t)$ uniquely for all $t > 0$. The number of product molecules can be recovered as $s_0 - (s + c)$. For a given s_0 and e_0, a computer solution is constrained not only by run time but also by the amount of computer memory that would be required just to store the current values of the function $p_{s,c}(t)$ on the 2-dimensional integer lattice space of the variables S and C. The master equation may be solved exactly only when s_0 and e_0 are small. For this model, there were two interacting populations ($m = 2$), and three ($m_\circ = 3$, in equation 9.46) intensity functions. Since only one particle from each population was implied in these intensity functions, all $\psi_{l,i}$ exponents were equal to one.

The possible states for substrate are 11 and 6 for the complex. **R** is a 66-dimensional matrix and the initial condition for the master equation is $p_{10,0}(0) = 1$. Figures 9.27 and 9.28 show the associated probabilities for each state as functions of time for the substrate and the complex, respectively. As previously, the full markers are the expected values and the solid lines the solution of the deterministic model. Notably, the expectation of the stochastic model does not follow the time profile of the deterministic system. This is the main characteristic of nonlinear systems.

Cumulant Generating Functions

To illustrate how to proceed using the cumulant generating functions, the well-known two-compartment model and the enzymatic reaction will be presented as examples of linear and nonlinear systems, respectively. In these examples, there are two interacting populations ($m = 2$) and the cumulant generating function is

$$\mathcal{K}(\theta_1, \theta_2, t) = \sum_{i,j \geq 0} \kappa_{ij}(t) \frac{\theta_1^i \theta_2^j}{i!j!} \tag{9.48}$$

Initial conditions for the compartmental model and the enzymatic reaction were set to $\underline{n}_0^T = [100\ 50]$, and $s_0 = 100$, $e_0 = 50$, and $c_0 = 0$, respectively. These values are higher than those used previously and they are more likely to resemble experimental reality.

Two-Compartment Model The model assumptions in (9.45) were substituted directly into operator equation (9.39), which was transformed via (9.37) to yield

$$\begin{aligned} \frac{\partial \mathcal{K}}{\partial t} &= \{h_{10}[\exp(-\theta_1) - 1] + h_{12}[\exp(-\theta_1 + \theta_2) - 1]\} \frac{\partial \mathcal{K}}{\partial \theta_1} \\ &\quad + \{h_{20}[\exp(-\theta_2) - 1] + h_{21}[\exp(\theta_1 - \theta_2) - 1]\} \frac{\partial \mathcal{K}}{\partial \theta_2}. \end{aligned}$$

Upon substituting the series expansion (9.48) into the previous equation and equating coefficients of θ_1 and θ_2, one has the following differential equations

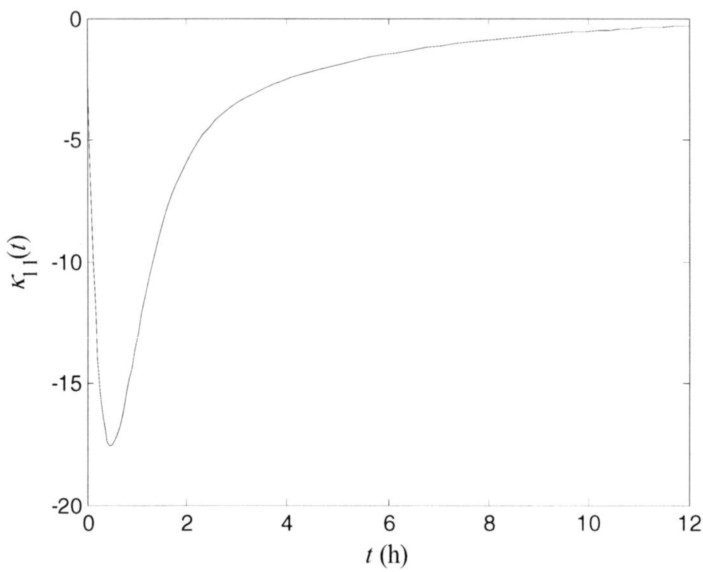

Figure 9.29: Cumulant $\kappa_{11}(t)$ profile expressing the statistical dependence of the population sizes for the compartmental model.

for the cumulant functions:

$$\begin{aligned}\dot{\kappa}_{10} &= -(h_{10}+h_{12})\kappa_{10} + h_{21}\kappa_{01},\\ \dot{\kappa}_{01} &= h_{12}\kappa_{10} - (h_{20}+h_{21})\kappa_{01},\end{aligned}$$

which are stochastic analogues of the deterministic formulation (8.4) and of the probabilistic transfer model (9.4). The equations for higher-order cumulants were obtained by equating coefficients of second- and third-order terms of $\theta_1^i\theta_2^j$. Especially for the second cumulants, the equations are

$$\begin{aligned}\dot{\kappa}_{20} &= -2(h_{10}+h_{12})\kappa_{20} + 2h_{21}\kappa_{11} + (h_{10}+h_{12})\kappa_{10} + h_{21}\kappa_{01},\\ \dot{\kappa}_{02} &= -2(h_{20}+h_{21})\kappa_{02} + 2h_{12}\kappa_{11} + (h_{20}+h_{21})\kappa_{01} + h_{12}\kappa_{10},\\ \dot{\kappa}_{11} &= -(h_{10}+h_{12}+h_{20}+h_{21})\kappa_{11} + h_{12}(\kappa_{20}-\kappa_{10}) + h_{21}(\kappa_{02}-\kappa_{01}).\end{aligned}$$

Since $p_{n_{01},n_{02}}(0) = 1$ and $p_{n_1,n_2}(0) = 0$ for $n_1 \neq n_{01}$ and $n_2 \neq n_{02}$, from (9.36) and (9.38) one has $\mathcal{M}(\theta_1,\theta_2,t) = \exp(\theta_1 n_{01} + \theta_2 n_{02})$ and $\mathcal{K}(\theta_1,\theta_2,t) = \theta_1 n_{01} + \theta_2 n_{02}$, respectively. Initial conditions for the cumulant differential equations are obtained by equating $\mathcal{K}(\theta_1,\theta_2,t)$ with the terms of the power expansion in (9.48): $\kappa_{ij}(0) = 0$ except for $\kappa_{10}(0) = n_{01}$ and $\kappa_{01}(0) = n_{02}$.

Simulations of these equations confirm the probabilistic behavior and the time profile of the distribution of particles that were already shown in Section

9.3.5 in Figures 9.15 and 9.16. Through the $\kappa_{11}(t)$ profile, this analysis reveals the statistical independence of the population sizes. Since the system has no entry, the two variables are negatively linked with an extreme value about 0.5 h as shown in Figure 9.29. This link is stronger when h_{12} and h_{21} are high compared to h_{10} and h_{20}.

Enzymatic Reaction The intensity functions (9.46) were substituted directly into operator equation (9.39), which was transformed by (9.37) to yield

$$\begin{aligned}\frac{\partial \mathcal{K}}{\partial t} &= k_{+1}\left[\exp\left(-\theta_1+\theta_2\right)-1\right]\left[e_0\frac{\partial \mathcal{K}}{\partial \theta_1}-\left(\frac{\partial^2 \mathcal{K}}{\partial \theta_1 \partial \theta_2}+\frac{\partial \mathcal{K}}{\partial \theta_1}\frac{\partial \mathcal{K}}{\partial \theta_2}\right)\right]\\ &\quad+\left\{k_{-1}\left[\exp\left(\theta_1-\theta_2\right)-1\right]+k_{+2}\left[\exp\left(-\theta_2\right)-1\right]\right\}\frac{\partial \mathcal{K}}{\partial \theta_2}.\end{aligned}$$

Upon substituting the series expansion (9.48) into the previous equation, and equating the coefficients of θ_1 and θ_2, one has the following differential equations for the expected value functions:

$$\begin{aligned}\dot{\kappa}_{10} &= -k_{+1}\left(e_0-\kappa_{01}\right)\kappa_{10}+k_{-1}\kappa_{01}+k_{+1}\kappa_{11},\\ \dot{\kappa}_{01} &= k_{+1}\left(e_0-\kappa_{01}\right)\kappa_{10}-\left(k_{-1}+k_{+2}\right)\kappa_{01}-k_{+1}\kappa_{11}.\end{aligned}$$

These equations are not equivalent to the deterministic formulation given by (8.8). The last term $k_{+1}\kappa_{11}$ involving the stochastic interaction in the previous equations expresses the main difference between deterministic and stochastic solutions for a nonlinear system.

The up to third order cumulant differential equations are

$$\begin{aligned}\dot{\kappa}_{20} &= k_{+1}\left(A-2B\right)+k_{-1}\left(\kappa_{01}+2\kappa_{11}\right),\\ \dot{\kappa}_{02} &= k_{+1}\left(A+2C\right)+k_0\left(\kappa_{01}-2\kappa_{02}\right),\\ \dot{\kappa}_{11} &= k_{+1}\left(B-A-C\right)-k_{-1}\left(\kappa_{01}-\kappa_{02}\right)-k_0\kappa_{11},\\ \dot{\kappa}_{30} &= k_{+1}\left(3B-A-3D\right)+k_{-1}\left(\kappa_{01}+3\kappa_{11}+3\kappa_{21}\right),\\ \dot{\kappa}_{21} &= k_{+1}\left(A+C+D-2B-2F\right)+k_{-1}\left(\kappa_{02}-\kappa_{01}+2\kappa_{12}-2\kappa_{11}\right)-k_0\kappa_{21},\\ \dot{\kappa}_{12} &= k_{+1}\left(B-A-2C-E+2F\right)+k_{-1}\left(\kappa_{03}+\kappa_{01}-2\kappa_{02}\right)+k_0\left(\kappa_{11}-2\kappa_{12}\right),\\ \dot{\kappa}_{03} &= k_{+1}\left(A+3C+3E\right)+k_0\left(\kappa_{01}+3\kappa_{02}-3\kappa_{03}\right),\end{aligned}$$

with

$$\begin{aligned}A &= \left(e_0-\kappa_{01}\right)\kappa_{10}-\kappa_{11},\\ B &= \left(e_0-\kappa_{01}\right)\kappa_{20}-\left(\kappa_{21}+\kappa_{10}\kappa_{11}\right),\\ C &= \left(e_0-\kappa_{01}\right)\kappa_{11}-\left(\kappa_{12}+\kappa_{10}\kappa_{02}\right),\\ D &= \left(e_0-\kappa_{01}\right)\kappa_{30}-\left(2\kappa_{20}\kappa_{11}+\kappa_{10}\kappa_{21}\right),\\ E &= \left(e_0-\kappa_{01}\right)\kappa_{12}-\left(2\kappa_{02}\kappa_{11}+\kappa_{10}\kappa_{03}\right),\\ F &= \left(e_0-\kappa_{01}\right)\kappa_{21}-\left(\kappa_{20}\kappa_{02}+\kappa_{11}^2+\kappa_{10}\kappa_{12}\right),\\ k_0 &= k_{-1}+k_{+2}.\end{aligned}$$

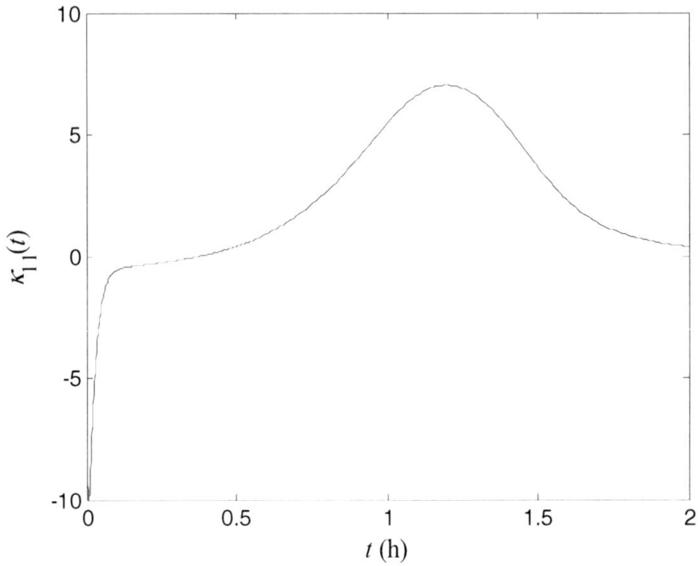

Figure 9.30: Cumulant $\kappa_{11}(t)$ profile expressing the statistical dependence of the population sizes for the enzymatic model.

In these equations, the contributions of the fourth- and higher-order cumulants are neglected.

From the above, we remark again as for the first-order cumulant that the differential equations for the second-order cumulants κ_{20}, κ_{02}, and κ_{11} imply the third-order cumulants κ_{12} and κ_{21} and so on. This can be generalized by noting that the differential equation for the jth cumulant function for a ψ-degree power in the intensity function model involves terms up to the $(j + \psi)$th cumulant. Obviously, this fact rules out exact solutions, such as those previously found for the linear kinetic model, for the present equations. A standard approach to this problem has been to assume that the population size variable follows a Gaussian distribution, and set to 0 all cumulants of order 3 or higher. One can also intend [385] to find approximating cumulant functions using a "cumulant truncation" procedure. In this approach, one approximates the cumulant functions of any specific order, say j, of a ψ-degree power model by solving a system of up to the first $(j + \psi)$ cumulant functions with all higher-order cumulants set to 0.

Initial conditions for the cumulant differential equations are $\kappa_{ij}(0) = 0$ except for $\kappa_{10}(0) = s_0$. Setting to 0 all cumulants of order 4 or higher, simulations of these equations confirm the expected behaviors and the associated confidence intervals. Through the $\kappa_{11}(t)$ profile shown in Figure 9.30, this analysis reveals the statistical independence of the population sizes. Moreover, $\kappa_{11}(t)$ magnified by k_{+1} evaluates the discrepancy between the deterministic and the stochastic solution: the substrate is overestimated at the early time of reaction by the deterministic model and underestimated over 0.5 h with a maximum about 1.2 h.

Stochastic Simulation Algorithm

As previously, initial conditions for the compartmental model and the enzymatic reaction were set to $\underline{n}_0^T = [100\ 50]$, and $s_0 = 100$, $e_0 = 50$, and $c_0 = 0$, respectively. Figures 9.31 and 9.32 show the deterministic prediction, a typical run, and the average and confidence corridor for 100 runs from the stochastic simulation algorithm for the compartmental system and the enzyme reaction, respectively. Figures 9.33 and 9.34 show the coefficient of variation for the number of particles in compartment 1 and for the substrate particles, respectively.

"On average," the solutions supplied by the deterministic system and the stochastic method are in close agreement, but the stochastic approach captures the fluctuations in the system. In comparing Figures 9.33 and 9.34, it is clear that when the number of molecules is large, the fluctuations might take the appearance of noise. But when there are small numbers of molecules, the fluctuations may in fact no longer be just noise but a significant part of the signal. Whether these fluctuations make a difference in the basic behavior of the system depends on the characteristics of that particular system. It may also be the case that the system moves between situations in which the fluctuations do and do not matter. However, when it is known that the system contains small numbers of molecules and the network is nonlinear, the stochastic approach appears to be a more appropriate method, because both of these situations will magnify any fluctuations that already exist in the system.

9.6 Fractals and Stochastic Modeling

In the classical book [4], the distinct models dealing with ion channel kinetics are extensively discussed. One of the important results is the connection established between fractal scaling and stochastic modeling. Based on experimental data, Liebovitch et al. [394] assessed the dependence of the effective kinetic constant $k_\circ$ on the sufficient time scale for detection $t_\circ$ by a fractal scaling relationship:

$$\log k_\circ (t_\circ) = \log \alpha + (1 - d_f) \log t_\circ, \tag{9.49}$$

where α is a constant and d_f is the fractal dimension. Moreover, the effective kinetic constant $k_\circ (t_\circ)$ can be considered as the conditional probability per unit time that the channel changes state (open vs. closed), i.e., $k_\circ (t_\circ)$ is considered as the hazard function $h (t_\circ)$ defined by (9.6). In that case, the survival function $\mathcal{S} (t_\circ)$ is the cumulative probability $\Pr [T_\circ > t_\circ]$ that the duration of the open (or closed) state $T_\circ$ is greater than $t_\circ$. Solving (9.6) and using the fractal scaling relationship (9.49), we obtain

$$\mathcal{S} (t_\circ) = \exp \left[-\frac{\alpha}{2 - d_f} (t_\circ)^{2-d_f} \right],$$

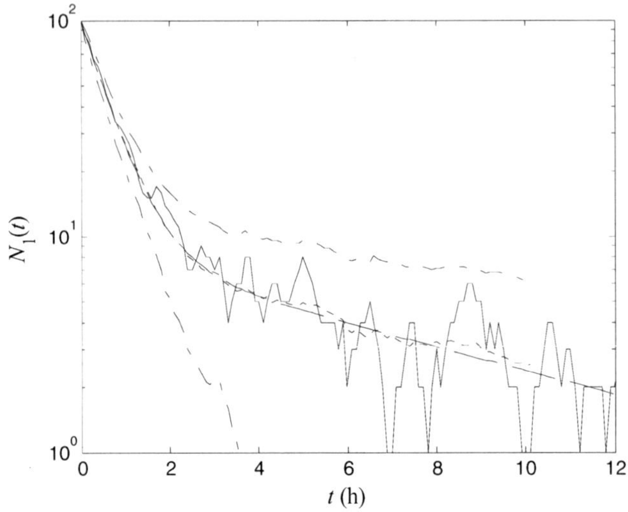

Figure 9.31: The deterministic profile (*dashed line*), typical run (*solid line*), average (*dotted line*), and confidence corridor (*dashed-dotted line*) in compartment 1.

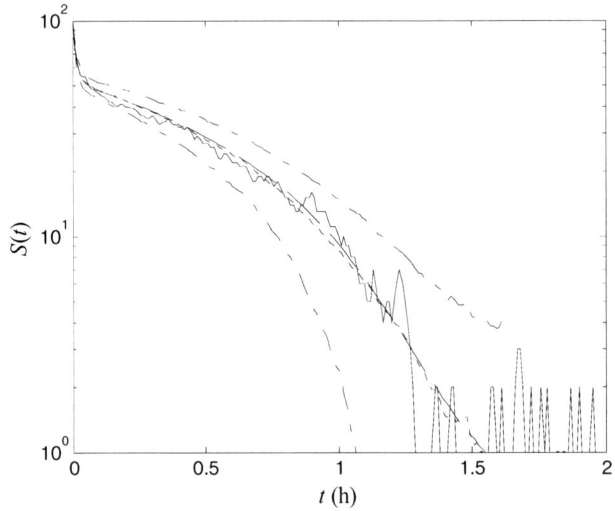

Figure 9.32: The deterministic profile (*dashed line*), typical run (*solid line*), average (*dotted line*), and confidence corridor (*dashed-dotted line*) for substrate particles.

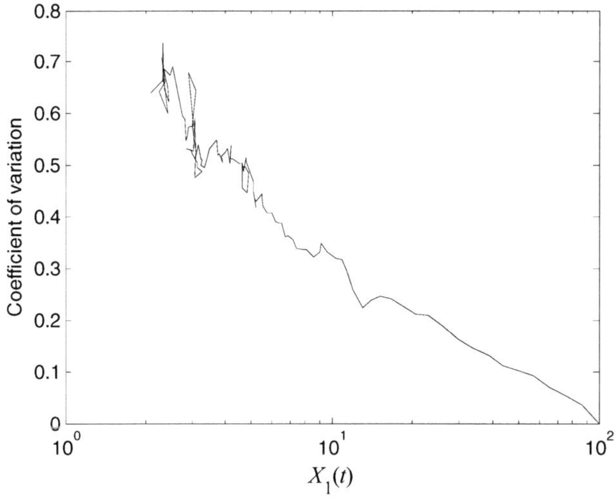

Figure 9.33: Coefficient of variation for the particles in compartment 1.

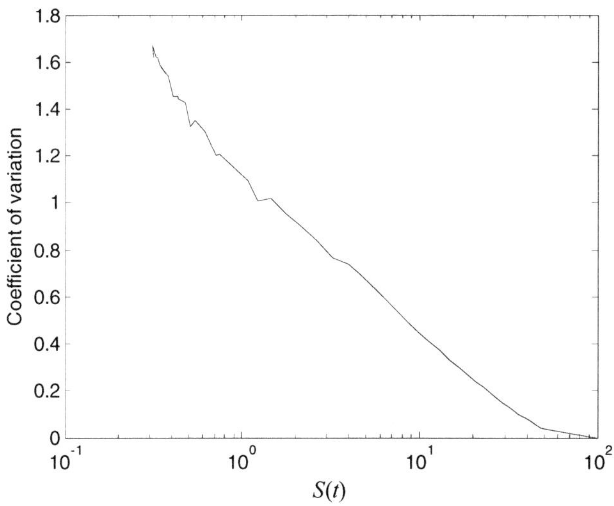

Figure 9.34: Coefficient of variation for the substrate particles.

which is the Weibull *survival* function already mentioned in Table 9.1. When the fractal dimension is close to 2, the previous equation takes the form of a power-law of time:

$$\mathcal{S}(t_\circ) = g(2 - d_f)(t_\circ)^{-\alpha},$$

where $g(2 - d_f)$ is a function of the fractal dimension. This form is equivalent to (2.8) in Chapter 2. From this development, we note the correspondence between the time scale sufficient for detection, $t_\circ$, and the age a of particles in a given compartment. This short presentation illustrates how fractality could be incorporated in retention-time distributions.

All the stochastic models presented here may include multiple compartments, age-varying rates, and heterogeneous particles with random rate coefficients, and their mathematical solutions tend to have various forms, e.g., exponential form, power function, damped oscillatory regimens, etc. This formulation concerns systems that are discrete in space, i.e., the particle can be located in one of a number of discrete compartments, and continuous in time; i.e., the particle is located continuously in one compartment until a transition occurs that discretely moves it to another compartment. In fact, stochastic models, and especially semi-Markov models, are tools for analyzing data when the response of interest is the time up to the occurrence of some event. Such events are generically referred to as *failures*, although the event may, for instance, be the ability of a power system to supply energy on demand without local failures or large-scale blackouts, the operating hours of replaced parts in equipment already in field use, industrial product testing, or the change of residence in a demographic study.

Certainly at the beginning, stochastic modeling had applications in the field of reliability, a relatively new field whose conception is primarily due to the complexity, sophistication, and automation inherent in modern technology. The problems of maintenance, repair, and field failures became severe for the military equipment used in World War II. In the late 1940s and early 1950s reliability engineering appeared on the scene [395–397]. For instance, the analysis of a process with operative and failure states can be based on the model presented in Figure 9.2 C. Compartments 1 and 2 correspond to states in which the process is operable and failed, respectively. A_1 corresponds to the time before failure, and A_2 to the time needed to repair. Lastly, ω and $1 - \omega$ correspond to the probabilities of entering reparable and irreparable failure states, respectively.

Recently, pharmacodynamicists have become interested in stochastic modeling for analyzing failure time data associated with pharmacological treatments [398]. Despite the unquestionable erudition of stochastic modeling, only a few of the stochastic models proposed to account for the observed biological data enjoy widespread use. The main reasons are that parameter estimation of stochastic processes in biology is a relatively recent enterprise and that a number of models involve the application of fairly advanced statistics that typically lie beyond the scope and knowledge of experimental biologists.

9.7 Stochastic vs. Deterministic Models

In many cases and with an acceptable degree of accuracy, the time evolution of a dynamic system can be treated as a continuous, deterministic process. For deterministic processes the law of mass conservation is well grounded in experiments and also leads to equations that can be readily solved. Besides the great importance of the differential equation approach for either compartmental analysis or analysis of reactions involved in a living system, we should not lose sight of the fact that the physical basis for this method leaves something to be desired. The approach evidently assumes that the time evolution of a real process is both *continuous* and *deterministic*. However, time evolution of such a system is not a continuous process, because particle population sizes can obviously change only in discrete integer amounts. Moreover, the time evolution is not a deterministic process. Even if we put aside quantum considerations and regard particle motions as governed by the equations of classical mechanics, it is impossible to predict the exact particle population size at some future time unless we take into account the precise positions and velocities of all the particles in the system.

This criticism has been supported by several recent experimental results that strongly suggest that several processes, like ecological systems, microscopic biological systems, and nonlinear systems driven to conditions of instability, in fact behave stochastically. So it was not until the early 1950s that it became clear that in small systems the law of mass conservation breaks down and that even small fluctuations in the number of molecules may be a significant factor in the behavior of the system [399]. Therefore, the equations obtained by using the law of mass conservation to describe fluctuations in the particle population sizes can be a serious shortcoming. Implicit in using the law of mass conservation are the key assumptions of continuity and determinism that could be warranted when there is a large number of the molecules of interest. These assumptions are reasonable for some systems of reactants, like a flask in the chemistry lab, but they are questionable when it comes to small living systems like cells and neurological synapses.

For instance, it turns out that inside a cell the situation is not continuous and deterministic, and that random fluctuations drive many of the reactions. With regard to the continuity assumption, it is important to note that the individual genes are often present only in one or two copies per cell and that the regulatory molecules are typically produced in low quantities [400]. The low number of molecules may compromise the notion of continuity and consequently that of homogeneity. As for determinism, the rates of some of these reactions are so slow that many minutes may pass before, for instance, the start of mRNA transcription after the necessary molecules are present. This may call into question the notion of deterministic change due to the fluctuations in the timing of cellular events. As a consequence, two regulatory systems having the same initial conditions might ultimately settle into different states, a phenomenon strengthened by the small numbers of molecules involved. This phenomenon is already reported as *sensitivity to initial conditions* (cf. Section

3.4) and it is characteristic of a nonlinear system exhibiting chaotic behavior. Thus, heterogeneity may be at the origin of fluctuations, and fluctuations are the prelude of instability and chaotic behavior.

Consequently, the observed process uncertainty may actually be an important part of the system and the expression of a structural heterogeneity. When the fluctuations in the system are small, it is possible to use the traditional deterministic approach. But when fluctuations are not negligibly small, the obtained differential equations will give results that are at best misleading, and possibly very wrong if the fluctuations can give rise to important effects. With these concerns in mind, it seems only natural to investigate an approach that incorporates the small volumes and small number of particle populations and may actually play an important part.

However, research along these lines is relatively scarce. The mathematical biology community continues to produce work that ignores the fact that there is a very different world inside a small biological system where topological heterogeneity prevails over homogeneity. So we turned to methods that are better able to capture the inherent stochastic nature of the system like the previously developed probabilistic transfer model, which expresses a structural heterogeneity and generates the process uncertainty corresponding to the observed fluctuations in the real process.

Aside from the continuity assumption and the discrete reality discussed above, deterministic models have been used to describe only those processes whose operation is fully understood. This implies a perfect understanding of all direct variables in the process and also, since every process is part of a larger universe, a complete comprehension of how all the other variables of the universe interact with the operation of the particular subprocess under study. Even if one were to find a real-world deterministic process, the number of interrelated variables and the number of unknown parameters are likely to be so large that the complete mathematical analysis would probably be so intractable that one might prefer to use a simpler stochastic representation. A small, simple stochastic model can often be substituted for a large, complex deterministic model since the need for the detailed causal mechanism of the latter is supplanted by the probabilistic variation of the former. In other words, one may deliberately introduce simplifications or "errors in the equations" to yield an analytically tractable stochastic model from which valid statistical inferences can be made, in principle, on the operation of the complex deterministic process.

For modeling purposes, the complexity pictured by heterogeneity undoubtedly requires more much knowledge than homogeneity conditions. If homogeneity prevails over heterogeneity, deterministic models may be good candidates to describe the real process. Conversely, the huge amount of knowledge needed to describe heterogeneity could be summarized only by the statistical concepts provided by stochastic modeling approaches.

Stochastic models have much to offer at the present time in strengthening the theoretical foundation and in extending the practical utility of the widespread deterministic models. After all, in a mathematical sense, the deterministic model is a special limiting case of a stochastic model.

The stochastic formulation was proposed to account for the heterogeneity in biological media since it supplies tractable forms to fit the data. These forms involve time-varying parameters in the dynamic modeling. But it is unlikely to have parameters depending on time through a single maturation or age dependence. We believe that internal dynamic states of the process are involved in these time-dependencies (cf. Appendix C). Introduction of these states leads to nonlinear dynamic modeling associated with various levels of stability. Naturally occurring, the nonlinear model may exhibit chaotic behavior. Thus, one must frequently expect chaotic-like behavior when the process is heterogeneous. In contrast, it is impossible to expect chaotic properties with homogeneous processes.

Part IV

MODELING IN PHARMACODYNAMICS

Pharmacodynamics has been described as "what the drug does to the body," in contrast to pharmacokinetics, which tells "what the body does to the drug." The simplest pharmacodynamic study is a phase I clinical trial, in which the purpose is to relate dose to toxicity. In a more general context, pharmacodynamics may concern efficacy as well as toxicity, or both. Such studies complement the more frequently performed pharmacokinetic studies, and pharmacokinetic-dynamic studies are becoming increasingly popular as a means of encompassing complete organism responses to pharmacological interventions. They can synthesize and integrate knowledge gained from multiple experiments, in vitro and in vivo, and can be used for forecasting or confirming (validation) the results of drug administration or of clinical trials.

In comparison to pharmacokinetic modeling, the task of defining appropriate structural and statistical models to represent and forecast pharmacodynamic outcomes is more complex. This complexity first relies on the difficulty of proper selection of the outcome measures in order to relate concentration to quantifiable effect. However, investigators tend to focus on those efficacy measures that are most reproducible or convenient for quantification, but these may bear little or no relationship to the therapeutically relevant effect. Hence, the importance of the appropriate choice of short-term markers predictive of long-term clinical outcomes has repeatedly been emphasized [401].

Efficacy measures can be categorized as *biomarkers*, *surrogate markers*, or *clinical outcomes*. While the clinical outcome is the ultimate efficacy measure quantifying the direct benefit to a patient, it is often difficult to measure. Instead, clinical outcomes are often predicted from surrogate markers that can readily and sooner be observed and can easily be quantified. The essential feature of a surrogate marker is that it predicts clinical outcome. Before surrogate endpoints can be used to predict clinical outcomes, however, it is crucial to validate them and prove their relevance. Biomarkers are measurable physiological or biochemical parameters that reflect some pharmacodynamic activity of the investigated drug, even if they are not directly related to clinical outcome. They may be useful to get insight into the overall pharmacodynamic behavior of a compound of interest. Surrogate markers are clinically validated biomarkers.

The complexity in pharmacodynamic studies also arises from the wide variety of pharmacological responses that are often discontinuous. These responses are classified as dichotomous, categorical, time-to-event, or counts:

- When the observed response is *dichotomous*, the probability of response is related to the level of the pharmacodynamic stimulus.

- *Categorical* responses can be regarded as generalized dichotomous responses: several distinct responses are possible, not just one [402].

- *Time-to-event* data record the time elapsing from some natural origin of the time scale to an event [398].

- The *count model* describes data that report the number of events in a volume of space or time.

Despite the problems associated with constructing appropriate efficacy measures, a plethora of classical pharmacodynamic models have been developed. Accordingly, in this last part of the book, we first present in Chapter 10 the classical modeling approaches used to analyze the observed pharmacodynamic effects. In Chapter 11, the last of the book, we present concepts and applications of nonlinear dynamics to pharmacodynamics, placing particular emphasis on chemotherapy and drugs affecting the central nervous system, the cardiovascular system, and the endocrine function.

10

Classical Pharmacodynamics

The master of the oracle at Delphi does not say anything and does not conceal anything, only hints.

Heraclitus of Ephesus (544-483 BC)

Receptors are the most important targets for therapeutic drugs [403]. Therefore, it is important to explore the mechanisms of receptor modulation and drug action in intact in vivo systems. Also, the need for a more mechanism-based approach in pharmacokinetic-dynamic modeling has been increasingly recognized [404, 405]. Hill [406] made the first explicit mathematical model of simulated drug action to account for the time courses and concentration–effect curves obtained when nicotine was used to provoke contraction of the frog rectus abdominis muscle.

Simple mathematical calculations by the first pharmacologists in the 1930s indicated that structurally specific drugs exert their action in very small doses and do not act on all molecules of the body but only on certain ones, those that constitute the drug receptors. For example, Clark [407] calculated that ouabain applied to the cells of the heart ventricle, isolated from the toad, would cover only 2.5% of the cellular surface. These observations prompted Clark [407, 408] to apply the mathematical approaches used in enzyme kinetics to the effects of chemicals on tissues, and this formed the basis of the occupancy theory for drug–receptor interaction. Thus, pharmacological receptor models preceded accurate knowledge of receptors by many years.

10.1 Occupancy Theory in Pharmacology

According to the occupancy theory, which has evolved chronologically from the original work of Clark [407, 408], the drug effect is a function of two processes:

- binding of drug to the receptor and drug-induced activation of the receptor, and
- propagation of this initial receptor activation into the observed pharmacological effect, where the intensity of the pharmacological effect is proportional to the number of receptor sites occupied by drug.

Therefore, the drug–receptor interaction follows the law of mass action and may be represented by the equation

$$\begin{gathered}\gamma\,[\text{drug molecules}] + [\text{receptor}] \underset{k_{-1}}{\overset{k_{+1}}{\rightleftarrows}} [\text{drug–receptor complex}] \\ [\text{drug–receptor complex}] \overset{k_2}{\Longrightarrow} [\text{pharmacological effect}]\,,\end{gathered} \tag{10.1}$$

where γ molecules of drug activate a receptor and give an activated receptor usually called the drug–receptor complex. Although γ is defined as the number of molecules interacting with one receptor, it is in practice merely used to provide better data fits. Rate constants k_{+1}, k_{-1} characterize the association and dissociation of the complex, respectively. The ratio k_{-1}/k_{+1} is defined in pharmacology as the dissociation constant k_D of the complex. The proportionality constant k_2 relates the drug–receptor complex concentration $\upsilon(t)$ with the pharmacological effect $E(t)$, through the equation

$$E(t) = k_2 \upsilon(t)\,. \tag{10.2}$$

When the total number of receptors r_0 is occupied, the effect will be maximal:

$$E_{\max} = k_2 r_0. \tag{10.3}$$

For drug concentration $c(t)$ and a total receptor concentration r_0 we thus have

$$\dot{\upsilon}(t) = k_{+1} c^{\gamma}(t)\left[r_0 - \upsilon(t)\right] - k_{-1}\upsilon(t)\,, \qquad \upsilon(0) = 0. \tag{10.4}$$

In the equilibrium state ($\dot{\upsilon}(t) = 0$ assumption **H1**) we have

$$\upsilon^* = \frac{r_0 c^{*\gamma}}{k_D + c^{*\gamma}}, \tag{10.5}$$

where c^*, υ^* are the drug and drug–receptor complex concentrations in the equilibrium, respectively. By combining the last equation with (10.2) and (10.3), we obtain the working equation for the so-called *sigmoid* $E_{\max}$ *model*:

$$E^* = \frac{E_{\max} c^{*\gamma}}{k_D + c^{*\gamma}}, \tag{10.6}$$

where E^* is the pharmacological effect at equilibrium. From the last equation, it can be seen that the dissociation constant k_D expresses also the γ-power of drug concentration needed to induce half maximal effect ($E_{\max}/2$). When γ is set to 1, the model is called the *basic* $E_{\max}$ *model*, but this model offers less flexibility in the shape of the function compared to the sigmoid $E_{\max}$ model.

Assuming relatively rapid drug–receptor equilibrium with respect to $c(t)$ variations, then $c^* \approx c(t)$ (assumption **H2**), so the previous equation becomes

$$E^*(t) = \frac{E_{\max} c^{\gamma}(t)}{k_D + c^{\gamma}(t)}, \tag{10.7}$$

where $E^*(t)$ indicates that the effect is driven by the pharmacokinetic time.

With $\gamma = 1$, (10.6) has been used extensively in pharmacology to describe the effect of chemicals on tissues in the modified form:

$$E^* = \frac{\varepsilon r_0 c^*}{k_D + c^*},$$

where ε is the intrinsic efficacy (inherent ability of the chemical to induce a physiological response). In other words, ε is the proportionality constant k_2 relating the receptor density r_0 with the maximal effect $E_{\max}$ (10.3). In order to avoid the use of the efficacy term (due to its ad hoc nature), Black and Leff [409] introduced in 1983 the operational model of drug action

$$E^* = \frac{\rho E_{\max} c^*}{k_D + (\rho + 1) c^*},$$

where ρ is equal to the ratio of the receptor density over the concentration of the complex that produces 50% of the maximal tissue response. In reality, this constant ratio characterizes the propensity of a given chemical–tissue system to yield a response.

Since the development of the occupancy theory, the mathematical models used to explain the action of ligands at receptors have been subject to continuous development prompted by new experimental observations. Currently, pharmacological studies deal with drug–receptor or drug–tissue interactions to get estimates for receptor (tissue) affinity and capacity. Thus, the operational model enjoys widespread application in the field of functional receptor pharmacology [410]. Although this model is routinely applied to in vitro studies, the estimates for receptor affinity and capacity can be used for prediction of the effect in vivo. In principle, k_D should be of the same order as the unbound Ec_{50}^{γ}, where Ec_{50} is the concentration at half maximal effect in vivo. In this context, Visser et al. [411] correlated the in vitro measurements with in vivo observations in rats when studying the effect of γ-aminobutyric acid receptor modulators on the electroencephalogram.

10.2 Empirical Pharmacodynamic Models

Combined pharmacokinetic-dynamic studies seek to characterize the time course of drug effects through the application of mathematical modeling to dose–effect–time data. This definition places particular emphasis on the time course of drug action. Pharmacodynamics is intrinsically related to pharmacokinetics, which encompasses the study of movement of drugs into, through, and out of the body. The term *pharmacodynamic models* exclusively refers to those models that relate drug concentration with the pharmacological effect.

The most common function used to relate drug concentration c with effect is the $E_{\max}$ model:

$$E = \frac{E_{\max} c^{\gamma}}{Ec_{50}^{\gamma} + c^{\gamma}}, \tag{10.8}$$

where $E_{\max}$ is the maximum effect and Ec_{50} is the concentration at half the maximal observable in vivo effect. Equation (10.8) corresponds to (10.6) with Ec_{50}^{γ} substituting k_D. It is also clear that (10.8) is a static nonlinear model in which c corresponds to the equilibrium point c^*. If we consider c as a time course $c(t)$, we must implicitly assume that equilibrium is achieved rapidly throughout $c(t)$, so $c^* \equiv c(t)$ (assumption **H2**).

If a baseline E_0 is introduced to the previous equation,

$$E = E_0 \pm \frac{E_{\max} c^{\gamma}}{Ec_{50}^{\gamma} + c^{\gamma}},$$

we obtain the $E_{\max}$ model describing either stimulation or inhibition of the effect by the concentration of the drug. Parameters $E_{\max}$, Ec_{50}, and γ are assumed constant and independent of the drug dose as well as the drug and receptor concentrations.

Other simpler empirical models have also been used since the early days of pharmacodynamics [412,413] to describe the drug concentration–effect relationship. The linear model relies on a linear relationship between E and c:

$$E = \alpha c + \beta, \tag{10.9}$$

where α is the slope indicating the sensitivity of the effect to concentration changes. The intercept β can be viewed as the baseline effect. Equation (10.9) reveals that the linearity between c and E is unlimited, and this feature is undoubtedly a drawback of the model. Besides, a log-linear model between E and c can also be considered:

$$E = \alpha \log(c) + \beta. \tag{10.10}$$

Due to the logarithmic expression of concentration in this model a larger concentration range is related "linearly" with the effect. As a rule of thumb, 20 to 80% of the concentration range of the $E_{\max}$ model can be approximately described with (10.10).

Although these empirical approaches may quantify and fit the data well, they do not offer a physical interpretation of the results.

10.3 Pharmacokinetic-Dynamic Modeling

In the mid-1960s, G. Levy [412, 413] was the first to relate the pharmacokinetic characteristics with the in vivo pharmacological response of drug using the above-mentioned linear models. In fact, as the pharmacological responses $E(t)$ and the drug concentration $c(t)$ can be observed simultaneously and repeatedly as a function of time, a combined pharmacokinetic-dynamic model is needed to describe these time courses. From the simple models, the discipline of pharmacokinetic-dynamic modeling emerged gradually, and in actuality even complex physiological processes controlling drug response can be modeled. The key mechanisms intrinsic to pharmacokinetic-dynamic models are the following:

- the processes may take place under either equilibrium or nonequilibrium conditions for the pharmacodynamic part,
- the binding of drug with the receptor may either be reversible or irreversible, and
- the bound drug may induce its effect directly or indirectly.

A general scheme for the basic components of pharmacokinetic-dynamic models is depicted in Figure 10.1. According to this scheme, the drug at the prereceptor phase is considered to distribute to an effect compartment; then it reacts with the receptors under equilibrium (direct link, assumption **H3**) or nonequilibrium (indirect link) conditions, and finally, at the postreceptor phase, the activated receptors can either produce the response directly (direct response, assumption **H4**) through the transducer function $\mathcal{T}$ (which is usually a proportionality constant like k_2 in equation 10.1) or they can interfere with an endogenous or already existing process that produces the final response (indirect response). In fact, all the processes of the general model depicted in Figure 10.1 are not necessarily incorporated in the final model used in practice. Almost always, one of these steps is considered to be the limiting one, and the model reduces to one of the basic models described below.

10.3.1 Link Models

During the first decades of the development of pharmacokinetic science, a lag time in pharmacological response after intravenous administration was often treated by applying a compartmental approach. If the plasma concentration declined in a biexponential manner, the observed pharmacodynamic effect was fitted to plasma or "tissue" compartment concentrations. Due to the lag time of effects, a successful fit was sometimes obtained between effect and tissue drug level [414]. However, there is no a priori reason to assume that the time course of a drug concentration at the effect site must be related to the kinetics in tissues that mainly cause the multiexponential behavior of the plasma time–concentration course. A lag time between drug levels and dynamic effects can also occur for drugs described by a one-compartment model.

Segre [415] was the first author to consider the possibility that the time course of pharmacological effect could itself be used to describe the transfer rate of a drug to the biophase. Thus, the lag time of the effect was modeled by

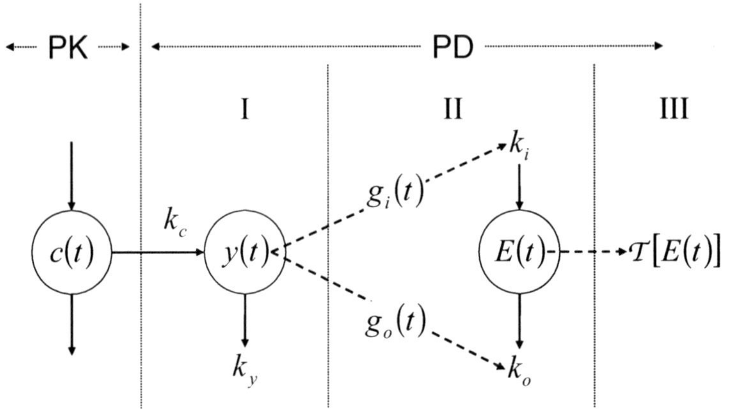

Figure 10.1: Schematic of the basic processes involved in pharmacokinetic (**PK**) - dynamic (**PD**) models. The phases I, II, and III refer to processes that take place in the prereceptor, receptor, and postreceptor proximity, respectively. The symbols are defined in the text.

including two hypothetical tissue compartments between the plasma compartment and the pharmacodynamic response compartment.

The idea of Segre was further developed, in an elegant way, by Sheiner and associates [416, 417] by linking the effect compartment to a kinetic model. This approach has since been called the *link model.* The time course of the drug in the effect site is determined by the rates of transfer of material into and from the effect compartment; the lag time of the effect site concentration is controlled by the elimination rate constant of the effect compartment. The beauty of this approach is that instead of relating the pharmacodynamic response to drug concentrations in some more or less well defined tissue, it is related to the plasma drug level, which in clinical practice is of great importance.

Direct Link

Strictly speaking, pharmacodynamic models are employed to relate the receptor site drug concentration to pharmacological response at any given time using data mainly from in vivo experiments. However, the receptor site drug concentration normally cannot be measured directly. Thus, the simplest pharmacokinetic-dynamic mechanistic model arises from assuming that the drug concentration in the blood, $c(t)$ (far left compartment of Figure 10.1) is the same at the receptor site, $y(t)$. Strictly speaking, this assumption expresses a prereceptor equilibrium (**H3**) and the resulting model does not utilize concentrations at the effect site.

Further, under the equilibrium conditions **H1**, we can use (10.6) to relate the pharmacological effect E^* with the drug concentrations c^*, or in addition, use (10.7) to relate the time courses $E^*(t)$ and $c^*(t)$ under the supplementary assumption **H2**. Thus, the simplest mechanistic models are once again the basic and the sigmoid $E_{\max}$ models, but now they have a specific physical interpretation in terms of drug–receptor reaction kinetics.

As is implicit from all the above, the measured concentration in plasma is directly linked to the observed effect for these simple mechanistic, pharmacokinetic-dynamic models. Accordingly, these models are called *direct-link models* since the concentrations in plasma can be used directly in (10.6) and (10.7) for the description of the observed effects. Under the assumptions of the direct-link model, plasma concentration and effect maxima will occur at the same time, that is, no temporal dissociation between the time courses of concentration and effect is observed. An example of this can be seen in the direct-link sigmoid $E_{\max}$ model of Racine-Poon et al. [418], which relates the serum concentration of the anti-immunglobulin E antibody CGP 51901, used in patients for the treatment of seasonal allergic rhinitis, with the reduction of free anti-immunglobulin E.

Under the assumptions of the direct-link model, neither a counterclockwise (Figure 10.2) nor a clockwise hysteresis loop (Figure 10.4) will be recorded in an effect vs. concentration plot. In principle, the shape of the effect vs. concentration plot for an ideal direct-link model will be a curve identical to the specific pharmacodynamic model, relating effect with concentration, e.g., linear for a linear pharmacodynamic model, sigmoid for the sigmoid $E_{\max}$ model (cf. Table 10.1 and following paragraphs and sections), etc.

Indirect Link: The Effect-Compartment Model

In the direct-link model, concentration–effect relationships are established without accounting for the intrinsic pharmacodynamic temporal behavior, and the relationships are valid only under the assumption of effect site, prereceptor equilibrium **H3**. In contrast, indirect-link models are required if there is a temporal dissociation between the time courses of concentration and effect, and the observed delay in the concentration–effect relationship is most likely caused by a functional delay between the concentrations in the plasma and at the effect site.

When a lag time of $E(t)$ is observed with respect to the $c(t)$ time course, the use of a combined pharmacokinetic-dynamic model, the indirect-link model, is needed to relate the drug concentration $c(t)$ to the receptor site drug concentration $y(t)$ (which cannot be measured directly) and the $y(t)$ to the pharmacological response $E(t)$.[1]

The effect–compartment model relaxes the assumption **H3** and it stems from the assumption of prereceptor nonequilibrium between drug concentration in the blood or plasma $c(t)$ and the receptor site $y(t)$. According to this model, an additional compartment is considered, the effect (or biophase) compartment, and

[1] In the classical pharmacokinetic-pharmacodynamic literature, the effect site concentration and the effect site elimination rate constant are denoted by c_E and k_{E0}, respectively. Here, the symbols $y(t)$ and k_y are used instead.

it is the concentration $y(t)$ in that compartment that reacts with the receptors, Figure 10.1.

Notation:

- V_c and V_y denote the apparent volumes of distribution of the plasma and effect compartments, respectively.
- k_c and k_y denote the first-order rate constants for the drug transfer from plasma to effect site and for drug elimination from the effect site, respectively.

Then assuming that the mass-flux equality holds for the effect compartment, i.e., $V_c k_c = V_y k_y$, the drug concentration $y(t)$ in the effect compartment can be described by the linear differential equation

$$\dot{y}(t) = k_y [c(t) - y(t)], \qquad y(0) = 0. \tag{10.11}$$

This equation can be solved by applying the Laplace transformation and convolution principles (cf. Appendix E):

$$y(t) = k_y \widetilde{y}(t), \tag{10.12}$$

where $\widetilde{y}(t)$ is defined as the *apparent effect site* drug concentration and it is given by

$$\widetilde{y}(t) = \int_0^t c(t') \exp[-k_y (t - t')] \, dt'.$$

The time symbols t', t denote the temporal dissociation between the time courses of concentration and effect, respectively. For various types of drug administration, the function $c(t)$ is known and therefore analytic solutions for $\widetilde{y}(t)$ have been obtained using the integral defined above. Substituting (10.12) into (10.7), we obtain the fundamental equation for the $E_{\max}$ indirect-link model:

$$E^*(t) = \frac{E_{\max}^{\gamma} \widetilde{y}^{\gamma}(t)}{\widetilde{y}_{50}^{\gamma} + \widetilde{y}^{\gamma}(t)}, \tag{10.13}$$

where $\widetilde{y}_{50}$ is the apparent effect site drug concentration producing 50% of the maximum effect.

In this model, the rate constant k_y was originally considered to reflect a distributional delay of drug from plasma to the effect compartment. However, it can also be regarded as a constant producing the delay in effects in relation to plasma, irrespective of whether this is caused by distributional factors, receptor events, production of a mediator of any kind, etc.

The basic feature of the indirect-link model is the counterclockwise hysteresis loop that is obtained from plotting the observed values of the effect vs. the observed plasma drug concentration values, Figure 10.2. In other words, the effect is delayed compared to the plasma drug concentration and this is reflected in the effect–concentration state space.

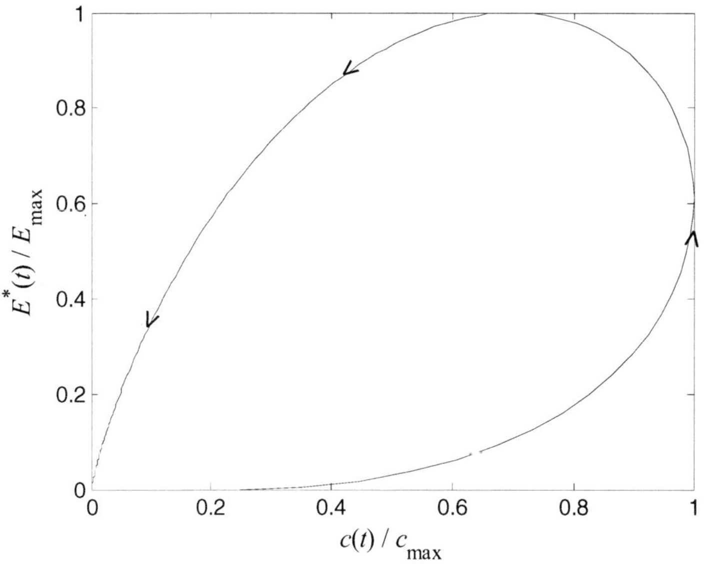

Figure 10.2: Normalized effect–plasma drug concentration state space for the indirect link model. As time flows (indicated by arrows) a counterclockwise hysteresis loop is formed. The rate constant for drug removal from the effect compartment k_y characterizes the temporal delay, that is, the degree of hysteresis.

Numerous applications of pharmacokinetic-dynamic models incorporating a biophase (or effect) compartment for a variety of drugs that belong to miscellaneous pharmacological classes, e.g., anesthetic agents [419], opioid analgesics [420–422], barbiturates [423, 424], benzodiazepines [425], antiarrhythmics [426], have been published. The reader can refer to a handbook [427] or recent reviews [405] for a complete list of the applications of the biophase distribution model.

In actual practice, nonlinear regression is used to fit a suitable pharmacokinetic model described by the function $c(t)$ to time–concentration data. Then, the estimated parameters are used as constants in the pharmacodynamic model to estimate the pharmacodynamic parameters. Alternatively, simultaneous fitting of the model to the concentration–effect–time data can be performed. This is recommended as $c(t)$ and $E(t)$ time courses are simultaneously observed.

Example 10 *Bolus Intravenous Injection*

An example of the indirect-link model after bolus intravenous injection can be seen in Figure 10.3. The arrow indicates the time flow. Each point represents

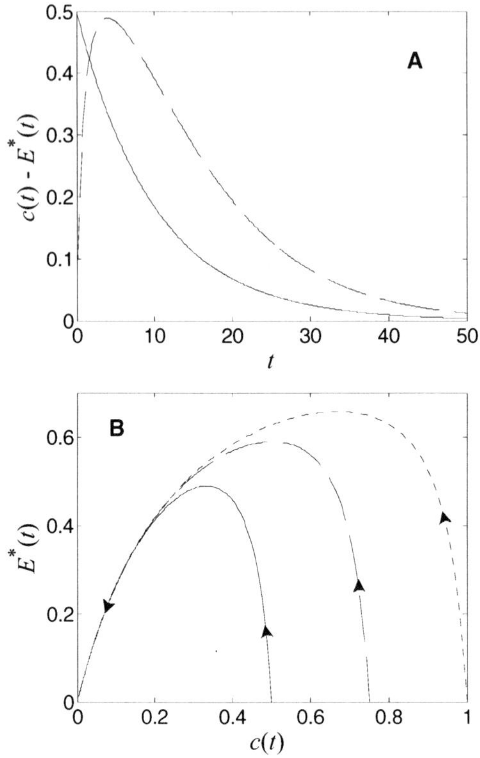

Figure 10.3: Indirect link model with bolus intravenous injection. (**A**) The classical time profiles of the two variables $c(t)$ (*solid line*) and $E^*(t)$ (*dashed line*) for dose $q_0 = 0.5$. (**B**) A two-dimensional phase space for the concentration $c(t)$ vs. effect $E^*(t)$ plot using three doses 0.5, 0.75, and 1 (*solid*, *dashed*, and *dotted lines*, respectively).

a uniquely defined state and only one trajectory may pass from it. The state space has a point attractor, i.e., a steady state, which is obviously the point $(c = 0,\ E^* = 0)$ reached at theoretically infinite time. Three different initial conditions of the form $c(0) = q_0/V_c$, $E^*(0) = 0$, are used to generate three different trajectories, all of which end up at the point attractor. The integrated equations of the system are

$$\begin{aligned} c(t) &= \frac{q_0}{V_c} \exp(-kt), \\ \widetilde{y}(t) &= \frac{q_0}{V_c} \frac{[\exp(-kt) - \exp(-k_y t)]}{k_y - k}, \\ E^*(t) &= \frac{E_{\max} \widetilde{y}(t)}{\widetilde{y}_{50} + \widetilde{y}(t)}, \end{aligned}$$

where q_0 is the dose, V_c and k are the volume of distribution and the elimination rate constant for pharmacokinetics, k_y is the effect site elimination rate constant, $E_{\max}$ is the maximum effect, and $\widetilde{y}_{50}$ is the concentration at which 50% of the maximum effect is observed. Parameter values were set to

$$\begin{array}{lll} V_c = 1, & k = 0.1, & \\ k_y = 0.5, & E_{\max} = 1, & \widetilde{y}_{50} = 0.7. \end{array}$$

where all units are arbitrary. ■

10.3.2 Response Models

Time is not an independent variable in the presented models. Dynamic behavior is either a consequence of the pharmacokinetics or the observed lag time by means of the effect compartment. Dynamic models from the occupancy theory and described by differential equations, such as (10.4), are scarce [428, 429].

Neglecting dynamic models in pharmacodynamics [430] is perhaps due to the fact in that instant equilibrium relationships between concentration and effect appear to occur for most drugs. For some drugs, such as cytotoxic agents, this delay is often extremely long, and attempts to model it are seldom made. One can describe these relationships as time-dissociated or nondynamic because the temporal aspects of the effect are not linked to the time–concentration profile.

In recent years, new models overcoming these defaults have been developed as the *indirect physiological models* introduced by Jusko and associates [431]. According to this last type of model, an endogenous substance or a receptor protein is formed at a constant rate and lost with a first-order rate constant. The drug concentration in plasma produces an effect by either stimulating or inhibiting the synthesis or removal of the endogenous substance leading to a change in the observed pharmacodynamic effect described by a suitable pharmacodynamic model.

Direct Response

The standard effect–compartment model, usually characterized as an atypical indirect-link model, also constitutes an example of what we will call a *direct-response* model in contrast to the *indirect-response* models. Globally, the standard direct-response models are models in which $c(t)$ affects all dynamic processes only linearly.

Indirect Response

Ariens [432] was the first to describe drug action through indirect mechanisms. Later on, Nagashima et al. [433] introduced the indirect response concept to pharmacokinetic-dynamic modeling with their work on the kinetics of the anticoagulant effect of warfarin, which is controlled by the change in the prothrombin complex synthesis rate. Today, indirect-response modeling finds extensive applications especially when endogenous substances are involved in the expression of the observed response.

From a modeling point of view, the last equilibrium assumption that can be relaxed, for the processes depicted in Figure 10.1, is **H4**, between the activated receptors (υ variable in the occupancy model) and the response E. Instead of the activated receptors directly producing the response, they interfere with some other process, which in turn produces the response E. This mechanism is usually described mathematically with a transducer function $\mathcal{T}$ which is no longer linear (cf. Section 10.4.1). This type of pharmacodynamic model is called *indirect response* and includes modeling of the response process usually through a linear differential equation of the form

$$\dot{E}(t) = k_i g_i(t) - k_o g_o(t) E(t), \qquad E(0) = k_i/k_o, \tag{10.14}$$

where k_o is a first-order rate constant, and k_i represents an apparent zero-order production rate of the response. Stationarity conditions set the initial response value $E(0)$ at the ratio k_i/k_o. Functions $g_i(t)$ and $g_o(t)$ depend on the drug concentration through $E_{\max}$ functions and can produce either stimulation or inhibition, respectively:

$$g(t) = 1 + \frac{S_{\max} c(t)}{Sc_{50} + c(t)} \qquad \text{or} \qquad g(t) = 1 - \frac{I_{\max} c(t)}{Ic_{50} + c(t)}. \tag{10.15}$$

In these expressions, $g(t)$ is either $g_i(t)$ or $g_o(t)$, $S_{\max}$ is maximum stimulation rate, $I_{\max}$ is maximum inhibition rate, Sc_{50} and Ic_{50} are the drug concentrations at which $g(t) = 1 + (S_{\max}/2)$ and $g(t) = 1 - (I_{\max}/2)$, respectively. Consequently, four basic models are formulated: inhibition of k_i, inhibition of k_o, stimulation of k_i, and stimulation of k_o, Figure 10.1.

This family of the four basic indirect response models has been proven to characterize diverse types of pharmacodynamic effects and it constitutes the current approach for pharmacokinetic-dynamic modeling of responses generated by indirect mechanisms. Thus, indirect response models have been used to interpret the anticoagulant effect of warfarin, adrenal suppression by corticosteroids, cell trafficking effects of corticosteroids, the antipyretic effect of ibuprofen, aldose reductase inhibition, etc. [434]. Basically, the indirect response concept is appropriate for modeling the pharmacodynamics of drugs that act through inhibition or stimulation of the production or loss of endogenous substances or mediators.

However, although the general model described above is considered to be mechanistic as opposed to the completely empirical approach, being based on a general physiological process like receptor activation, it is still too general and abstract to describe complicated drug processes. Stimulation and inhibition of a baseline through the saturable $E_{\max}$ function is often not enough, since drugs interplay with complicated physiological processes. Thus, during the last ten years Jusko's group and other investigators have expanded the application of indirect response mechanisms to real mechanistic pharmacodynamic modeling and have included detailed modeling of the underlying physiology and then modeled the effect of drugs on it. These models are called *extended indirect response models* [435] and they have been used to describe tolerance and rebound phenomena [436], time-dependency of the initial response [437, 438], cell trafficking dynamics [439], etc.

It is rather obvious that an indirect response mechanism, whatever the detailed processes involved, results in a counterclockwise hysteresis loop for the effect–concentration relationship, Figure 10.2. Here, however, the elaboration of the observed response is usually secondary to a previous time-consuming synthesis or degradation of an endogenous substance(s) or mediator(s). Since both the indirect-link and indirect response models have counterclockwise hysteresis effect–concentration plots, an approach based on the time of the maximum effect has been applied to furosemide data [440] for indirect (link or response) model selection.

When one looks into the basic functions of the link and indirect response models, it is clear that one of the differences resides in the input functions to the effect and the receptor protein site, respectively. For the link model a linear input operates in contrast to the indirect model, where a nonlinear function operates. For the link model the time is not directly present and the pharmacological time course is exclusively dictated by the pharmacokinetic time, whereas the indirect model has its own time expressed by the differential equation describing the dynamics of the integrated response.

10.4 Other Pharmacodynamic Models

A number of other pharmacodynamic approaches focusing either on prereceptor or postreceptor events have been proposed in the literature and are discussed below.

10.4.1 The Receptor–Transducer Model

First, mention can be made of cases in which the measured effect instead of being proportional to the activated receptors, follows a more general function $E = \mathcal{T}(v)$. This model is called *receptor–transducer* and was introduced by Black and Leff [409]. The function $\mathcal{T}$ is called a *transducer function* and its most common form is yet again the $E_{\max}$ function, which when replaced in (10.5) results in an $E_{\max}$ model but with different shape parameters called an *operational model* [441].

10.4.2 Irreversible Models

All the above-mentioned pharmacokinetic-dynamic models are characterized by reversibility of the drug–receptor interaction. In several cases, however, drug action relies on an irreversible bimolecular interaction; thus, enzyme inhibitors and chemotherapeutic agents exert their action through irreversible bimolecular interactions with enzymes and cells (bacteria, parasites, viruses), respectively.

The irreversible inactivation of endogenous enzymes caused by drugs, e.g., the antiplatelet effect of aspirin after oral administration [442], the 5α-reductase inhibition by a new nonsteroidal inhibitor [443], and the H^+, K^+-ATPase inactivation by proton pump inhibitors [444], is modeled with turnover models. The simplest model [442] includes terms for the production rate k_i and loss rate k_o of the response E, coupled with a function $g(c)$ representing the change of plasma or effect–compartment drug concentration:

$$\dot{E}(t) = k_i - [k_o + g(c)] E(t),$$

where k_i and k_o have the same meaning as defined for (10.14) while the function $g(c)$ is either linear or of Michaelian type.

The models used for the irreversible effects of chemotherapeutic agents quantify the response $E(t)$ in terms of the cell number since irreversible inactivation leads to cell killing. In these models, the function of the natural proliferation of cells $r(E)$ is combined with the cell-killing function $g(c)$, which again represents the change of plasma or effect–compartment drug concentration:

$$\dot{E}(t) = r(E) - g(c) E(t).$$

The function $r(E)$ can take various forms describing the natural growth of the cell population in the absence of drug [372, 445], while $g(c)$ can be either linear or nonlinear [435, 446, 447]. Due to the competitive character of the functions $r(E)$ and $g(c)$, the cell number vs. time plots are usually biphasic with the minimum effective concentration of drug being the major determinant for the killing or regrowth phases of the plot.

10.4.3 Time-Variant Models

Contrary to the already mentioned models, which include constant parameters, pharmacodynamic models may include time-varying parameters as well. Typical examples include models of drug tolerance or sensitization, where the parameters vary as a function of the dosing history. Other examples concern modeling of circadian rhythms where parameters depend explicitly on time through biological clocks, e.g., the baseline of a pharmacological response, and it is necessary to include periodicity in the pharmacokinetic-dynamic modeling. This is usually done by empirical periodic functions directly on the baseline, such as trigonometric functions, for example. An example is the effect of fluticasone propionate on cortisol [438]. All models associated with these phenomena are called time-variant.

Drug Tolerance

This phenomenon is characterized by a reduction in effect intensity after repeated drug administration. The explanation for the diminution of the effect as a function of time is attributed either to a decrease in receptor affinity or a

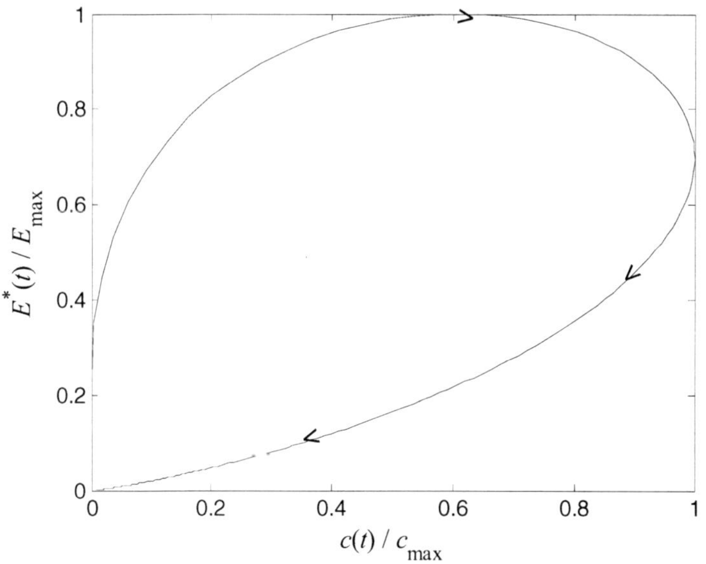

Figure 10.4: Normalized effect–plasma drug concentration state space for tolerance phenomena. As time flows (indicated by arrows) a clockwise hysteresis loop is formed.

decrease in the number of receptors. These changes result in a clockwise hysteresis loop when the effect is plotted vs. the plasma concentration, Figure 10.4. Usually, tolerance phenomena are discussed with respect to the $E_{\max}$ model. In this case, tolerance is associated with either a decrease in $E_{\max}$ over time or an increase in Ec_{50}^{γ} over time (10.8). An example of this kind of time dependency is the work of Meibohm et al. [448] on the suppression of cortisol by triamcinolone acetonide during prolonged therapy.

Apart from the decrease in the number or affinity of the receptors, more complex mechanisms have been proposed for tolerance phenomena. In the so-called *counterregulation* models, the development of tolerance is driven by the primary effect of drug perhaps via an intermediary transduction step. This mechanism was postulated by Bauer and Fung [449] for hemodynamic tolerance to nitroglycerine. According to these authors, initial nitroglycerin-induced vasodilatation controls the counterregulatory vasoconstrictive effect. Moreover, the *desensitization of receptors* can reduce the drug effects on prolonged exposure. The receptor-inactivation theory [450] can be used to model this mechanism.

Drug Sensitization

This term is used to describe the increase in pharmacological response with time to the same drug concentration. The up-regulation of receptors is considered to be the primary cause for sensitization. This phenomenon is observed when the negative feedback of an agonist is removed. A clinical example of sensitization is the chronic administration of beta-blockers, which induce up-regulation of beta-adrenoreceptors. This leads to increased adenyl cyclase activity and hypersensitivity to catecholamines after sudden withdrawal of the antagonist [451]. Due to the increase of the effect over time in sensitization phenomena, the effect–plasma concentration plots have a counterclockwise hysteresis loop, Figure 10.2.

10.4.4 Dynamic Nonlinear Models

Using the approach of Sheiner and Verotta [452], a large number of pharmacodynamic models can be considered as hierarchical models composed of a series of submodels. These submodels are linear or nonlinear, static or dynamic input–output, elementary models. Several possible combinations of such submodels have been considered, but they have systematically associated the linear with dynamic features, and the nonlinear with static ones. Is there hesitation or fear of using nonlinear dynamics in the traditional pharmacokinetic-dynamic modeling context?

The most interesting case arises by removing assumption **H1**, i.e., when the reaction between drug and receptor is not at equilibrium [428]. This happens when relatively slow rates of association and dissociation of the complex are observed. Under these conditions, a slow dynamic receptor-binding model is most applicable. By maintaining the proportionality between the effect and the concentration of the drug–receptor complex, (10.4) can be written in terms of the effect

$$\dot{E}(t) = k_{+1}c^{\gamma}(t)\left[E_{\max} - E(t)\right] - k_{-1}E(t), \qquad E(0) = 0. \tag{10.16}$$

This equation is a nonlinear differential equation describing the time course of the effect and using an intrinsic pharmacodynamic time. An application of this model can be found in the work of Shimada et al. [429], who applied the drug–receptor nonequilibrium assumption to model the pharmacodynamics of eight calcium channel-blocking agents in hypertensive patients on the basis of their in vitro binding data. This model is rarely used because it produces profiles similar to the indirect-link model described above. However, the drug–receptor nonequilibrium model has more theoretical and practical interest since more complex solutions are also feasible by adding a feedback component to the effect of the drug [453]. The resulting model has a very rich dynamic behavior and is the essence of Chapter 11.

10.5 Unification of Pharmacodynamic Models

Historically, delays between drug exposure and effect have been described with the so-called effect–compartment model, first described by Segre [415] and popularized by Sheiner and coworkers [416, 417]. Recently, Dayneka [431] focused attention on a set of indirect-effect models to introduce intrinsic pharmacodynamic time. The relevance of combined pharmacokinetic-dynamic modeling has been largely recognized [454, 455]. The discussion in Section 10.3 indicates that the development of the various pharmacokinetic-dynamic models was based on the dominating assumption for one of the drug processes depicted in Figure 10.1. Thus, the pharmacokinetic-dynamic models can be classified kinetically on the basis of the assumptions associated with:

- the prereceptor equilibrium,
- the drug–receptor interaction, and
- the postreceptor equilibrium.

A very general scheme for relating effects to concentration, of which both the effect–compartment and the indirect-effect models are special cases, was outlined by Sheiner and Verotta [452]. The models presented in the study can be considered to be a special case of that unified scheme. As judiciously presented by these authors, both direct-response and indirect-response models are composed of one nonlinear static submodel and one dynamic submodel, but the placement of the submodels in the global model differs:

- In a direct-response model, the output of a linear dynamic model (the link model) with input $c(t)$ drives a nonlinear static model (usually the $E_{\max}$ model) to produce the observed response.
- In an indirect-response model, the above order of models is reversed and now the static model precedes the dynamic one. The dynamic model describes the formation and loss of the response variable through a linear differential equation whose parameters are nonlinear saturable forms of the driving concentration $c(t)$.

All these models introducing the prereceptor and postreceptor events have an interesting appeal with respect to physiologically implied mechanisms. Sheiner and Verotta [452] pointed out the importance of knowing where the rate-limiting step is located in a series of events from pre- to postreceptor drug interactions.

The fundamental assumption and equations governing the effect–concentration relationship for each one of the models considered are listed in Table 10.1. The presence or not of an hysteresis loop in the effect–plasma concentration plot of each model is also quoted in Table 10.1. At present, the methodology for performing efficient pharmacokinetic-dynamic modeling is well established [405, 456, 457].

Table 10.1: Assumptions and operable equations for the pharmacokinetic-dynamic models. The hysteresis column "Hyster" refers to the graph of the effect–concentration plot.

Model	Prereceptor	Receptor	Postreceptor	Hyster
Empirical $E_{\max}$	None	10.8	None	No
Indirect link	10.11	10.13	None	Yes
Indirect response	Equilibrium	10.15	10.14	Yes
Transducer	None	None	$E(t) = \mathcal{T}(v)$	-
Nonlinear	Equilibrium	10.16	None	Yes

10.6 The Population Approach

The goal of pharmacokinetic and pharmacodynamic investigations is to establish a rational basis for the therapeutic use of a drug. Specifically, clinical trials aim at determining the dose and the dosage regimen of the new drug that will produce therapeutic benefit in patients while minimizing the inconvenience of side effects and risks of adverse drug reactions. This is particularly true in the clinical evaluation of new chemical and biological entities during drug development [458].

Data destined for pharmacokinetic analysis consist of one or more drug concentration vs. time observations, while pharmacodynamic data consist of specific concentration levels corresponding to a specific therapeutic effect or its validated biomarker. One distinguishes two types of data:

- *Experimental data* arise from studies carried out specifically for pharmacokinetic investigations, under controlled conditions of drug dosing and extensive blood sampling.

- *Observational data* are collected as a supplement in a study designed and carried out for another purpose. These data are characterized by lack of control and few design restrictions: the amount of kinetic data collected from each individual is variable, the timing of blood sampling differs and the number of blood samples per patient is small, typically from 1 to 5.

It should be emphasized that in the collected data, several responses may be measured (e.g., drug plasma concentration, arterial blood pressure), and diverse administration schedules (single dose and chronic dosing) may be considered.

10.6.1 Inter- and Intraindividual Variability

The population approach is a new point of view in clinical drug evaluation and therapy. It emphasizes the estimation of parameters describing the dose–concentration–response relationship both between and within patients (including average behavior and variability). The population approach recognizes variability as an important feature that should be identified and measured during drug evaluation [459]. Indeed:

- We need to know something about the distributions of the deviations of individual patient pharmacokinetic-dynamic model parameters from their population average values, and how these deviations correlate with one another. The deviations are population parameters of a different type: random individual effect parameters; random because individual deviations are regarded as occurring according to chance mechanisms.

- One may ask how much drug outcome (concentration/effect) varies across a modeling cycle within an individual. To answer this question, yet other random-effect population parameters are needed: the variance of the combined random intraindividual and measurement error; random because outcome fluctuations and measurement errors are also regarded as occurring according to chance mechanisms.

- One may immediately imagine further subdividing the last type of variability. For example, one might wish to distinguish intraindividual variability due to different aspects of kinetics and separate all such variability from that due to measurement error. The problem with so doing is only that most data are insufficiently detailed and complete to allow these components of variance to be estimated separately. The two-way division we have proposed appears to suffice for most applications and data sets.

According to the population approach, the analysis of collected data requires an explicit mathematical model, including parameters quantifying population mean profiles, interindividual variability, and residual variability including intraindividual variability and measurement error [460].

10.6.2 Models and Software

Nonlinear mixed-effects modeling methods as applied to pharmacokinetic-dynamic data are operational tools able to perform population analyses [461]. In the basic formulation of the model, it is recognized that the overall variability in the measured response in a sample of individuals, which cannot be explained by the pharmacokinetic-dynamic model, reflects both interindividual dispersion in kinetics and residual variation, the latter including intraindividual variability and measurement error. The observed response of an individual within the framework of a population nonlinear mixed-effects regression model can be described as

$$y_{ij} = g\left(\underline{\theta}_i, t_{ij}\right) + \varepsilon_{ij},$$

where y_{ij} for $j = 1, \ldots, n_i$ are the observed data at time points t_{ij} of the ith individual. An appropriate model of this type is defined for all $i = 1, \ldots, m$, where m is the number of individuals in the sample. The function $g\left(\underline{\theta}, t\right)$ is a specific function for predicting the response, $\underline{\theta}_i$ is the vector of unknown individual-specific parameters, and ε_{ij} accounts for the error between the unknown value and the corresponding measurement.

The sample of individuals is assumed to represent the patient population at large, sharing the same pathophysiological and pharmacokinetic-dynamic parameter distributions. The individual parameter $\underline{\theta}$ is assumed to arise from some multivariate probability distribution $\underline{\Theta} \sim f(\underline{\Psi})$, where $\underline{\Psi}$ is the vector of so-called hyperparameters or "population characteristics." In the mixed-effects formulation, the collection of $\underline{\Psi}$ is composed of population "typical values" (generally the mean vector) and of population "variability values" (generally the variance–covariance matrix). Mean and variance characterize the location and dispersion of the probability distribution of $\underline{\Theta}$ in statistical terms.

Then, given a model for data from a specific drug in a sample from a population, mixed-effect modeling produces estimates for the complete statistical distribution of the pharmacokinetic-dynamic parameters in the population. Especially, the variance in the pharmacokinetic-dynamic parameter distributions is a measure of the extent of inherent interindividual variability for the particular drug in that population (adults, neonates, etc.). The distribution of residual errors in the observations, with respect to the "mean" pharmacokinetic or pharmacodynamic model, reflects measurement or assay error, model misspecification, and, more rarely, temporal dependence of the parameters.

Population modeling software varies in the number of assumptions made regarding the statistical distributions of the pharmacokinetic-dynamic parameters, the within-individual or residual error, and, particularly, the interindividual variability (random effect). They take either a parametric approach with strong assumptions, typically of a Gaussian distribution [462, 463] or Bayesian approaches [464], a semiparametric view with relaxed assumptions [465], or a nonparametric, no assumptions approach [466, 467]. NONMEM (NONlinear Mixed Effect Modeling, NONMEM Project Group, University of California at San Francisco, CA [462]) and NPEM2 (NonParametric Expectation Maximization, Laboratory of Applied Pharmacokinetics, University of Southern California, Los Angeles, CA [467]) are examples of parametric and nonparametric population modeling packages respectively. Recently, Aarons [468] reviewed some of the software currently available for performing nonlinear mixed-effects modeling.

However, the Bayesian analysis using Gibbs sampling (BUGS) requires special mention since is as a general program for performing analysis for a wide range of statistical problems and is available on PC and Unix platforms and also in a PC Windows version. The Bayesian analysis is based on complex statistical models using Markov chain Monte Carlo methods [469–471].

10.6.3 Covariates

In the initial stage of the analysis, the pharmacokinetic or pharmacodynamic observations are blind with respect to the patients, i.e., no patient-specific demographic or physiological covariates are included, other than the dose. Both parametric and nonparametric population methods then, in this first stage, produce a base model for the centering and spread of the parameters in the population, which can then be used in subsequent steps in various ways.

However, the base model provides inadequate individualization, and to assist clinical decision-making, it is important to relate differences among individuals to readily identifiable and routinely measurable individual attributes or covariates, such as demographic (e.g., age), pathophysiological (e.g., serum creatinine, renal, or hepatic function) or genotypic (e.g., CYP2D6 polymorphism) data. Knowing the value of an influential covariate in a new patient before starting therapy increases the predictive power and therefore makes the choice of dose more reliable.

Explanation of parameter variability using covariates can be achieved:

- by simple regression of the individual empirical Bayes parameters from the base model with the covariates and
- within the population fitting process, estimating the covariate term coefficients jointly with the pharmacokinetic parameters.

Parametric population methods also obtain estimates of the standard error of the coefficients, providing consistent significance tests for all proposed models. A hierarchy of successive joint runs, improving an objective criterion, leads to a "final" covariate model for the pharmacokinetic parameters. The latter step reduces the unexplained interindividual randomness in the parameters, achieving an extension of the deterministic component of the pharmacokinetic model at the expense of the random effects. Recently used individual empirical Bayes estimations exhibit more success in targeting a specific individual concentration after the same dose.

10.6.4 Applications

The knowledge of population kinetic parameters has been proved important, and up to the present, the population approach has had a wide spectrum of applications:

- It is currently accepted medical practice to measure a few drug levels after dosage has progressed for some time. In order to make useful the measured drug levels, one should estimate individual parameters using Bayesian estimation techniques. They consist in combining a few or even a single individual drug level measurement with the probability distribution function expressing interindividual variability. Once the individual parameters are obtained, the time-dependent pharmacokinetic model can be used for forecasting and predictive exploration of dosing regimens.
- Most decisions regarding drug regulation involve knowledge of the typical or average behavior of a drug in a population. To the extent that pharmacokinetic aspects of drugs are of interest to drug regulatory agencies, population pharmacokinetics will also be of interest.

- Although intraindividual kinetic variability has only been regarded as a nuisance, the typical degree of intraindividual kinetic variability from all causes can be used to fix rational limits on the increments for tablet dosage, and on permissible tablet-to-tablet and lot-to-lot variability.

- Finally, in drug development or evaluation phase studies, logistical trade-offs of pharmacokinetic-dynamic data may lead to reduced samples per patient (sparse data) and/or reduced patient group sizes, as well as noisy data (e.g., unknown variability in the dose strategy, noncompliance) (phase IV).

The ability to handle, in a statistically rigorous explanatory and predictive framework, large datasets of drug-related pharmacokinetic-dynamic clinical observations is of increasing importance to the industry, regulatory agencies, and patients, in order to reduce human and budgetary risks.

Beyond pharmacokinetics and pharmacodynamics, population modeling and parameter estimation are applications of a statistical model that has general validity, the nonlinear mixed effects model. The model has wide applicability in all areas, in the biomedical science and elsewhere, where a parametric functional relationship between some input and some response is studied and where random variability across individuals is of concern [458].

11

Nonclassical Pharmacodynamics

The whole is more than its parts.

Aristotle (384-322 BC)
Metaphysics

Whereas the concentration of a drug depends on the administration protocol and on intrinsic regulation mechanisms, endogenous substances are certainly controlled. For example, the neurotransmitter norepinephrine is released from sympathetic nerve endings and its concentration is regulated by enzymes and by a mechanism for reuptake of this catecholamine into nerve endings. Deficiencies in the control of such important chemicals may result in vasospasm, spasticity, and a variety of behavioral abnormalities. Such observations strongly suggest the existence of control systems represented by negative feedback mechanisms. By means of those mechanisms, the dynamic system controls the local concentration of critical endogenous chemicals that interact with receptors according to the mass-action law. Indeed, the biomedical literature, particularly that of functional and biochemical pharmacology, is rich with detailed descriptive mechanisms of control and its modification induced by an extensive list of drugs and chemicals. However, mathematical analysis of such control is virtually nonexistent in the pharmacological literature. In contrast, there has been a steady evolution of concepts of control theory and dynamic modeling in many areas of physiology, elegantly traced by Glass and Mackey [31], with an extension by these authors to physiopathological states [48].

As stated in Chapter 10, when the drug–receptor interaction involves feedback, the system becomes more complex. Hence, we will first present modeling and associated mathematical analysis of two typical processes. This will be followed by several examples involving drug pharmacodynamics organized around pharmacotherapy with drugs affecting the endocrine, central nervous, and cardiovascular systems.

11.1 Nonlinear Concepts in Pharmacodynamics

Here, we provide new insights that may aid in understanding the variety of oscillations displayed in biological systems and how they may be related to the maintenance or loss of control in such systems. Examples of periodic phenomena abound in biological systems, in many cases due to fluctuations of ligand interacting with a receptor.

Homeostatic regulation will be studied as represented by negative feedback mechanisms. First, it will be shown how the properties of negative feedback are related to the geometric properties of the binding and control curves in a ligand–receptor interaction and, further, how changes in their geometry affect the system's response to variations of the ligand release. Second, in the analysis of the hemopoietic chain, the negative feedback is supplemented by a lag time that leads to bifurcations and oscillatory, chaotic behavior.

For both analyses, the procedure is to use dimensionless parameters in the set of differential equations describing the model, look for the steady state, investigate the linear stability, and determine the conditions for instability. Near the bifurcation values of the parameters, which initiate an oscillatory growing solution, a perturbation analysis provides an estimate for the period of the ensuing limit cycle behavior.

11.1.1 Negative Feedback

The interaction of a drug or an endogenous ligand with a specific receptor is most often modeled as a bimolecular reversible reaction

$$[\text{ligand}] + [\text{receptor}] \underset{k_{-1}}{\overset{k_{+1}}{\rightleftarrows}} [\text{ligand-receptor complex}]\,,$$

to which the mass-action law applies. This is the classic model presented elsewhere (8.8), (10.1). It is the basis of most studies aimed at quantitatively characterizing receptors with specific radioligands as well as in functional studies where the effect is related to receptor occupancy [450].

For a concentration $c(t)$ of the drug or ligand and a total receptor concentration r_0, we thus have

$$\dot{\upsilon}(t) = k_{+1} c(t) [r_0 - \upsilon(t)] - k_{-1} \upsilon(t)\,, \qquad \upsilon(0) = 0, \tag{11.1}$$

where $\upsilon(t)$ is the concentration of the ligand–receptor complex, and k_{+1} and k_{-1} are the forward and reverse rate constants, respectively, of the reaction. The main features for the ligand model are that ligand is continuously released at rate $r(t)$, and eliminated exponentially with a rate constant k, and that there exists a negative feedback control function $\Phi(\upsilon)$ that depends on the concentration of occupied receptor $\upsilon(t)$ that modulates the release; thus

$$\dot{c}(t) = -kc(t) + \Phi(\upsilon) + r(t)\,, \qquad c(0) = 0, \tag{11.2}$$

with $\Phi(\upsilon) \geq 0$ and $\mathrm{d}\Phi(\upsilon)/\mathrm{d}\upsilon \triangleq \Phi'(\upsilon) \leq 0$. The model is based on evidence, obtained largely from studies of the release of neurotransmitters, that the quantity of ligand released per unit time is modulated by the nerve terminal itself as a result of stimulation by the neurotransmitter of a subset of the receptors termed "autoreceptors" [450]. Thus, receptor stimulation not only produces effects but also inhibits or augments release, thereby maintaining a basal level of the ligand. The feedback signal may originate at a site other than the occupied receptor; however, it is functionally related to $\upsilon(t)$.

We make the variables of the above equations dimensionless:

$$\begin{aligned} \lambda &= \frac{k_{+1}}{k}, \qquad \mu = \frac{k_{-1}}{k}, \qquad \kappa = \frac{k_{-1}}{k_{+1}}, \\ y(\tau) &= \frac{\upsilon(\tau)}{r_0}, \qquad \phi(y) = \frac{\Phi(y)}{k}, \qquad \rho(\tau) = \frac{r(\tau)}{k}, \end{aligned}$$

with $\tau = kt$. The set of differential equations becomes

$$\begin{aligned} \dot{y}(\tau) &= \lambda c(\tau)[1 - y(\tau)] - \mu y(\tau), \qquad y(\tau) = 0, \\ \dot{c}(\tau) &= -c(\tau) + \phi(y) + \rho(\tau), \qquad c(\tau) = 0. \end{aligned} \tag{11.3}$$

Equilibrium points (y^*, c^*) of the system are those for which $\dot{y}(\tau) = \dot{c}(\tau) = 0$.

- Equating $\dot{y}(\tau)$ to zero, we have the *binding curve*. The binding curve and its slope are given by

$$c_B = \kappa y/(1-y) \quad \text{and} \quad \mathrm{d}c_B/\mathrm{d}y = \kappa/(1-y)^2 \tag{11.4}$$

  respectively.

- Equating $\dot{c}(\tau)$ to zero, we have the *feedback curve*. The feedback curve and its slope are given by

$$c_F = \phi(y) + \rho(\tau) \quad \text{and} \quad \mathrm{d}c_F/\mathrm{d}y = \phi'(y) \tag{11.5}$$

  respectively.

Equilibrium points of the system are the intersections of the binding curve with the feedback curve, i.e., $c^* = c_B = c_F$. Their location in the state space depends on κ and $\rho(\tau)$ (equations 11.4 and 11.5).

The stability of equilibrium points is determined by standard stability analysis (cf. Appendix A). The Jacobian matrix of the linearized system,

$$\mathbf{A}(y) = \begin{bmatrix} -(\lambda c + \mu) & \lambda(1-y) \\ \phi'(y) & -1 \end{bmatrix},$$

supplies the eigenvalues ζ_1 and ζ_2. Given (11.4) and (11.5), these eigenvalues are

$$\zeta_{1,2} = \frac{1}{2}\left[-(\nu+1) \pm \sqrt{(\nu+1)^2 + 4\lambda(1-y)\left(\phi'(y) - \frac{\mathrm{d}c_B}{\mathrm{d}y}\right)}\right], \tag{11.6}$$

where

$$\nu = \lambda c + \mu = \frac{\mu}{1-y}.$$

Since $y < 1$ at any equilibrium point, it follows that a negative feedback $\phi'(y) \leq 0$ ensures that the second term under the radical is negative, so that the eigenvalues are real and negative or complex with a negative real part; hence such an equilibrium point is stable. For a large negative value of $\phi'(y)$ the eigenvalues are complex and the point is a stable focus. A shallow negative slope gives two real negative eigenvalues and thus a stable node. In the previous equation it is seen that the eigenvalues do not depend on the normalized ligand input rate $\rho(\tau)$.

In simple negative feedback, $\phi(y)$ is a monotone decreasing function and the equilibrium point is unique. However, due to a variety of factors, it is expected that at very low ligand–receptor numbers, $\phi(y)$ becomes an increasing function, implying a positive feedback. The feedback becomes a mixture of positive and negative feedbacks, called *mixed feedback*, and it has been reported elsewhere [472]. Positive slopes in $\phi(y)$ generate other equilibrium points [453, 473, 474]. The characterization of the eigenvalues in these new equilibrium points, and hence the stability of the system, follows from the application of the following theorem (cf. Appendix H): "The derivatives of two successive intersection points between two continuous functions, one of which is monotone, have opposite signs."

Application of this theorem permits analysis of the equilibrium points of the system with a monotone binding curve. If in the equilibrium point $P_1 = (y_1^*, c_1^*)$ we have $\phi'(y_1^*) < 0$, the system is stable. If the feedback curve is assumed to be continuous over a domain of permissible values of receptor occupancy $y(\tau)$, in the nearest equilibrium point $P_2 = (y_2^*, c_2^*)$, we will have $\phi'(y_2^*) > 0$. This condition is necessary but not sufficient for the instability of the system. But if moreover $\phi'(y_2^*) > \mathrm{d}c_B^*/\mathrm{d}y^*$, one eigenvalue from (11.6) is positive and the system becomes unstable at P_2, which is an unstable saddle point or repellor.

Example 11 *Stability with Feedback*

We use a feedback curve similar to the Weibull distribution with both scale and shape parameters equal to 5; for the binding curve we set $\kappa = 20$ and $\rho(\tau) = 0.2$. In the state space, Figure 11.1 illustrates the position of the three equilibrium points, $P_1 = (0.2304, 5.9878)$, $P_2 = (0.1119, 2.5204)$, and $P_3 = (0.0099, 0.2002)$ (Figure 11.1 A). The graphic slope analysis determines $y_A = 0.2947$, $y_B = 0.1901$, and $y_C = 0.0182$ (Figure 11.1 B), which are threshold values for $\phi'(y)$ and $\mathrm{d}c_B/\mathrm{d}y$ comparison. Thus, according to (11.6), we note that for:

- $y^* < y_C$, the equilibrium is stable since $\phi'(y) < \mathrm{d}c_B/\mathrm{d}y$; since $y_3^* < y_C$, P_3 is stable;
- $y_C \leq y^* < y_B$, the equilibrium is unstable since $\phi'(y) > \mathrm{d}c_B/\mathrm{d}y$; since $y_C < y_2^* < y_B$, P_2 is unstable;

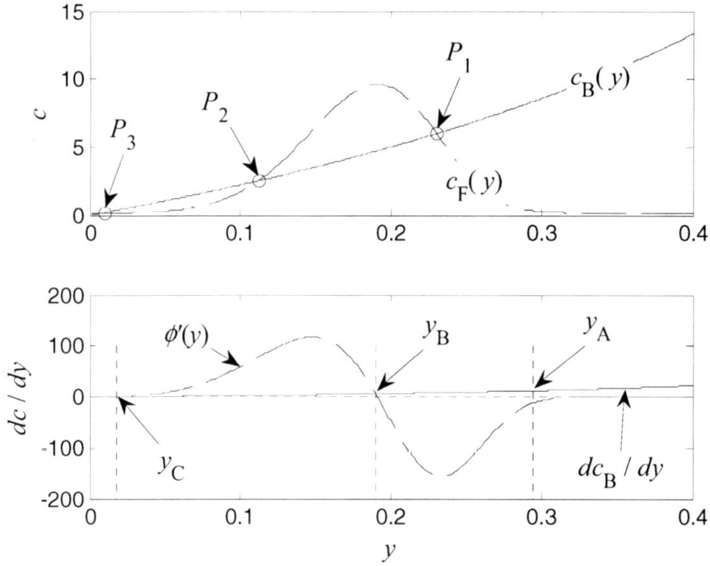

Figure 11.1: Upper panel: Binding curve (*solid line*) and the intersections P_1, P_2, and P_3, with the feedback curve (*dashed line*) to give the equilibrium points. Lower panel: the slope graphical analysis determines y_A, y_B, and y_C, which are intersections of slopes $\phi'(y)$ (*dashed line*) and dc_B/dy (*solid line*) helping us to analyze the stability graphically.

- $y_B \leq y^* < y_A$, the equilibrium is a stable focus since $|\phi'(y)| \gg \mathrm{d}c_B/\mathrm{d}y$; since $y_B \leq y_1^* < y_A$, P_1 is a stable focus; and
- $y_A \leq y^*$, the equilibrium is stable since $|\phi'(y)| < \mathrm{d}c_B/\mathrm{d}y$.

These results are supported by the standard stability analysis of Figure 11.2, where λ is set to 0.1 and $\mu = 2$ $(\mu = \kappa\lambda)$. The eigenvalues computed by (11.6) are plotted as functions of y. In this figure, unstable and stable equilibrium points are clearly separated by an interval, $[0.1974; 0.2790]$, where eigenvalues are complex, leading to a stable focus. With increasing λ, this interval becomes narrower and for $\lambda > 0.65$, the eigenvalues have only real parts.

Finally, Figure 11.3 illustrates the dynamics of the model described by (11.3) when a different initialization is used. The unstable P_2 point is actually a repellor, P_1 is a stable focus, and P_3 is stable. Depending on the magnitude of the release of ligand from intracellular storage sites or exogenous administration, and the location of the unstable point, the state vector will either return to the operating point or be propelled to extreme states. Such propulsion means loss of control in the effector system governed by occupation of the receptor. The

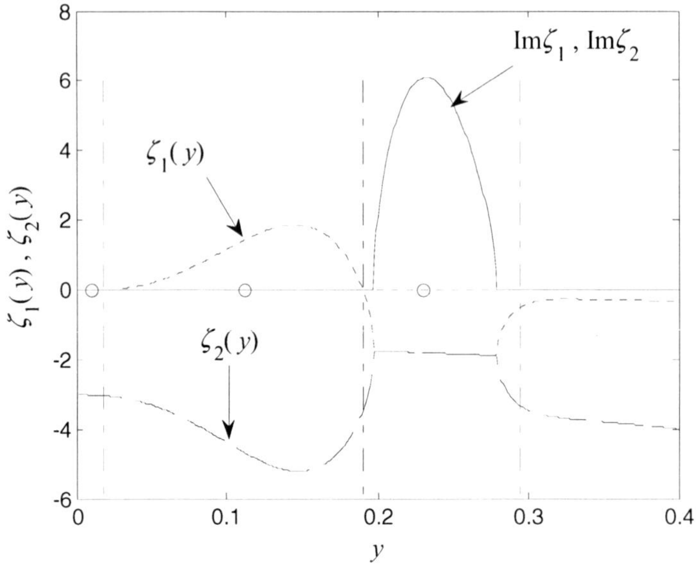

Figure 11.2: Eigenvalues computed via (11.6). (◯) indicate the positions of equilibrium points P_1, P_2, and P_3, and vertical *dash-dotted lines*, the positions of y_A, y_B, and y_C. The *solid line* represents the imaginary parts, and *dashed* and *dotted lines* represent the real parts of eigenvalues.

distance between the stable and the unstable points is crucial in this regard. Equally important are the eigenvalues at the control point. When these are complex, the recovery follows a path that is closer to the unstable point.

It is also interesting to comment on the influence of the parameters in the system behavior:

1. The elimination rate constant k is a time-scale parameter and it does not affect the present stability analysis.

2. The forward and reverse rate constants, k_{+1}, and k_{-1} respectively, govern the affinity. A change in affinity, with the same receptor density, also affects stability. A decrease in affinity (decrease in λ or increase in κ expressing the dissociation constant) results in an increase in distance between P_1, P_2, and P_3. The parameter κ affects exclusively the binding curve.

3. The steep negative slope $\phi'(y) \leq 0$ results in complex eigenvalues. The frequency of the oscillation increases with the steepness. The operating point in such cases is a stable focus. In contrast, shallow negative slopes

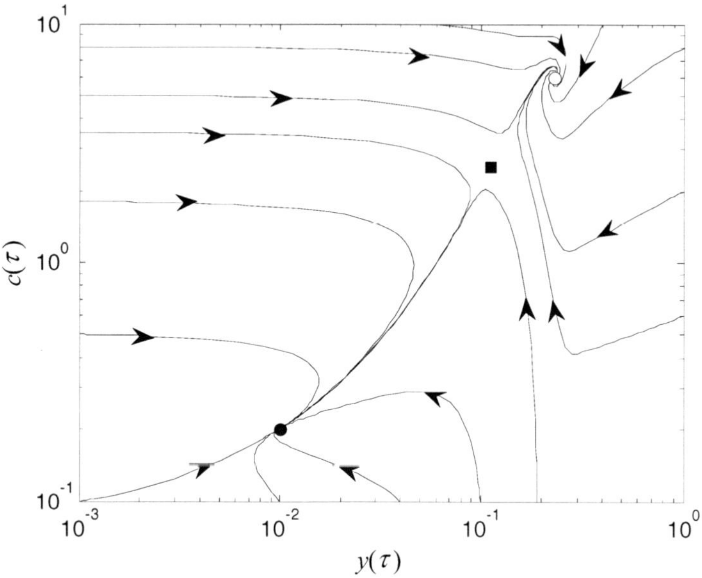

Figure 11.3: State space for different initial conditions. The equilibrium point P_1 (○) is a stable focus, P_2 (■) is an unstable saddle point, and P_3 (•) is stable.

are indicative of a nonoscillating operating point or stable node. Since eigenvalues do not depend on $\rho(\tau)$, figures like Figure 11.2 are useful for following the positions of equilibrium points in the state space when $\rho(\tau)$ varies.

4. When fixed, the number of receptors r_0 is a scale parameter and it does not affect the stability. When r_0 is not fixed but changes as a result of pharmacological interventions or pathological states, the operating point will, of course, change.

When $r(t)$ and r_0 are varying and the other parameters are fixed, simulations (not presented here) with (11.1) and (11.2) reveal that a decrease in $r(t)$ results in a decrease in distance between P_2 and P_3. Conversely, an increase in r_0 results in a decrease in distance between P_1 and P_2. ■

The above-mentioned theorem allows speculation about a monotone feedback curve and a nonlinear binding curve. Their intersections will have derivatives with alternate signs, and therefore, they lead to stable and unstable equilibrium points. In this sense, Tallarida [474] used a U-shaped feedback curve to analyze experiments involving neurotransmitter norepinephrine systems [475].

Analysis on the state space proved to be very useful and demonstrated how possible changes in receptor affinity or receptor number affect the distance between the operating point and the unstable equilibrium point, and thus the ability of the system to return to the operating point after a perturbation such as endogenous release. The new information reported here pertains to the geometry on the state space, which allows us to predict both the stability of equilibrium points and the characteristic frequency from the slopes of both curves at their intersection. The relationship between the slope and the frequency of the system is especially important in the further development of models for particular receptor systems, since examples of rhythmic phenomena abound in biological systems.

As we proceed it will be seen that the most important results do not depend on a particular assumption regarding the form of the feedback function. Thus far we have not located the equilibrium points for the system under study because the function $\Phi(v)$ was kept general. The model we have used is applicable to both endogenous and exogenous substances.

In a series of contributions, Tallarida also studied the control of an endogenous ligand in the presence of a second compound (agonist or antagonist) that interacts with the same receptor [475] or under periodic release of the ligand [476]. That author showed by computer simulation how the parameters of the model affect the time course of released ligand resulting from administration of an antagonist and the suppression of such release when the second compound is an agonist.

A new quantitative concept that describes the feedback control of the dopaminergic system was also introduced, the control curve. Once known, the ligand's control curve has predictive value that may be useful in the design of efficient drug tests. These theoretical results were confirmed experimentally on numerous cases as for neurotransmitters, hormones, peptides, etc., whose concentrations in the various organs and tissues remain bounded. For example, the control of dopamine release by negative feedback was confirmed in the rat striatum [477]. A consequence of this model is that competitive antagonists augment dopamine release, whereas competing agonists reduce such release.

These findings may be of general importance since baseline parameters are crucial in determining pharmacodynamic responses [478], while feedback mechanisms are frequently involved in physiological processes, e.g., the secretion of hormones and the recurrent inhibitory pathway for γ-aminobutyric acid (GABA) in the hippocampus, which has been described in almost every type of neural tissue ranging from the lowest invertebrates through humans [479], and the production of biotech products in humans [480].

11.1.2 Delayed Negative Feedback

An interesting case is that of a delay mechanism inserted in a closed loop process with negative feedback. The typical process is the hemopoietic process that incorporates some control elements that regulate homeostatically the rates of release of marrow cells to proliferation, maturation, and to the blood.

The dynamic response of the process of neutrophil granulocyte production to perturbations has been studied in a number of ways. In leukopheresis, neutrophils are removed from the blood artificially over a short span of time. Following such an acute depletion of the neutrophil blood count, referred to as a state of *neutropenia*, neutrophils rapidly enter the blood from the marrow and produce an abnormally large number of neutrophils in the blood, or a state of *neutrophilia* [481]. The magnitude of the neutrophil blood count seen in such a state is about 2–3 times normal. Such observations suggested the presence of some mechanisms for regulating the release of marrow neutrophils in response to the number of neutrophils circulating in the blood [482].

The most notable feature of the dynamic response of the process to large perturbations in the number of blood cells is that the system "rings," displaying an oscillatory behavior in the number of cells in the blood and other compartments of the system, as a function of time. Such large perturbations are produced by leukopheresis or exposure of the system to disease, which depletes the number of blood cells, and in total body irradiation experiments or some drug treatments as chemotherapy, which deplete the total number of cells in the production process of neutrophil granulocytes.

Besides these perturbations on the hemopoietic process, there are some diseases, collectively referred to as the *periodic diseases*, in which symptoms recur on a regular basis of days to months. The most common of these disorders are cyclic neutropenia (also known as periodic hemopoiesis) [483], and cyclic thrombocytopenia [484]. It has long been suspected that periodic hematologic diseases arise because of abnormalities in the feedback mechanisms that regulate blood cell number [485–487]. But in a dynamic feedback process such as hemopoiesis it is difficult to distinguish between cause and effect. Oscillations occurring in one cell stage may induce cycling in other stages via feedback regulation. The mechanisms regulating neutrophil production are not as well understood. The important role of the cytokine granylocyte colony-stimulating factor (G-CSF) for the in vivo control of neutrophil production was demonstrated by Lieschke [488, 489]. Several studies have shown an inverse relation between circulating neutrophil density and serum levels of G-CSF [490]. Coupled with the in vivo dependency of neutrophil production on G-CSF, this inverse relationship suggests that the neutrophils would regulate their own production through negative feedback, in which an increase (decrease) in the number of circulating neutrophils would induce a decrease (increase) in the production of neutrophils through the adjustment of G-CSF levels. G-CSF has synergetic effects on the entry into cycling of dormant hemopoietic stem cells.

These observations have provided impetus for mathematicians to determine the conditions for the observed oscillations. Thus far, there have been two surprising discoveries [47, 48]:

- qualitative changes can occur in blood cell dynamics as quantitative changes are made in feedback control; and
- under appropriate conditions, these feedback mechanisms can produce aperiodic, irregular fluctuations, which could easily be mistaken for noise and/or experimental error [31, 491, 492].

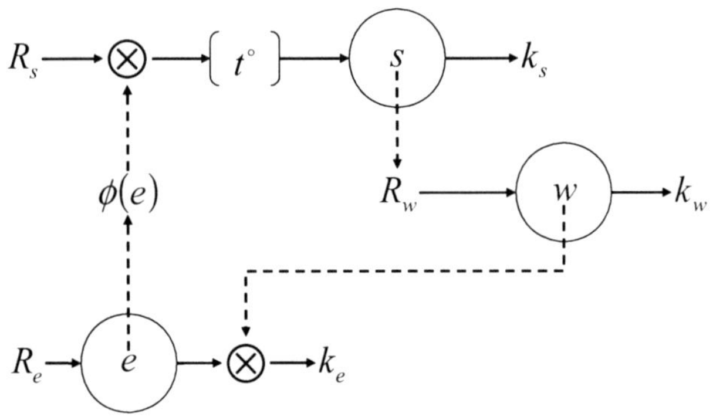

Figure 11.4: The organization of normal hemopoiesis. Symbols are defined in the text.

In the following, we will examine some theoretical developments and discuss their implications in a pharmacodynamic context. A simple, physiologically realistic mathematical model of neutrophil lineage is first proposed, including homeostatic regulation by means of cytokine G-CSF. Next, investigation of the properties of the model by stability analysis shows that this variety of clinical outcome can be described mainly from the dynamics of neutrophil counts governed by feedbacks. The sharpness of the feedback signals is essentially determined by the stability of the oscillatory behavior.

Modeling of Neutrophil Regulation

The organization of normal hemopoiesis is shown in Figure 11.4. It is generally believed that there exists a self-maintaining pluripotent stem cell population capable of producing committed stem cells specialized for the erythroid, myeloid, or thromboid cell lines [493]. The lineage studied here is the myeloid, ending with the neutrophils in the bloodstream. Three differential equations describe the mechanisms implied in this hemopoietic scheme:

1. The influx R_s of cells from the pluripotent stem cell population to the committed stem cell lines is assumed mainly regulated by long-range humoral mechanisms $\phi(e)$, implicating the cytokine G-CSF, $e(t)$. An intrinsic property of the hemopoietic chain is the presence of a time delay t° that arises because of finite cell maturation times and cell replication times for the neutrophil myelocytes, $s(t)$. In fact, it is important to remember that once a cell from the pluripotent stem cell population is committed to the neutrophil lineage, it undergoes a series of nuclear divisions and enters a maturational phase for a period of time ($t^\circ \approx 5$–$7\,\mathrm{d}$) before release into circulation. The production function R_s has not only to be amplified, but

also to be delayed as described by the feedback component $\phi(e(t-t^\circ))$, because a change in the blood neutrophil numbers can only augment or decrease the influx into the circulation after a period of time t° has elapsed. Thus, changes that occur at time t were actually initiated at a time $t-t^\circ$ in the past. For the sake of simplicity, we will use the notation $e^\circ(t)$ instead of $e(t-t^\circ)$. We can describe these dynamics by

$$\dot{s}(t) = R_s\phi(e^\circ(t)) - k_s s(t),$$

where k_s is the loss rate for $s(t)$. The previous equation is a differential-delay equation. In contrast to ordinary differential equations, for which we need only to specify an initial condition as a particular point in the state space for a given time, for differential-delay equations we have to specify an initial condition in the form of a function, usually called the *history function*, $\psi(t)$ and defined for a period of time equal to the duration of the time delay. Thus, we will select

$$s(t) = \psi(t), \qquad -t^\circ \leq t \leq 0.$$

We will consider only initial functions that are constant, i.e., $s(t) = s_0$.

2. Mature neutrophil myelocytes $s(t)$ are now controlling the input rate R_w of neutrophils $w(t)$ that disappear from the blood with a rate constant k_w. The input function R_w in its simplest form can be considered as an amplification of $s(t)$, expressing the proliferation of neutrophil myelocytes, i.e., $R_w = \alpha s(t)$:

$$\dot{w}(t) = \alpha s(t) - k_w w(t), \qquad w(0) = w_0.$$

3. At the physiological equilibrium state, cytokine G-CSF, $e(t)$, is delivered at the rate R_e and cleared by mechanisms characterized by rate constant k_e. A fall in circulating neutrophil numbers $w(t)$ leads to an acceleration of the G-CSF clearance, which has as consequence a decrease in $e(t)$ levels. This decrease in turn triggers the production of committed stem cells, which increases cellular efflux of neutrophil precursors, and ultimately augments $w(t)$ (i.e., negative feedback). This regulated behavior can be implemented by means of the $e(t)$ clearance depending on $w(t)$ levels and the $\phi(e^\circ(t))$ function [494]. The differential equation for $e(t)$ is expressed by

$$\dot{e}(t) = R_e - k_e w(t)\, e(t), \qquad e(0) = e_0.$$

4. Of primary importance is the form of feedback mechanism implying the previous differential equation, where the $w(t)$ level controls the $e(t)$ clearance, and the function $\phi(e^\circ(t))$ that modulates the committed stem cell production. More specifically

$$\phi(e^\circ(t)) = \vartheta\left\{1 - \exp\left[\log\left(\frac{\vartheta-1}{\vartheta}\right)\left(\frac{e^\circ(t)}{e_0}\right)^{\theta}\right]\right\},$$

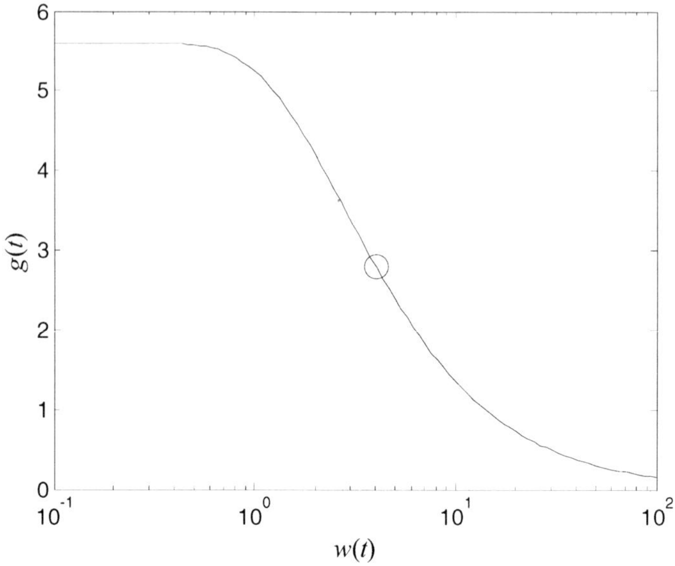

Figure 11.5: Homeostatic control for regulation of neutrophil production. Parameters are $w_0 = 4$ and $e_0 = 1$ cells$\times 10^6$ ml^{-1}, $\vartheta = 2$ and $\theta = 1$, and $k_w = 0.7$ d^{-1}. $\circ$ indicates the position of the equilibrium point.

a monotone increasing Weibull-like function with $\phi(e_0) = 1$. In the present model, parameters ϑ and θ express the amplification and the sharpness of the G-CSF effect.

In the above equations, s_0, w_0, and e_0 denote history and initial conditions corresponding to the undisturbed state of the process at equilibrium.

Thus, we propose to study a model with delayed feedback, as done by several investigators [47, 48, 485–487, 495–499].

Normally, the equilibrium behavior of the process requires that the production rate equal the disappearance rate. These conditions and the introduction of the variable transformation $g(t) = \alpha s(t)$ allows us to determine the input rates R_s and R_e:

$$\begin{aligned}
\dot{g}(t) &= k_s k_w w_0 \phi\left(e^\circ(t)\right) - k_s g(t), & g\left(-t^\circ \leq t \leq 0\right) &= k_w w_0, \\
\dot{w}(t) &= g(t) - k_w w(t), & w(0) &= w_0, \\
\dot{e}(t) &= k_e \left[w_0 e_0 - w(t)\, e(t)\right], & e(0) &= e_0.
\end{aligned} \tag{11.7}$$

Figure 11.5 illustrates the function regulating neutrophil production depending on the circulating neutrophil numbers.

We make the variables of the above equations dimensionless:

$$g(\tau) = k_w w_0 x(\tau), \qquad w(\tau) = w_0 y(\tau), \qquad e(\tau) = e_0 z(\tau),$$

with $\tau = k_e w_0 t$, and we set

$$\tau^\circ = k_e w_0 t^\circ, \qquad \gamma = k_s / (k_e w_0), \qquad \lambda = k_w / (k_e w_0).$$

The set of differential equations becomes

$$\begin{aligned} \dot{x}(\tau) &= \gamma [\phi(z^\circ(\tau)) - x(\tau)], & x(-\tau^\circ \leq \tau \leq 0) &= 1, \\ \dot{y}(\tau) &= \lambda [x(\tau) - y(\tau)], & y(0) &= 1, \\ \dot{z}(\tau) &= 1 - y(\tau) z(\tau), & z(0) &= 1, \end{aligned} \tag{11.8}$$

with $z^\circ(\tau) = z(\tau - \tau^\circ)$ and

$$\phi(z^\circ(\tau)) = \vartheta \left\{ 1 - \exp \left[\log \left(\frac{\vartheta - 1}{\vartheta} \right) (z^\circ(\tau))^\theta \right] \right\}. \tag{11.9}$$

Current analytic and numerical work determine the time-dependent changes in blood cell number as certain quantities, referred to as control parameters, are varied. Examples of control parameters in the regulation of hemopoiesis are the dimensionless maturation time τ° and the peripheral destruction rates γ and λ.

It is well established that under appropriate circumstances, delayed negative feedback mechanisms can produce oscillations. To illustrate this point we continue with the stability analysis.

Stability Analysis

Equilibrium points (x^*, y^*, z^*) of the system are those for which $\dot{x}(\tau) = \dot{y}(\tau) = \dot{z}(\tau) = 0$. As previously defined, the equilibrium state of the process leads to the single equilibrium point $(x^*, y^*, z^*) = (1, 1, 1)$. Since $z(\tau)$ is not changing with time, we have also $(z^\circ)^* = z^* = 1$. We would now like to know what conditions on the parameters of our model are required to warrant stability, and even further, what happens in the case of instability.

Because the model (11.8) that describes this physiological process is nonlinear, we cannot answer these questions in total generality. Rather, we must be content with understanding what happens when we make a small perturbation on the states x, y, and z away from the equilibrium. The fact that we are assuming that the perturbation is small allows us to carry out what is known as linear stability analysis of the equilibrium state.

The nonlinearity of (11.8) comes from the terms $\phi(z^\circ(\tau))$ and $y(\tau) z(\tau)$ involved in the nonlinear negative feedback regulation. What we want to do is replace these nonlinear terms by a linear function in the vicinity of the equilibrium state (x^*, y^*, z^*). This involves writing the Jacobian matrix of the linearized system (cf. Appendix A):

$$\mathbf{A} = \begin{bmatrix} -\gamma & 0 & \gamma \phi'(1) (\mathrm{d}z^\circ / \mathrm{d}z) \\ \lambda & -\lambda & 0 \\ 0 & -1 & -1 \end{bmatrix}.$$

To analyze stability of the linearized model, we have to examine the eigenvalues that are solutions of the characteristic equation of $\mathbf{A}$. Usually the eigenvalue is a complex number $\zeta = \mu + \mathrm{i}\omega$. If $\mu = \mathrm{Re}\,\zeta < 0$, then the solution is a decaying oscillating function of time, so we have a stable situation. If $\mu = \mathrm{Re}\,\zeta > 0$ on the other hand, then the solution diverges in an oscillatory fashion and the solution is unstable. The boundary between these two situations, where $\mu = \mathrm{Re}\,\zeta = 0$, defines a Hopf bifurcation in which an eigenvalue crosses from the left-hand to the right-hand complex plane.

The usual procedure to obtain solutions of the characteristic equation of $\mathbf{A}$ is to assume that the solution of $z(\tau)$ has the form

$$z(\tau) \propto \exp(\zeta\tau)$$

and find out the requirements on the parameters of the equation so that there is an eigenvalue ζ allowing $z(\tau)$ to be written in this form. Under this assumption,

$$\mathrm{d}z^\circ/\mathrm{d}z = \exp(-\zeta\tau^\circ),$$

and the eigenvalues are given as solutions of the characteristic equation

$$(\zeta+\gamma)(\zeta+\lambda)(\zeta+1) + \gamma\lambda\phi'(1)\exp(-\zeta\tau^\circ) = 0. \tag{11.10}$$

In contrast to systems without delay, the previous equation has generally an infinite number of roots. Nevertheless, there are only a finite number of roots with real parts [500]. Figure 11.6 illustrates solutions of (11.10) with $\gamma = 0.025$, $\lambda = 0.175$, $\phi'(1) = 2$, and $\tau^\circ = 0, 10, 100$. We note that:

- all roots are complex conjugates, except for $\tau^\circ = 0$, where one root is real,
- only for $\tau^\circ = 100$ do we have one pair of roots with positive real part, $\mu = \mathrm{Re}\,\zeta \approx 0.002947$,
- in the complex plane, the roots' density near the origin is higher for $\tau^\circ = 100$ than the density for the other τ° values.

Let us find the critical delay value $\tau^\bullet$ at which the characteristic roots intersect the stability boundary, i.e., the imaginary axis $\mu = 0$, thus rendering the system unstable with $\zeta = \mathrm{i}\omega^\bullet$. We substitute this into the previous equation, and after separating real and imaginary parts, we have

$$\begin{aligned} \gamma\lambda\left[1+\phi'(1)\cos\omega^\bullet\tau^\bullet\right] &= (\gamma+\lambda+1)\,\omega^{\bullet 2}, \\ \gamma\lambda\phi'(1)\sin\omega^\bullet\tau^\bullet &= (\gamma\lambda+\gamma+\lambda)\,\omega^\bullet - \omega^{\bullet 3}. \end{aligned} \tag{11.11}$$

Note that if $\mathrm{i}\omega^\bullet$ is a characteristic root of (11.10), then $-\mathrm{i}\omega^\bullet$ is also a characteristic root. Then, we can assume that $\omega^\bullet > 0$. Squaring and adding the above two equations defines a polynomial equation in $\omega^{\bullet 6}$ with only even powers. If we set $\chi = \omega^{\bullet 2} > 0$, this equation becomes a third-order polynomial equation in χ:

$$\chi^3 + \beta_2\chi^2 + \beta_1\chi + \beta_0 = 0$$

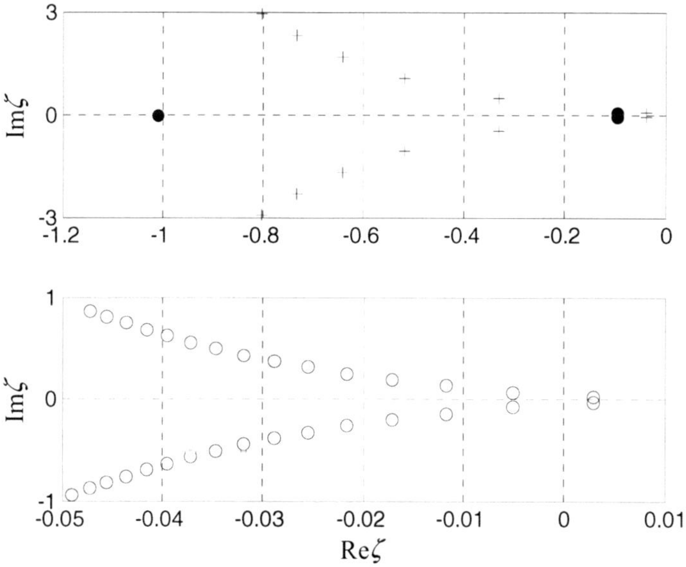

Figure 11.6: Characteristic roots for different τ° values: $\tau^{\circ} = 0$ ($\bullet$), $\tau^{\circ} = 10$ (+), and below for $\tau^{\circ} = 100$ ($\circ$).

with

$$\beta_2 = \gamma^2 + \lambda^2 + \gamma + \lambda + 1,$$
$$\beta_1 = \gamma^2\lambda^2 + \gamma^2 + \lambda^2,$$
$$\beta_0 = (\gamma\lambda)^2 \left[1 - \phi'(1)^2\right].$$

According to the Descartes rule of signs and since β_1 and β_2 are positive, the inequality $\beta_0 < 0$ or $\phi'(1) > 1$ is a necessary condition to have the unique positive solution $\chi > 0$.

After evaluating this solution, we obtain $\omega^{\bullet}$, and then, from one of (11.11), we calculate the critical value of $\tau^{\bullet}$. Figure 11.7 shows the frequency $\omega^{\bullet}$ (upper panel) and the critical value of $\tau^{\bullet}$ (lower panel) as functions of $\phi'(1)$. We can say that the real parts of ζ will be positive, and thus (11.8) will be unstable, if and only if the actual delay τ° is greater than $\tau^{\bullet}$. For example, for $\phi'(1) = 2$ and the set γ and λ values, $\tau^{\bullet} \approx 44.049$ and any $\tau^{\circ} > \tau^{\bullet}$ triggers periodic oscillations following some perturbation in the system. Otherwise, the system is locally stable.

The period T of the periodic solution can be obtained by noting that $\omega = 2\pi/T$. In general, the period of an oscillation produced by a delayed negative feedback mechanism is at least twice the delay [485]. For our model of neutrophil production, the functional relationship between $\tau^{\bullet}$ and T is shown in

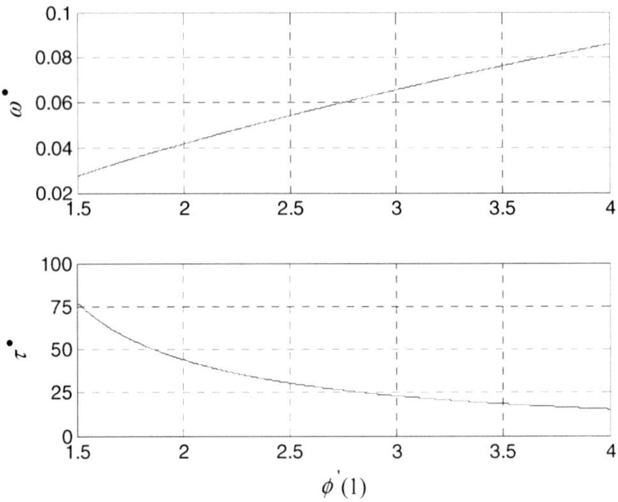

Figure 11.7: Critical frequency $\omega^{\bullet}$ and delay $\tau^{\bullet}$ as functions of $\phi'(1)$.

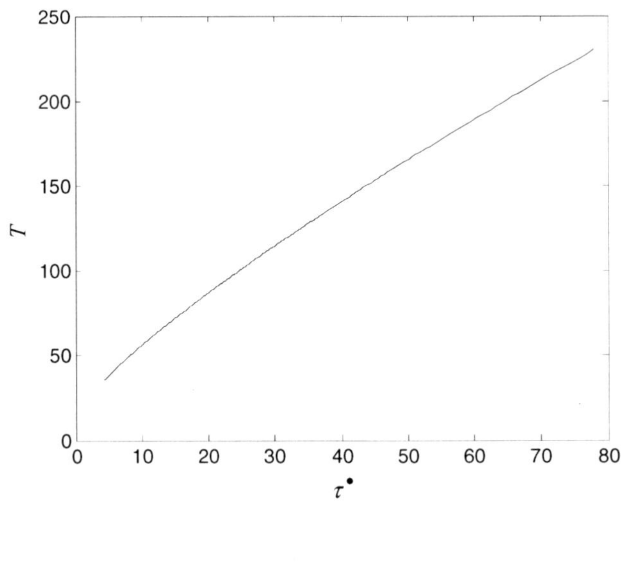

Figure 11.8: The period T of oscillations as a function of $\tau^{\bullet}$.

Figure 11.8. We note that the period of the oscillation should be four times the delay as reported by Mackey [486] and as also concluded by another simpler model without the neutrophil myelocytes $s(t)$ (results not shown). Since the maturational delay for neutrophil production is $\tau \approx 5$–$7\,\mathrm{d}$, we would expect to see oscillations in neutrophil numbers with periods of about 3–4 weeks.

Chemotherapy

Neutropenic episodes place patients at increased risk for infective processes (e.g., abscesses, pneumonia, septicemia), and up to 20% of patients may die during these episodes. For example, in patients undergoing cancer chemotherapy, neutropenia is frequently a dose-limiting side effect. The importance of the time profile of hematologic effects in analyzing properties of anticancer agents has been recently recognized [501, 502]. Different models of the entire time course of responses have been proposed. They can be classified as either mechanistic or empirical. The latter models postulate explicit relationships between the effect and pharmacodynamic and pharmacokinetic parameters [430, 501], whereas the mechanistic models describe the biological processes controlling the change of the affected cells [502, 503].

Here, a new model of hematologic toxicity of anticancer agents is introduced. The postulated mechanisms that influence the response variable (e.g., neutrophil count) are:

- an indirect mechanism, where by means of a logistic function, the cell production rate of neutrophil myelocytes $s(t)$ is modulated by the blood concentrations of the anticancer drug, and
- direct toxicity of the anticancer drug levels, according to which the killing rate of neutrophil myelocytes is proportional to $s(t) \times$ [drug levels] [504].

This model was identified from data gathered in a clinical study [505] aiming to define a regular and tolerable dose of the epirubicin-docetaxel combination in first-line chemotherapy on 65 patients with metastatic breast cancer. Following the analysis of these data, parameters were set to

$$\begin{array}{lll} k_w = 0.7\,\mathrm{d}^{-1}, & k_e = 1\,\mathrm{d}^{-1}, & k_s = 0.1\,\mathrm{d}^{-1}, \\ w_0 = 4000 \text{ cells} \times \mathrm{mm}^{-3}, & e_0 = 1000 \text{ cells} \times \mathrm{mm}^{-3}, & \\ \vartheta = 2, & \theta = 2.8854, & \end{array}$$

leading to the dimensionless parameters

$$\gamma = 0.025, \quad \lambda = 0.175, \quad \phi'(1) = 2.$$

Two different delays were also assessed, $t^\circ = 5$ and $15\,\mathrm{d}$, corresponding to $\tau^\circ = 20$ and 60, respectively. The simulation of the neutrophil count is presented in the Figure 11.9. We observe that:

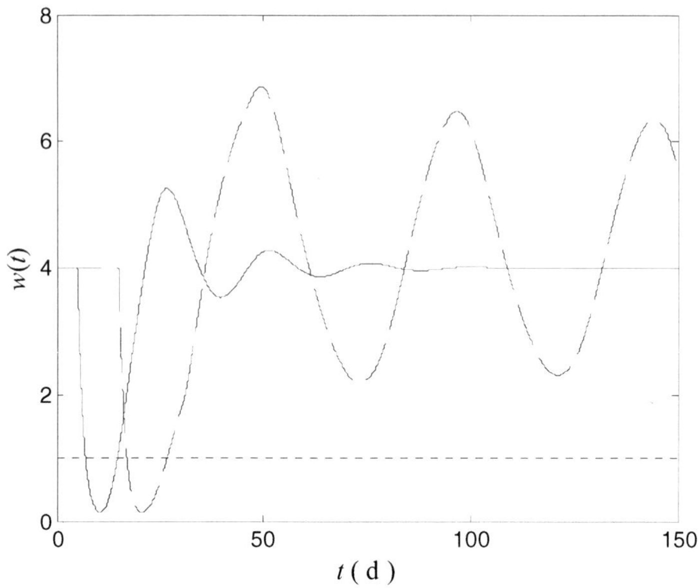

Figure 11.9: Simulation of the neutrophil count kinetics for $t^{\circ} = 5$ (*solid line*) and 15 d (*dashed line*). The *dotted line* indicates the minimum allowed neutrophil level.

- while in the first month following initial delay the two kinetic patterns look the same, their behavior has been differentiated for the subsequent time leading either to oscillatory or dampening behavior corresponding to $t^{\circ} = 15$ and $t^{\circ} = 5$ d delays, respectively, and

- the periods of oscillations in both cases are 4-fold higher than the initial delay.

The clinical significance of the previous analysis is that it may be possible to develop new therapeutic strategies with agents shortening the period of chemotherapy-associated neutropenia such as lenograstim [506]. Such agents may reduce incidence or duration of serious infections and enable greater dose-intensification. In the long run, quantitative modeling may support the design of chemotherapy or growth factor drug regimens based on manipulation of feedback [31, 47–49]. Alternatively, the model can be used to identify more specifically the effects of drugs in the hemopoietic system.

Mixed Feedback

Recently, Bernard et al. [499] studied oscillations in cyclical neutropenia, a rare disorder characterized by oscillatory production of blood cells. As above, they developed a physiologically realistic model including a second homeostatic control on the production of the committed stem cells that undergo apoptosis at their proliferative phase. By using the same approach, they found a local supercritical Hopf bifurcation and a saddle-node bifurcation of limit cycles as critical parameters (i.e., the amplification parameter) are varied. Numerical simulations are consistent with experimental data and they indicate that regulated apoptosis may be a powerful control mechanism for the production of blood cells. The loss of control over apoptosis can have significant negative effects on the dynamic properties of hemopoiesis.

In the previous analysis, delayed negative feedback mechanisms were considered only for neutrophil regulation. However, if over a wide range of circulating neutrophil levels, the neutrophil production rate decreases as the number of neutrophils increases (i.e., negative feedback), in the range of low neutrophil numbers the production rate must increase as neutrophil number increases (i.e., positive feedback). This type of feedback was reported as *mixed feedback* [472].

In order to contrast the dynamics that arise in delayed negative and mixed feedback mechanisms, Mackey et al. [507] considered periodic chronic myelogenous leukemia in which peripheral neutrophil numbers oscillate around elevated levels with a period of 30–70 d even in the absence of clinical interventions [508, 509]. On closer inspection it can be seen that the number of days between successive maximum numbers of neutrophils is not constant, but varies by a few days. Moreover, the morphology of each waveform differs slightly and there are shoulders on some of them. Mackey et al. [47, 48] have explored the possibility that these irregularities are intrinsic properties of the underlying control mechanism. These studies indicate that the dynamics of mixed feedback are much richer than for the simple negative feedback model. Increases in t° are of particular interest since a prolongation of the neutrophil maturation time is inferred in patients with chronic myelogenous leukemia [510]. As t° is increased an initially stable equilibrium becomes unstable and stable periodic solutions appear. Further increases in t° lead to a sequence of period-doubling bifurcations, which ultimately culminates in an apparently chaotic or aperiodic regime. Here, the model predicts that levels of circulating neutrophils are random simply as a consequence of their own deterministic evolution.

The observations in these notes emphasize that an intact control mechanism for the regulation of blood cell numbers is capable of producing behaviors ranging from no oscillation to periodic oscillations to more complex irregular fluctuations, i.e., chaos. The type of behavior produced depends on the nature of the feedback, i.e., negative or mixed, and on the value of certain underlying control parameters, e.g., peripheral destruction rates γ and λ or maturation times τ°. Pathological alterations in these parameters can lead to periodic hematologic disorders. The observation that periodic hematologic diseases have periods that are multiples of 7 may simply be a consequence of the combination of delayed feedback mechanisms with maturation times that are on the order of 5–7 d. Thus it is not necessary to search for elusive and mystical entities [511], such as circadian rhythms, to explain the periodicity of these disorders.

The realization that physiological control mechanisms can generate exceedingly complex oscillations, such as chaos, is a subject of great interest [31,47,48, 491,492]. It is quite possible that both interesting and relevant dynamic changes are often observed, but their significance is wrongly ascribed to environmental noise and/or experimental error. Careful attention to these dynamic behaviors may eventually provide important insights into the properties of the underlying control mechanisms.

Periodic and Dynamical Diseases

The first explicit description of the concept of periodic diseases, where the disease process itself may flare or recur on a regular basis of days to months, was provided over 40 years ago by H. Reimann [512]. That author described and catalogued a number of periodic disease states ranging from certain forms of arthritis to some mental illnesses and hereditary diseases such as familial Mediterranean fever. As an extension to the concept of periodic diseases introduced by Reimann and to encompass irregular physiologic dynamics thought possibly to represent deterministic chaos, the term *dynamical disease* has been introduced [31,47–49]. A dynamical disease is defined as a disease that occurs in an otherwise intact physiological control system but operates within a range of control parameters that leads to abnormal dynamics. Clearly the hope is that it may eventually be possible to identify these altered parameters and then readjust them to values associated with healthy behaviors.

11.2 Pharmacodynamic Applications

During the last fifteen years many investigators have expanded traditional pharmacodynamic modeling (law of mass action at equilibrium) to mechanistic pharmacodynamic modeling including detailed modeling of the underlying physiology and then modeling the effect of drugs on it. On the other hand, as just pointed out, deterministic chaos is typically the recorded behavior of complex physiological systems implicating feedback regulations and nonlinear elements. In the next paragraphs, three major fields of physiological systems with great importance in pharmacotherapy, namely cardiovascular, central nervous, and endocrine systems, where tools and concepts from nonlinear dynamics have been applied, will be discussed.

11.2.1 Drugs Affecting Endocrine Function

It is widely appreciated that hormone secretion is characterized by pulsatility. The first experimental studies of the pulsatile nature of hormone secretion started more than thirty years ago. Hellman et al. reported in 1970 [513] that "cortisol is secreted episodically by normal man." It was also realized that this pulsatility was not due to noise, but was actually associated with physiological processes. Indeed, the circadian clock, the interaction between hormones through feedback mechanisms, and the interaction of hormones with central and autonomic nervous systems are some of the reasons for this behavior. It has been apparent that the theory of dynamic systems is the right field in which to find useful tools for the study of hormonal systems. This has been done along two directions:

- experimental studies using tools from time series analysis and
- modeling with differential equations.

A Dynamic System for Cortisol Kinetics

Although the detailed features of the interactions involved in cortisol secretion are still unknown, some observations indicate that the irregular behavior of cortisol levels originates from the underlying dynamics of the hypothalamic–pituitary–adrenal process. Indeed, Ilias et al. [514], using time series analysis, have shown that the reconstructed phase space of cortisol concentrations of healthy individuals has an attractor of fractal dimension $d_f = 2.65 \pm 0.03$. This value indicates that at least three state variables control cortisol secretion [515]. A nonlinear model of cortisol secretion with three state variables that takes into account the simultaneous changes of adrenocorticotropic hormone and corticotropin-releasing hormone has been proposed [516].

The Model These observations prompted us to model cortisol plasma levels [517] relying on the well-established erratic secretion rate [518] and the circadian rhythm, while other factors controlling cortisol secretion are also considered but not expressed explicitly:

- Cortisol concentration is described by a nonlinear time-delay differential equation [47, 519] with two terms, i.e., a secretion rate term that adheres to the negative feedback mechanism [520, 521] and drives the pulsatile secretion, and a first-order output term with rate constant k_o:

$$\dot{c}(t) = k_i \frac{g^{\gamma} c^{\circ}(t)}{g^{\gamma} + [c^{\circ}(t)]^{\gamma}} - k_o c(t), \qquad (11.12)$$

  where $c(t)$ is the cortisol concentration, $c^{\circ}(t)$ is the value of $c(t)$ at time $t - t^{\circ}$, γ is an exponent, and k_i and k_o are the input and output rate constants, respectively.

- The circadian rhythm of cortisol secretion is implemented phenomenologically by considering the parameter of the model as a simple cosine function of 24-hour period:

$$g(t) = \alpha \cos\left[(t - \varphi)\frac{2\pi}{1440}\right] + \beta, \qquad (11.13)$$

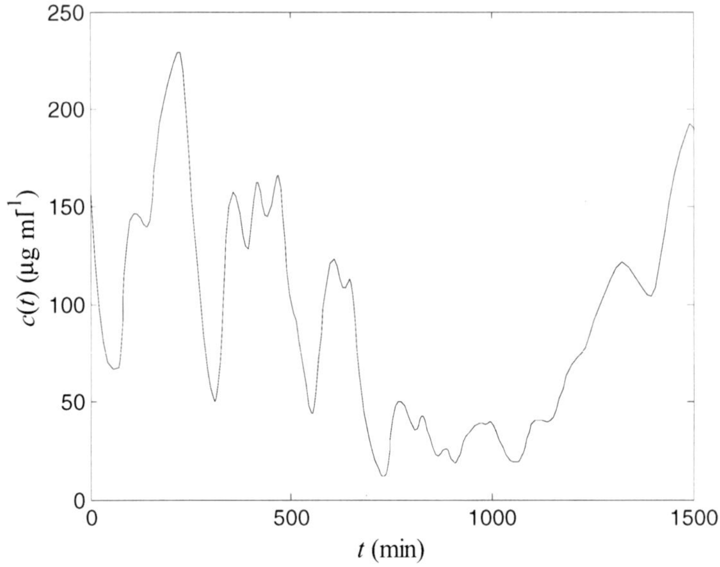

Figure 11.10: A 24-h simulated profile generated by the model of cortisol kinetics.

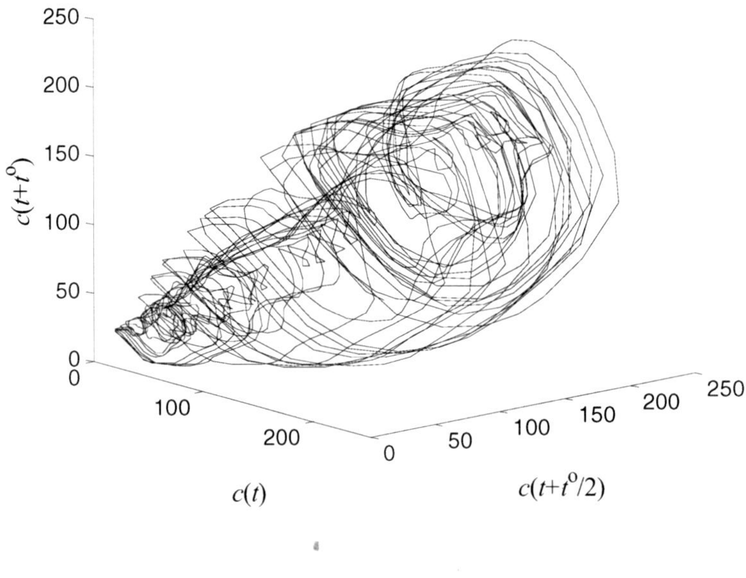

Figure 11.11: A pseudo-phase space for the model of cortisol kinetics. Variables $c(t)$, $c(t+t^\circ/2)$, $c(t+t^\circ)$ are expressed in μg ml^{-1}.

where α and β are constants with concentration units, φ is a constant with time units, and t is time in minutes. Similar approaches relying on simple periodic functions were used by Rohatagi et al. [522] to describe the secretion rate of cortisol.

Our dynamic model consists of (11.12) and (11.13). The physical meaning of the time delay in (11.12) is that the cortisol concentration $c(t)$ affects other physiological parameters of the hypothalamic–pituitary–adrenal process (not present in equation 11.12), which in turn affect, via the feedback mechanism, cortisol concentration; thereby cortisol controls its own secretion [438]. This cycle is postulated to last time t°, and that is how the concentration $c^\circ(t)$ at time $t - t^\circ$ arises.

The simulated profile generated by (11.12) and (11.13) is shown in Figure 11.10. Model parameters take the values

$$\begin{array}{llll} k_i = 0.0666\,\text{min}^{-1}, & k_o = 0.0333\,\text{min}^{-1}, & c(0) = 170\,\mu\text{g ml}^{-1}, & \gamma = 10, \\ \alpha = 70\,\mu\text{g ml}^{-1}, & \beta = 100\,\mu\text{g ml}^{-1}, & \varphi = 250\,\text{min}, & t^\circ = 70\,\text{min}. \end{array}$$

The value assigned to t° corresponds to about one cortisol secretion burst per hour in accordance with experimental observations [518]. The simulations were performed by a numerical solution of (11.12) and (11.13). This simulation exhibits the circadian rhythm, as well as the pulsatile nature of the cortisol secretion system.

Since (11.12) has an infinite number of degrees of freedom [523], we constructed a pseudophase space [4, 32] for the system of (11.12) and (11.13) using the model variables $c(t)$, $c(t + t^\circ/2)$, $c(t + t^\circ)$, Figure 11.11. The use of three dimensions is in accordance with the embedding dimension that Ilias et al. [514] have found. The attractor of our system is quite complicated geometrically, i.e., it is a strange attractor. The real phase space is of infinite dimension. However, trajectories may be considered to lie in a low-dimensional space (attractor). The model parameters take the same values as in Figure 11.10 and time runs for 10 days.

A Dynamic Perspective of Variability The model under study here offers an opportunity to refer to some implications of the existence of nonlinear dynamics. Apart from the jagged cortisol concentration profile, elements such as the sensitive dependence from the initial conditions (expressed by the positive Lyapunov exponent), as well as the system's parameters, play an important role and may explain the inter- and intraindividual variability observed in the secretion of cortisol. These implications, together with other features absent from classical models, are demonstrated in Figure 11.12.

In all plots the dashed line is generated from (11.12) and (11.13) using the above parameter values, while the sampling interval is fixed to 30 min. The solid lines correspond to the same set of parameter values applying a change only in one of them. This change, however, is enough to produce significant visual change in the profile: (A): k_o is set to $0.03\,\text{min}^{-1}$; (B): $c(0)$ is set to $160\,\mu\text{g ml}^{-1}$;

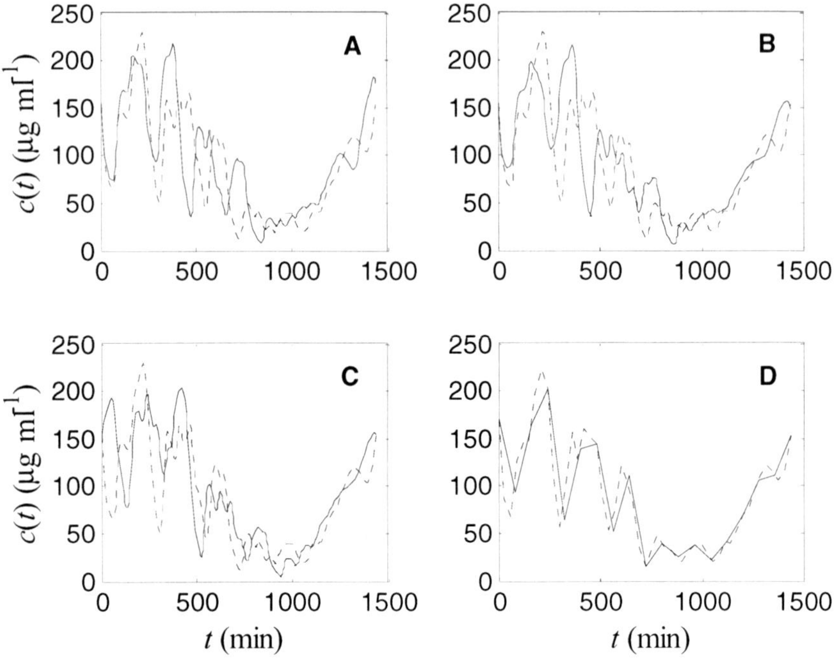

Figure 11.12: The dotted lines are generated from the model of cortisol kinetics using the same parameter values as for Fig. 11.10. The solid lines correspond to the same set of parameter values applying a change only in one of them: k_o (**A**), $c(0)$ (**B**), and observation sampling (**D**). In (**C**), the second-day profile is compared to the first-day profile.

(C): the second-day profile is compared to the first-day profile; (D): sampling is performed every 80 min instead of 30 min. The dashed and solid lines of plots C and D have identical values for the model parameters. Thus, a change in the initial conditions or the parameter values of (11.12) and (11.13) may be depicted in a relatively large change of the final profile, Figure 11.12 A and B. Also, the profiles corresponding to two successive days (Figure 11.12 C), or two different sampling designs (Figure 11.12 D) may differ remarkably, even though the exact same set of parameter values is used. Overall, our analysis based on nonlinear dynamics offers an alternative explanation for the fluctuation of cortisol levels. However, the most important implication of the presence of nonlinear dynamics in cortisol secretion processes is the limitation for long-term prediction, which makes practical application of the classical models questionable.

As we have already mentioned (cf. Chapter 3), one of the most important features of nonlinear dynamics is the sensitivity to initial conditions. A measure to verify the chaotic nature of a dynamic system is the Lyapunov exponent [32], which quantifies the sensitive dependence on initial conditions. In the present model we found [524] the largest Lyapunov exponent to have a positive value of around 0.00011 min^{-1}, which is a clear indication for chaotic behavior.

Cortisol Suppression by Corticosteroids The model presented here allows the consideration of external corticosteroid administration as a perturbation of the cortisol secretion system. As a matter of fact, corticosteroids cause a temporary diminution of plasma cortisol levels [522]. Assuming that the drug follows one-compartment model disposition with first-order input and output, the effect-site [525] concentration is described by the following equation [417]:

$$y(t) = \frac{F_a q_0}{V} \frac{k_a k_y}{k_a - k_e} \left[\frac{\exp(-k_y t) - \exp(-k_e t)}{k_e - k_y} - \frac{\exp(-k_y t) - \exp(-k_a t)}{k_a - k_y} \right], \tag{11.14}$$

where F_a is the bioavailable fraction of dose q_0, V is the volume of distribution of the pharmacokinetic compartment, k_a, k_e are the input and elimination first-order rate constants from the pharmacokinetic compartment, respectively, and k_y is the elimination rate constant from the effect compartment.

The effect-site concentration of the corticosteroids can be considered to affect one or more parameters of the model described by (11.12). This must be implemented so that the presence of $y(t)$ suppresses the cortisol secretion in accordance with the experimental data. Instead of $g(t)$, the parameter describing the circadian rhythm, a new parameter $\widetilde{g}(t)$ was introduced to include the effect of corticosteroid administration following a receptor reduction:

$$\widetilde{g}(t) = g(t) \left[1 - \frac{y(t)}{Ec_{50} + y(t)} \right], \tag{11.15}$$

where Ec_{50} is a coefficient that expresses the concentration of the drug when $\widetilde{g}(t) = g(t)/2$. In this simple way, and in the presence of external corticosteroid drug administration, realistic cortisol blood levels can be obtained as illustrated in Figure 11.13, for the case of fluticasone propionate [438].

The solid and dashed lines represent simulations for two "individuals" with significantly different profiles corresponding to different initial conditions, $c(0) = 90\,\mu\text{g ml}^{-1}$ (solid) and $c(0) = 150\,\mu\text{g ml}^{-1}$ (dashed). Parameter values were set to

$$\begin{array}{llll} k_i = 0.0666\,\text{min}^{-1}, & k_o = 0.0333\,\text{min}^{-1}, & \gamma = 10, & \\ \alpha = 90\,\mu\text{g ml}^{-1}, & \beta = 100\,\mu\text{g ml}^{-1}, & \varphi = 200\,\text{min}, & t^{\circ} = 70\,\text{min}. \\ k_a = 0.14\,\text{min}^{-1}, & k_e = 0.002\,\text{min}^{-1}, & k_y = 0.005\,\text{min}^{-1}, & \\ V = 22.2\,\text{l}, & F_a q_0 = 1\,\text{mg}, & Ec_{50} = 20\,\mu\text{g ml}^{-1}, & \end{array}$$

These values were selected in order to generate qualitatively similar profiles to the experimental data and were not optimized since fitting is not well established for chaotic systems. In parallel, the sensitive dependence of the detailed final profile from the exact values of the concentration $y(t)$ should be emphasized, since $y(t)$ directly affects one of the parameters of the chaotic oscillator (11.15).

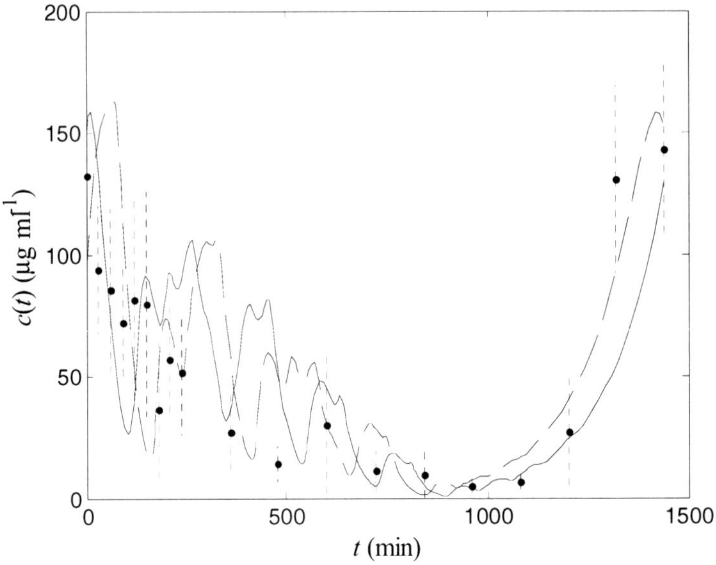

Figure 11.13: Diminution of cortisol blood levels in the presence of fluticasone propionate. *Circles* represent averaged experimental data of four volunteers after the administration of 1 mg of inhaled drug [438], while the *solid* and *dashed lines*, generated by the model of cortisol kinetics, represent simulated data for two "individuals" with different initial conditions.

Finally, experimental evidence indicates that fluctuations in cortisol secretion are not produced by random processes. In fact, the large inter- and intraindividual variability observed in studies dealing with the effect of fluticasone propionate on cortisol levels [526] may be partly explained with the erratic behavior of the system of (11.12) to (11.15).

Parametric Models

Numerous other experimental studies of hormonal systems utilize tools from nonlinear dynamic systems theory. Smith in 1980 [527] used a mathematical model of three interacting hormones, namely testosterone, luteinazing hormone, and luteinazing hormone-releasing hormone, to describe qualitatively their behavior. The initial model was improved later by Cartwright and Husain [528], introducing time-retarded terms of the three state variables to make the system more realistic, exhibiting limit cycle solutions. Further improvements of the model were studied by Liu and Deng [529] and also by Das et al. [530]. Apart from testosterone other efforts in the same context have been made to model the secretion of hormones. Examples are the work of Lenbury and Pacheenburawana [516] in the system of cortisol, adrenocorticotropic hormone, and corticotrophin-releasing hormone, the work of Topp et al. in the system of β-cell mass, insulin, and glucose [531], and also the work of Londergan and Peacock-Lopez [532]. The latter is a general model of hormone interaction description with negative feedback, exhibiting very rich dynamics and even chaotic behavior.

Many drugs affect the normal hormonal secretion, either as their primary target of action or as a side effect. Many studies in recent years have considered models of hormonal secretion together with the dominant pharmacokinetic-dynamic concepts of drug action. Examples include the effect of corticosteroids on cortisol by Chakraborty et al. [438]; the effect of the gonadotropin-releasing hormone antagonist on testosterone and luteinazing hormone by Fattinger et al. [533]; the effect of the dopaminomimetic drug DCN 203-922 on prolactin by Francheteau et al. [534]; the effect of the calcimimetic agent R-568 on parathyroid hormone by Lalonde et al. [535]; and the effect of ipamorelin on growth hormone by Gobburu et al. [536].

All the above studies share a common element. The hormone secretion modeling is kept to a minimum, usually consisting of a single differential equation or even an algebraic equation that gives a simple smooth hormone baseline. Then, the pharmacokinetic-dynamic models, such as direct or indirect link and response [405], relate the inhibition or the stimulation of the baseline with the drug concentration. In order to set the baseline, only the most obvious characteristics of the hormone profile are integrated, like a periodic circadian rhythm. The dynamic structure of the underlying physiology is practically ignored and so is pulsatility, which is considered to be noise. The only studies in which pulsatility is considered as a feature of the profile are the works of Francheteau et al. [534] for the effect of dopaminomimetic drug DCN 203-922 on prolactin and Chakraborty et al. [438] for the effect of fluticasone propionate on cortisol. However, even in these studies the pulsatility is integrated phenomenologically through spline terms or Fourier harmonics, respectively, and not through modeling of the dynamic origin of the pulsatility. It must be noted though that there are studies in which the pulsatility does not play an important role, like the study of Gobburu et al. [536] for the effect of ipamorelin on growth hormone, where the baseline of the hormone is reasonably considered zero due to the multifold amplification of the growth hormone levels after the administration of the drug.

A mathematical model of the insulin–glucose feedback regulation in man was proposed by Tolic et al. [537] to examine the effects of an oscillatory supply of insulin compared to a constant supply at the same average rate. The model analysis allowed them to interpret seemingly conflicting results of clinical studies in terms of their different experimental conditions with respect to hepatic glucose release. If this release is operating near an upper limit, an oscillatory insulin supply will be more efficient in lowering the blood glucose level than a constant supply. If the insulin level is high enough for the hepatic release of glucose to nearly vanish, the opposite effect is observed. For insulin concentrations close to

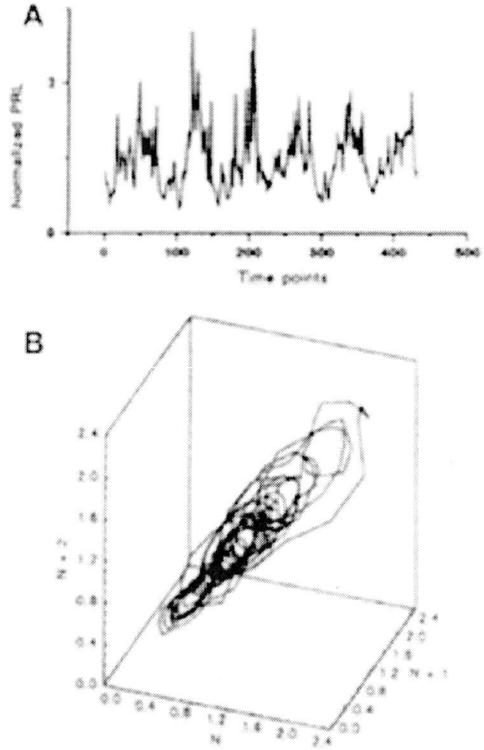

Figure 11.14: (**A**) The composite prolactin time series. (**B**) Sketch of the 3-dimensional attractor of prolactin generated by the data of plot **A**. Reprinted from [539] with permission from Blackwell.

the point of inflection of the insulin–glucose dose–response curve, an oscillatory and a constant insulin infusion produce similar effects.

Nonparametric Models

The phase space reconstruction approach, making use only of the hormone plasma profiles, was utilized in order to assess the dimensionality and thus expose the chaotic nature of the underlying dynamics of various hormones. In all these studies, the reconstruction of the phase space gave attractors of fractal dimension, evidence for the presence of nonlinear dynamics. Such examples are the work of Prank et al. [538] on parathyroid hormone, Ilias et al. [514] on cortisol and growth hormone, and Papavasiliou et al. [539] on prolactin.

By using methods of nonlinear dynamics, Papavasiliou et al. [539] analyzed the circadian profiles of prolactin, directly from the experimental data, by combining in a single time series (432 measurements), six individual 24-h prolactin profiles (72 measurements per profile, 20 min sampling interval), obtained from young healthy human volunteers, under basal conditions, Figure 11.14 A. Significant autocorrelation exists between any given point of the time series and a limited number of its successors. Fourier analysis showed a dominant frequency of 1 cycle$\times$ d^{-1}, without sub-24-h harmonics. Poincaré section indicated the presence of a fractal attractor, and a sketch of the attractor revealed a highly convoluted geometric structure with a conical contour. The box-counting dimension was found to be fractional, namely $d_f = 1.66$, indicating that diurnal prolactin secretion is governed by nonlinear dynamics. *Information dimension* and *correlation dimension* confirmed the above value of the attractor. The two dimensions did not differ significantly from each other, and exhibited saturation at an embedding dimension of 2. A 3-dimensional plot of the attractor is presented in Figure 11.14 B. The evidence taken together suggests that under basal conditions, the daily changes in the peripheral blood levels of prolactin are governed by nonlinear deterministic dynamics, with a dominant rhythm of 1 cycle$\times$ d^{-1} mixed with a higher-frequency, low-amplitude signal.

Pincus developed in 1991 a different method to quantify the hormone pulsatility, which is referred to as the approximate entropy algorithm [540] and is based on the concept of Lyapunov exponents. This method has been applied for several hormones such as adrenocorticotropic hormone, cortisol, prolactin, insulin, growth hormone, testosterone, and luteinazing hormone, quantifying the observed pulsatility and comparing it between different groups such as sick against healthy, different age groups, etc. ([541] and references therein). The experimental evidence of the chaotic nature of hormonal underlying dynamics clarifies the origin of the pulsatility and acts as a guide for proper modeling.

Serial data of glucose and insulin values of individual patients vary over short periods of time, since they are subject to biological variations. The classic homeostatic control model assumes that the physiological mechanisms maintaining the concentrations of glucose and insulin are linear. The only deviations over a short period of time one should observe are in relation to a glucose load or major hormonal disturbance. Otherwise, the values of glucose and insulin should be constant and any variations should be due to random disturbances. Kroll [542] investigated previously published serial data (three for glucose and one for insulin) with both linear and nonlinear techniques to evaluate the presence of deterministic components hidden within the biological (intraindividual) variation. Within the linear techniques, the power spectra failed to show dominant frequencies, but the autocorrelation functions showed significant correlation, consistent with a deterministic process. Within the nonlinear techniques, the correlation dimension was finite, around 4.0, and the first Lyapunov exponent was positive, indicative of a deterministic chaotic process. Furthermore, the phase portraits showed directional flow. Therefore, the short-term biological variation observed for glucose and insulin records arises from nonlinear, deterministic chaotic behavior instead of random variation.

From the above studies, it is evident that although significant progress has been made as far as the physiological modeling of hormonal systems is concerned, the relevant pharmacodynamic modeling, even in state-of-the-art studies dealing with the effect of drugs on hormonal levels, practically ignores these findings. It is a necessity to develop new pharmacodynamic models for drugs related to hormonal secretion, compatible with the physiological modeling and the experimental findings that suggest low-dimensional nonlinear dynamic behavior. This kind of modeling not only is more realistic but integrates a new rationale as well. The notions of the sensitivity from the initial conditions and the qualitatively different behavior for different, even slightly, values of the control parameters, surely play an important role and must be taken into account in modeling since their presence is suggested by experiments.

An important outcome of these studies is the opportunity that it offers to discuss the implications of the presence of nonlinear dynamics in processes such as the secretion of cortisol. Based on the aforementioned discussion it is evident that the concepts of deterministic nonlinear dynamics should be adopted in pharmacodynamic modeling when supported by experimental and physiologic data. This is valid not only for the sake of more detailed study, but mainly because nonlinear dynamics suggest a whole new rationale fundamentally different from the classical approach. Moreover, the clinical pharmacologist should be aware of the limitations of chaotic models for long-term prediction, which is contrary to the routine use of classical models.

If chaotic dynamics are present, the experimental errors do not originate exclusively from classical randomness. Thus, the measures of central tendency used to describe or treat experimental data are questionable, since averaging is inappropriate and masks important information in chaotic systems [234].

11.2.2 Central Nervous System Drugs

The application of nonlinear dynamics to brain electrical activity offered new information about the dynamics of the underlying neuronal networks and formulated the brain disorders on the basis of qualitatively different dynamics [479].

Parametric Models

Serotonin plays an active role in temperature regulation and in particular in the maintenance of the body's set point [543–545]. More recently, numerous pharmacological studies have suggested the involvement of homeostatic control mechanisms [544, 546] that are achieved through interplay between the 5-hydroxytryptamine (HT)1A and 5-HT2A/C receptor systems [545, 547, 548]. Administration of a 5-HT1A-receptor agonist that is used therapeutically as an antidepressant and antianxiety drug causes hypothermia [549, 550].

So far, only very few of these models incorporate complex regulatory behavior [534, 551, 552]. Specifically, no mathematical models have been developed to characterize the complex time behavior of the hypothermic response in a strict quantitative manner, and neither attempts to link existing temperature regulation models [553] to pharmacokinetic models describing the time course of the drug concentration in the body.

To characterize 5-HT1A-agonist-induced hypothermia, Zuideveld et al. [554] developed a mathematical model that describes the hypothermic effect on the basis of the concept of a set point and a general physiological response model [431,555]. The model was applied to characterize hypothermic response vs. time profiles after administration of different doses of the reference 5-HT1A receptor agonists R- and S-8-OH-DPAT.

Example 12 *Temperature Regulation*

The classical three-compartment model describes pharmacokinetics of 5-HT1A receptor agonists. By means of a sigmoidal function $E(c)$, the 5-HT1A agonist concentration $c(t)$ influences the set-point signal that dynamically interacts with the body temperature. By using $x(t)$ and $y(t)$ as dimensionless state variables for the set-point and temperature, respectively, the model is expressed by the set of two nonlinear differential equations:

$$\begin{aligned}
E(c) &= S_{\max} c^{n}(t)\left[Sc_{50}^{n} + c^{n}(t)\right]^{-1}, \\
\dot{x}(t) &= A\left[1 - E(c) - y(t)\right], \\
\dot{y}(t) &= B\left[1 - x^{-\gamma}(t)\, y(t)\right],
\end{aligned}$$

where the initial conditions are those at equilibrium $(x^{*}, y^{*}) = (1, 1)$. The symbols and the parameter values are as reported in [554]. Figures 11.15 and 11.16 simulate the dynamic behavior of the model for two dose levels, 200 and 1000 mg. We note that for the low dose, damped oscillations appear in the temperature $y(t)$ variable, whereas for the larger dose the perturbed temperature slowly reaches the reference value. These behaviors result from the type of the eigenvalues of the linearized model, complex conjugated for the low dose and real with negative part for the higher dose. ■

The developed model is able to reproduce the observed complex effect vs. time profile:

- When the model is not fully "pushed" into the maximal effect, a plateau phase appears. This plateau originates from damped oscillations that occur around the equilibrium point on returning to baseline. Hence, the observed plateau phase is an intrinsic part of the regulatory mechanism related to the oscillatory behavior found in many regulatory systems [556, 557].

- When the model is fully "pushed" into its maximal effect, such as in the case of a relatively high dose of a full agonist, the system becomes overdamped, thereby losing its oscillatory behavior.

The model described above has been successfully applied to characterize the in vivo concentration effect relationships of several 5-HT1A agonists including flesinoxan and buspirone [558, 559]. This model has also linked with the operational model of agonism into a full mechanism-based pharmacokinetic-dynamic model [560].

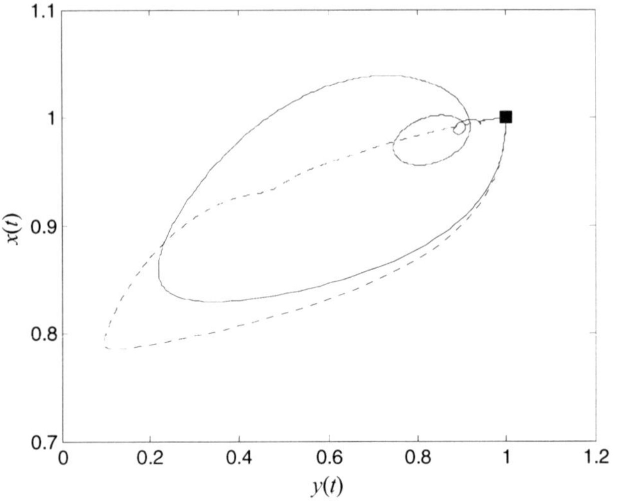

Figure 11.15: The state space of the dimensionless set-point and temperature variables $x(t)$ and $y(t)$, respectively. *Solid* and *dashed lines* correspond to the low and high doses, respectively. (■) represents the stable equilibrium point.

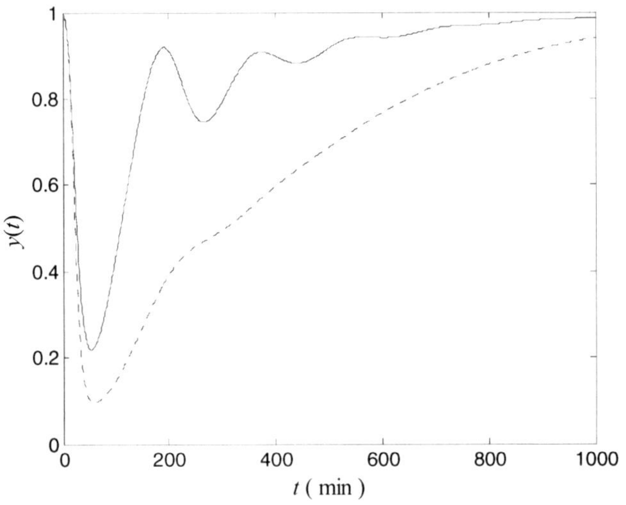

Figure 11.16: The perturbed dynamics of the dimensionless temperature variable $y(t)$. *Solid* and *dashed lines* correspond to the low and high doses, respectively.

Nonparametric Models

Once again, most studies applying nonlinear tools in this field are based on experimental electroencephalogram recordings and demonstrate the irregular behavior of the brain electrical activity. Various metrics have been used to assess the electroencephalogram variability, using phase space reconstruction techniques or even calculating the fractality of the electroencephalogram recording in real time [561]. These tools, apart from pointing out the obvious complexity of the brain electrical signals, offer supplemental information to the classical techniques, such as Fourier analysis, in order to distinguish qualitatively different electroencephalogram recordings, e.g., in epileptic seizures [562], in Parkinson's disease [563], or in schizophrenia [564]. In the same context, low doses of ethanol have been found to reduce the nonlinear structure of brain activity [565]. Most of the pharmacokinetic-dynamic studies of centrally acting drugs rely on quantitative measures of electroencephalogram parameters [566]. However, an ideal electroencephalogram parameter to characterize the central nervous system effect of drugs has not been found as yet. To the best of our knowledge, time series analysis of electroencephalogram data of pharmacodynamic studies with central nervous system drugs using techniques of nonlinear dynamics are limited. Examples include investigations of the influence of anticonvulsive [567] and antiepileptic [568] drugs in epilepsy, the study of sleep electroencephalogram under lorazepam medication [569], the study of the effects of pregnenolone sulfate and ethylestrenol on rat behavior [570], the investigation of the electrophysiological effects of the neurotoxin 5, 7-dihydroxytryptamine [571], and the study of epileptiform bursts in rats after administration of penicillin and K^+ ions [572].

However, the pharmacodynamic mixed-effects model for the effect of temazepam on sleep [573] requires special mention. The model is based on hypnogram recordings and describes the probability of changes in sleep stage as a function of time after drug intake. The model predictions were found to be consistent with the observations of the effect of temazepam on sleep electroencephalogram patterns. Also, the effect of temazepam on the sleep–wake status was interpreted in terms of known mechanisms for sleep generation and benzodiazepine pharmacology.

Modeling in the brain is mainly targeted to the general qualitative principles underlying various phenomena such as epileptic seizures [574], and not to quantitative assessment and forecasting as one would expect to achieve in simpler systems. For example, in [479], recurrent inhibition and epilepsy are studied and also penicillin is considered as a γ-aminobutyric acid inhibitor.

The analysis of brain activity using tools from chaos theory can provide important information regarding the underlying dynamics if one takes into consideration that the qualitative electroencephalogram changes, induced by centrally acting drugs, e.g., ketamine, thiopental, etomidate, propofol, fentanil, alfentanil, sulfentanil, and benzodiazepines, differ considerably [566]. This exercise can also unmask the sources of extremely high variability (the coefficient of variation for model pharmacodynamic parameters of benzodiazepines in humans ranges from 30 to 100%) [566]. A plausible interpretation for the extremely high variability of pharmacodynamic parameters of benzodiazepines may be associated with the dynamic behavior of the underlying system, i.e., the recurrent inhibitory pathway of γ-aminobutyric acid [479].

It is also worthy of mention the work on the pharmacodynamics of midazolam in rats of Cleton et al. [575]. These authors found that the rate of change in plasma concentration is an important determinant of midazolam pharmacodynamics. In addition, the relationship found between the rate of change of blood concentration and the values of the different pharmacodynamic parameters is rather complex. These findings indicate that in vivo a homeostatic control mechanism is operative that may modify the sensitivity to midazolam and whose activation is largely influenced by the rate of presentation of the drug in blood.

Keeping patients at a well-defined level of anesthesia is still a difficult problem in clinical practice. If anesthesia is too deep, a decompensation of the cardiovascular system is threatening. When anesthesia is too weak, the patient may wake up. Depth of anesthesia is expected to be reflected in the electroencephalogram. In current clinical practice, one or a few channels of electroencephalogram are routinely displayed during difficult anesthesias. Since the attending personnel have to monitor several critical parameters (blood pressure, heart rate, etc.), the vast amount of information contained in the electroencephalogram must be severely condensed in order to be useful. Only a few numbers may be monitored at a typical intervention time scale. Most pragmatically, a single number should be produced that indicates the instantaneous depth of anesthesia of the patient.

In that spirit, Widman et al. [576] adapted a prescription for an overall index of nonlinear coherence that has been found powerful for anticipating epileptic seizures from implanted electrode recordings. This index based on phase space reconstruction and correlation sums was called d^*, and it contains many ingredients familiar from the Grassberger–Procaccia algorithm for the correlation dimension [577].

Widman et al. [578] compared several indices measuring the depth of anesthesia from electroencephalogram data gathered from 17 patients undergoing elective surgery and anesthetized with sevoflurane. Two of these measures are based on the power spectrum, and the third is the bispectral index BIS [579]. The power spectrum measures are essentially useless and unreliable as indicators of depth of anesthesia in the investigated group of patients. While for both of the two nonlinear measures, bispectral index and d^*, such a relationship seems to exist, the correlation is strongest for d^*. Dimension d^* seems to be able to improve the quantification of depth of anesthesia from brain electrical activity, at least when sevoflurane is used as an anesthetic drug. To assess the depth of anesthesia of the patient, Bruhn et al. [580] recently proposed another index based on the Shannon entropy.

11.2.3 Cardiovascular Drugs

Numerous applications of nonlinear dynamics and chaos theory to cardiac physiology have been published [581]. Many techniques, either statistical, like spec-

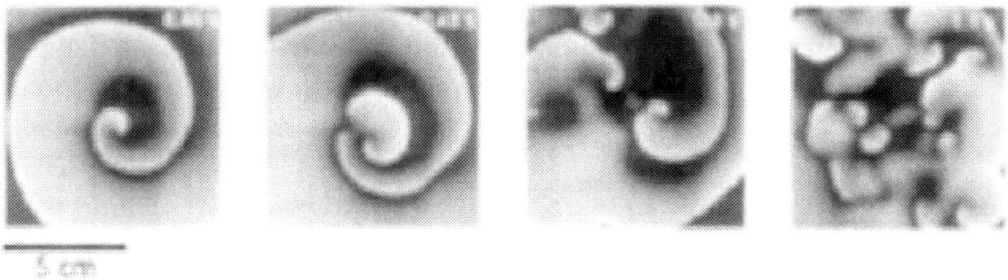

Figure 11.17: The four snapshots show the evolution and breakup of a spiral wave pattern in 2-dimensional simulated cardiac tissue (300×300 cells). The chaotic regime shown in the final snapshot corresponds to fibrillation. Reprinted from [587] with permission from Lippincott, Williams and Wilkins.

tral analysis, or dynamic, like phase space reconstruction, applied to electrocardiogram data clearly indicate that the frequency of the heartbeat is essentially irregular. The electrocardiogram was in fact, one of the first biological signals studied with the tools of nonlinear dynamics. Studies applying concepts from chaos theory to electrocardiogram data, regarding the effects of drugs on the dynamics of cardiac physiology, have also been published. Examples include the effect of atropine on cardiac interbeat intervals [582], the induction of cellular chaos during quinidine toxicity [583], the attempt to control cardiac chaos using ouabain [584], and the effect of anticholinergic drugs on heart rate variability [585].

Another very successful application of nonlinear dynamics to the heart is through mathematical modeling. An example in which a simple model based on coupled oscillators describes the dynamics of agonist induced vasomotion is in the work of de Brouwer et al. [586], where the route to chaos in the presence of verapamil, a class IV antiarrhythmic drug, is studied.

Undoubtedly, the most promising modeling of the cardiac dynamics is associated with the study of the spatial evolution of the cardiac electrical activity. The cardiac tissue is considered to be an excitable medium whose the electrical activity is described both in time and space by reaction–diffusion partial differential equations [519]. This kind of system is able to produce spiral waves, which are the precursors of chaotic behavior. This consideration explains the transition from normal heart rate to tachycardia, which corresponds to the appearance of spiral waves, and the following transition to fibrillation, which corresponds to the chaotic regime after the breaking up of the spiral waves, Figure 11.17. The transition from the spiral waves to chaos is often characterized as electrical turbulence due to its resemblance to the equivalent hydrodynamic phenomenon.

These concepts have been successfully applied to the effect of antiarrhythmic drugs as well. It is widely known that although class II antiarrhythmic drugs, like isoproterenole, have shown satisfactory results [588], class I and III agents, such as encainide, flecainide, and moricizine, have been shown even to increase sudden death rate caused by ventricular fibrillation [589]. Although it is unclear how to integrate the drug action in the excitable media models, successful attempts have been made to simulate, mainly, 2-dimensional cardiac tissue [590, 591]. Three-dimensional cardiac tissue has been simulated as well [592], where the 3-dimensional equivalent of spiral waves, the scroll waves, appear. These models explain how a drug can exhibit antiarrhythmic action in a single-cell system, which ignores the spatial evolution, while acting as proarrhythmic in a system of a whole cardiac tissue of spatial dimension 2 or 3. This has given rise to a new approach for antiarrhythmic drug evaluation based on the chaotic dynamics of transition from tachycardia to fibrillation [587,591,592], which is also supported by experimental evidence [592]. The results of these recent studies [587] indicate that the failure to predict long-term efficacy of class I and III antiarrhythmic agents in patients with ischemic heart disease [589] may be associated with the limitations of the classical approach, which is based only on the suppression of premature ventricular polarization on the electrocardiogram, i.e., the initiation of tachycardia. Sudden cardiac death resulting from ventricular fibrillation, however, is separated into two components: initiation of tachycardia and degeneration of tachycardia to fibrillation. These studies suggest that a new antiarrhythmic drug classification scheme must be adopted, which should incorporate the antifibrilatory profile based on results from excitable media modeling, together with the classical antitachycardiac profile (classes I to IV scheme). Also, the drug bretylium is proposed as a prototype for future development of antifibrillatory agents [592].

In the pharmaceutical literature [593] the pharmacodynamics of antiarrhythmic drugs are treated with the classical models, $E_{\max}$, indirect link with effect compartment, etc. Variability, wrong dosage scheme, narrow therapeutic index, and lack of individualization of treatment are the dominant interpretations for the failure of these drugs. Another factor held responsible for the failure in treatment with antiarrhythmics is the possible nonbioequivalency of the generics used [594]. However, classical bioequivalence studies are based only on the comparison of pharmacokinetic parameters of the formulations ($c_{\max}$, area under curve AUC). Although testing for therapeutic equivalence is implied, pharmacodynamics are not taken into account at all. Thus, classical bioequivalence studies may be inappropriate for assessing the effects of antiarrhythmic drugs if their mechanism of action arises from nonlinear dynamic processes.

11.2.4 Conclusion

These studies show that it is possible to predict the time course of drug effects in vivo in situations in which complex homeostatic control mechanisms are operative. As such, they form the basis for the development of an entirely new class of pharmacokinetic-dynamic models. These models are important for the development of new drugs and the application of such drugs in clinical practice. For example, on the basis of this kind of model, it becomes possible to predict whether withdrawal symptoms will occur on cessation of (chronic) drug treatment. Hence, these models may provide a scientific basis either for the selection of alternative drug candidates or the design of dosing regimens that show less-pronounced withdrawal phenomena. It is further anticipated that such models will provide a basis for pharmacokinetic-dynamic modeling with disease progression.

The time-dependent behavior of the examined data sets exhibits strong deterministic components. The deterministic components in each data set show considerable variation (chaotic behavior), and are the source of an important portion of the observed biological (intraindividual) variation. This information also changes the view of biological variation. In the past, it was thought that the source of variation was external to the internal workings of the organism, that the environment, such as temperature, food ingestion, immobilization, venous occlusion, were responsible for the short-term changes. The source of biological variation for glucose and insulin comes from within the organism itself; it is endogenous. Beyond that, the variation demonstrates chaotic behavior.

Are these phenomena unique, or are they typical of biological systems? From a mathematical perspective, enzyme systems fall into a class of nonlinear organization, and a chain of enzyme reactions with negative feedback easily can demonstrate oscillatory behavior [520]. Glass has noted that in general, any nonlinear system with multiple negative feedback may demonstrate oscillations that lead to chaotic behavior [595].

Nonlinear analysis requires the use of new techniques such as embedding of data, calculating correlation dimensions, Lyapunov exponents, eigenvalues of singular-valued matrices, and drawing trajectories in phase space. There are many excellent reviews and books that introduce the subject matter of nonlinear dynamics and chaos [515, 596–599].

Since most drugs are modifiers of physiological and biochemical states that are in some way abnormal, and are given to move the system toward normality, it follows that many concepts of modern nonlinear dynamic theory have potential application to pharmacology and drug development. Indeed, a growing body of biological problems is the subject of studies in journals and books on dynamic modeling. Some pharmacological problems are discussed by Riggs [308]; many others appear in a growing literature on nonlinear dynamics nicely summarized in recent monographs [3, 31, 32, 40].

Appendix A

Stability Analysis

Stability is determined by the eigenvalue analysis at an equilibrium point for flows and by the characteristic multiplier analysis of a periodic solution at a fixed point for maps [3].

- The *equilibrium point* $\underline{y}^*$ for flows is the solution of $\underline{g}\left(\underline{y}^*, t, \theta\right) = \underline{0}$. The local behavior of the flow near $\underline{y}^*$ is determined by linearizing $\underline{g}$ at $\underline{y}^*$; let $\mathbf{A}$ be the matrix formed by elements

$$a_{jk} = \left. \frac{\mathrm{d}g_j\left(\underline{y}\right)}{\mathrm{d}y_k} \right|_{\underline{y}=\underline{y}^*} .$$

  Let the eigenvalues of $\mathbf{A}$ be ζ_j with corresponding eigenvectors η_j. If ζ_j is real, the eigenvalue is the rate of contraction (if $\zeta_j < 0$) or expansion (if $\zeta_j > 0$) near $\underline{y}^*$ in the direction of η_j. If ζ_j are complex-conjugate pairs, the trajectory is a spiral in the phase space spanned by $\mathrm{Re}\left[\eta_j\right]$ and $\mathrm{Im}\left[\eta_j\right]$. The real part of ζ_j gives the rate of contraction (if $\mathrm{Re}\left[\zeta_j\right] < 0$) or expansion (if $\mathrm{Re}\left[\zeta_j\right] > 0$) of the spiral; the imaginary part of the eigenvalue is the frequency of rotation. Hence, one can conclude that if $\mathrm{Re}\left[\zeta_j\right] < 0$ for all ζ_j, then all sufficiently small perturbations tend toward 0 as $t \to \infty$, and $\underline{y}^*$ is asymptotically *stable*. If $\mathrm{Re}\left[\zeta_j\right] > 0$ for all ζ_j, then any perturbation grows with time, and $\underline{y}^*$ is *unstable*. If there exist j and k such that $\mathrm{Re}\left[\zeta_j\right] < 0$ and $\mathrm{Re}\left[\zeta_k\right] > 0$, then $\underline{y}^*$ is *unstable*. An unstable equilibrium point is often called a *saddle point*. A stable or unstable equilibrium point with no complex eigenvalues is often called a *node*.

- The *fixed point* $\underline{y}^*$ for maps is the solution of $\underline{y}^* = \underline{g}\left(\underline{y}^*, \theta\right)$. The local behavior of the map near $\underline{y}^*$ is determined by linearizing the map at $\underline{y}^*$; let $\mathbf{A}$ be the matrix formed by elements

$$a_{jk} = \left. \frac{\mathrm{d}g_j\left(\underline{y}\right)}{\mathrm{d}y_k} \right|_{\underline{y}=\underline{y}^*} .$$

Let the eigenvalues of $\mathbf{A}$ be ξ_j with corresponding eigenvectors η_j. The eigenvalues ξ_j are called *characteristic multipliers* and they are a generalization of the eigenvalues at an equilibrium point. The characteristic multipliers' position in the complex plane determines the stability of the fixed point. If ξ_j is real, the characteristic multiplier is the amount of contraction (if $\xi_j < 1$) or expansion (if $\xi_j > 1$) near $\underline{y}^*$ in the direction of η_j for one iteration of the map. If ξ_j are complex-conjugate pairs, the orbit is a spiral in the phase space spanned by Re $[\eta_j]$ and Im $[\eta_j]$. The magnitude of ξ_j gives the amount of expansion (if $|\xi_j| > 1$) or contraction (if $|\xi_j| < 1$) of the spiral for one iteration of the map; the angle of the characteristic multiplier is the frequency of rotation. Hence, one can conclude that if $|\xi_j| < 1$ for all ξ_j, then all sufficiently small perturbations tend toward 0 as $i \to \infty$, and $\underline{y}^*$ is asymptotically *stable* and is said to be an *attracting equilibrium.* If $|\xi_j| > 1$ for all ξ_j, then any perturbation grows with iterations, and $\underline{y}^*$ is *unstable.* If there exist j and k such that $|\xi_j| < 1$ and $|\xi_k| > 1$, then $\underline{y}^*$ is *unstable.* An unstable fixed point is often called a *saddle point*. The critical values $\xi = \pm 1$ are where the fixed point $\underline{y}^*$ changes its behavioral character. The case $\xi = 1$ is called a *tangent bifurcation* and the case $\xi = -1$ is called a *pitchfork bifurcation*.

Both equilibrium and fixed points are simply referenced as *steady states*. The matrix $\mathbf{A}$ of the linearized system is called the *Jacobian* of the system.

Appendix B

Monte Carlo Simulations in Drug Release

Models that, either naturally or through approximation, can be discretized are suitable for study using Monte Carlo simulations. As an example, we give a brief outline below of the simulations of drug release from cylinders assuming Fickian diffusion of drug and excluded volume interactions. This means that each molecule occupies a volume V where no other molecule can be at the same time.

First, a 3-dimensional lattice in the form of a cube with L^3 sites is constructed. Next, a cylinder inside this cubic lattice is defined. The cylinder can leak from its side, but not from its top or bottom. A site is uniquely defined by its 3 indices i, j, k (coordinates). The sites are labeled as follows (R is the radius of the cylinder):

- When for a site $(R-1)^2 \leq i^2 + j^2 \leq R^2$, it is considered to be a leak site and it is marked as such.
- If $i^2 + j^2 \leq (R-1)^2$, then it belongs to the interior of the cylinder and it can host drug molecules.
- If, on the other hand, $i^2 + j^2 > R^2$, then it is outside the cylinder, and it is marked as a restricted area, so that particles are not allowed to go there; cf. Figure B.1 for a schematic.

When spherical matrices are constructed, the sites with indices $i^2 + j^2 + k^2 > R^2$ are considered outside of the sphere with radius R and marked as a restricted area, while leak sites are those whose indices satisfy the inequalities $(R-1)^2 \leq i^2 + j^2 + k^2 \leq R^2$.

The simulation method proceeds as follows: a number of particles is placed randomly on the sites of the cylinder, according to the initial concentration, avoiding double occupancy. The diffusion process is simulated by selecting a particle at random and moving it to a randomly selected nearest-neighbor site.

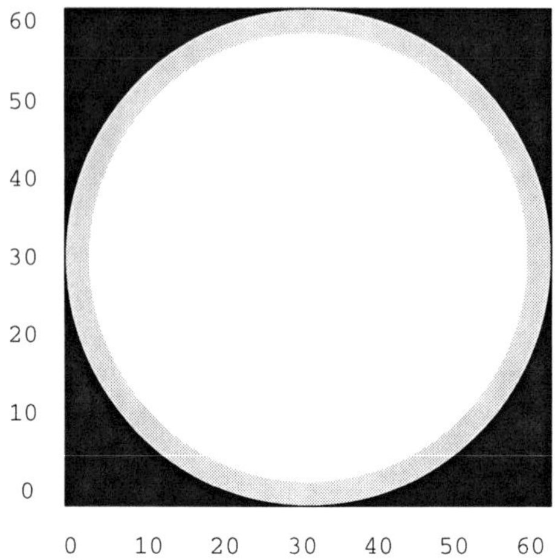

Figure B.1: A cylindrical cross section with radius $R = 30$ sites. The dark area is restricted to particles. The gray area indicates the leaking sites. The white area is where the drug particles are initially located. Each site in the white area can be either occupied or empty.

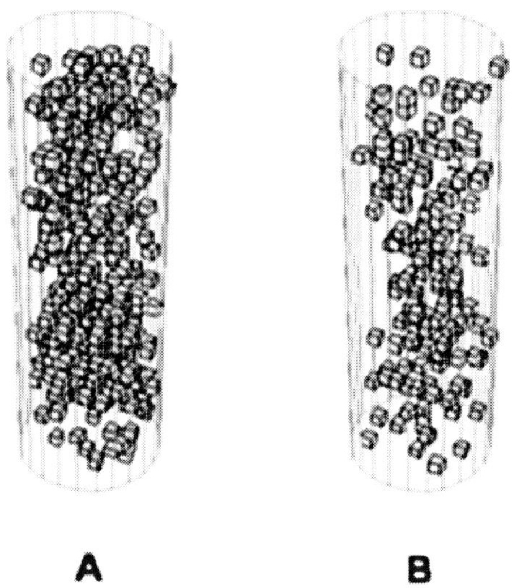

Figure B.2: (**A**) A Cylinder with radius 5 units and half-height 20 units initially contains 282 particles at completely random positions. Each particle is represented by a cuboid of volume 1 $(\text{unit})^3$. (**B**) A snapshot of the same cylinder during the release procedure. Now only 149 particles are left inside the cylinder. The positions of the particles are no longer completely random. On average a concentration gradient forms with fewer particles at the cylinder border.

If the new site is an empty site then the move is allowed and the particle is moved to this new site. If the new site is already occupied, the move is rejected (since excluded volume interactions are assumed).

A particle is removed from the lattice as soon as it migrates to a site lying within the leak area. After each particle move, time is incremented. The increment is chosen to be $1/n(t)$, where $n(t)$ is the number of particles remaining in the system. This is a typical approach in Monte Carlo simulations. The number of particles that are present inside the cylinder as a function of time until the cylinder is completely empty of particles is monitored. The results are averaged using different initial random configurations, but the same parameter. A pictorial view of particles in the cylinder at two different time points is presented in Figure B.2.

Appendix C

Time-Varying Models

The fact that some kinetic profiles are fitted by sums of exponentials, and others are fitted by power functions, suggests that different types of basic mechanisms are at work. In fact, as concluded in Chapter 7, while kinetics from homogeneous media can be fitted by sums of exponentials, heterogeneity shapes kinetic profiles best represented by empirical power-law models. Conversely, when power laws fit the observed data, they suggest that the rate at which a material leaves the site of a process is itself a function of time in the process, i.e., age of material in the process.

But any empirical model is of limited interest because it is able to describe only the observed data. In contrast, phenomenological models are more useful, allowing simulation, design, and control. In order to develop such phenomenological models, Marcus considered stochastic modeling for a summary description based on a single compartment model featuring the key mechanisms [300]. Thus, when exchange rates of material depend on time in the process, phenomenological models may be obtained through stochastic modeling techniques fully analyzed in Chapter 9.

The stochastic formulation would be the most appropriate choice to capture the structural and functional heterogeneity in these biological media. When the process is heterogeneous, one frequently observes chaos-like behaviors. Heterogeneity is at the origin of fluctuations, and fluctuations are the prelude of instability and chaotic behavior. Stochastic modeling is able to:

- generate process uncertainty,
- express process memory or the age of the material in the process, and
- supply tractable forms involving time-varying parameters.

The usual deterministic approach is incapable of accurately describing all these features. However:

- it is technically hard to conceive how to reproduce instability conditions by means of a model with time-varying parameters, and

- it is unlikely that this time-varying feature on the observed processes would be explained by a single functional relation with maturation or age.

More likely, we think that time-varying parameters are expressions of feedback regulation mechanisms involving the states of the process. This is our fundamental working hypothesis in the subsequent procedure. To unveil the dependence of the time-varying parameters on the states of the process, we propose the following procedure:

1. Start to describe the process by means of a phenomenological model according to the underlying physiological structure. For instance, use compartmental configuration to sketch the fundamental mechanisms. The parameters of this holistic description, e.g., exchange rates and volumes of distribution, will be allowed to vary across time. The issue is a state-space dynamic model described by a set of differential equations continuous in time:
$$\dot{\underline{y}}(t) = \mathbf{A}(t)\,\underline{y}(t) + \underline{b}(t)\,r(t), \tag{C.1}$$
where $\underline{y}(t)$ and $r(t)$ are the states and inputs, respectively, and $\mathbf{A}(t)$ and $\underline{b}(t)$ are the time-varying parameters. It is often easier to describe physical or biological processes in terms of continuous-time models. The reason is that most physical laws are expressed in continuous time as differential equations. However, as discussed in Section 9.4, the key problem is how to decide the partition of the dynamic behavior between the contribution of the basic phenomenological model, and the contribution of the time-varying parameters in the model. Several candidate models can be proposed and the final model should be validated and selected by the screening process involving model-selection criteria.

2. Given a set of experimental data, we look for the time profile of $\mathbf{A}(t)$ and $\underline{b}(t)$ parameters in (C.1). To perform this key operation in the procedure, it is necessary to estimate the model "on-line" at the same time as the input-output data are received [600]. Identification techniques that comply with this context are called *recursive identification* methods, since the measured input-output data are processed recursively (sequentially) as they become available. Other commonly used terms for such techniques are on-line or real-time identification, or sequential parameter estimation [352]. Using these techniques, it may be possible to investigate time variations in the process in a real-time context. However, tools for recursive estimation are available for discrete-time models. If the input $r(t)$ is piecewise constant over time intervals (this condition is fulfilled in our context), then the conversion of (C.1) to a discrete-time model is possible without any approximation or additional hypothesis. Most common discrete-time models are difference equation descriptions, such as the *A*uto-*R*egression with e*X*tra inputs (*ARX*) model. The basic relationship is the linear difference equation:
$$y(t) + a_1 y(t-1) + \cdots + a_n y(t-n) = b_1 r(t-1) + \cdots + b_m r(t-m),$$

which relates the current output $y(t)$ to a finite number of past outputs $y(t-k)$ and inputs $r(t-k)$. State-space and ARX models describe the functional relation between inputs and outputs. The order of the state-space model relates to the number of delayed inputs and outputs used in the corresponding difference ARX equation and they are rearranged so that only one delay is used in their expression.

3. Analyze the time profile of $\mathbf{A}(t)$ and $\underline{b}(t)$ against the states $\underline{y}(t)$. For instance, one looks at the dependence of $a_k(t)$ on $y_j(t)$ by plotting $a_k(t)$ or $\log a_k(t)$ as a function of $y_j(t)$ or $\log y_j(t)$. This dependence can be expressed by a second-level modeling of the form $\mathbf{A}\left[\underline{y}(t)\right]$ and $\underline{b}\left[\underline{y}(t)\right]$ that results in the nonlinear differential equation

$$\dot{\underline{y}}(t) = \mathbf{A}\left[\underline{y}(t)\right]\underline{y}(t) + \underline{b}\left[\underline{y}(t)\right] r(t).$$

This second-level modeling of the feedback mechanisms leads to nonlinear models for processes, which, under some experimental conditions, may exhibit chaotic behavior. The previous equation is termed *bilinear* because of the presence of the $\underline{b}\left[\underline{y}(t)\right] r(t)$ term and it is the general formalism for models in biology, ecology, industrial applications, and socioeconomic processes [601]. Bilinear mathematical models are useful to real-world dynamic behavior because of their variable structure. It has been shown that processes described by bilinear models are generally more controllable and offer better performance in control than linear systems. We emphasize that the unstable inherent character of chaotic systems fits exactly within the complete controllability principle discussed for bilinear mathematical models [601]; additive control may be used to steer the system to new equilibrium points, and multiplicative control, either to stabilize a chaotic behavior or to enlarge the attainable space. Then, bilinear systems are of extreme importance in the design and use of optimal control for chaotic behaviors. We can now understand the butterfly effect, i.e., the extreme sensitivity of chaotic systems to tiny perturbations described in Chapter 3.

The transformation procedure of a time-varying parameter model to a nonlinear one has already been applied in other contexts. For instance in a simple case, if it is possible to approximate $\log a(t)$ linearly at any logarithmically transformed state $\log y(t)$, one obtains $\log a(t) = \lambda + \mu \log y(t)$. In terms of the original variables, that gives a power-law approximation

$$a(t) = \lambda y^{\mu}(t).$$

This approximation is better over a wider range than the linearization in the space of the original states. Subsequently, the differential equation with time-varying parameters (C.1) is transformed into a differential equation of the form

$$\dot{y}(t) = \lambda y^{\mu+1}(t).$$

Table C.1: Classical and nonclassical considerations of the in vitro and in vivo drug processes.

	Media	
Models or features	**Homogeneous**	**Heterogeneous**
Empirical model	Sum of exponentials	Power law
Phenomenological model	Deterministic	Stochastic
Retention probability	Exponential	Weibull
Process memory	No	Yes
Process uncertainty	No	Yes

In the presence of multiple states, the right-hand-side term consists of sums, products, and nesting of elementary functions such as y^{μ}, $\log y$, $\exp y$, and trigonometric functions, called the S-system formalism [602]. Using it as a canonical form, special numerical methods were developed to integrate such systems [603]. The simple example of the diffusion-limited or dimensionally restricted homodimeric reaction was presented in Section 2.5.3. Equation 2.23 is the traditional rate law with concentration squared and time-varying time "constant" $k(t)$, whereas (2.22) is the power law $(c^{\gamma}(t))$ in the state differential equation with constant rate.

The presented procedures show how to expect chaotic behaviors with processes revealing uncertainty and which are described by models involving time-varying parameters. All these considerations oriented us to complete the initial Table 1 referenced in the preface by Table C.1.

Appendix D

Probability

D.1 Basic Properties

- *Poincaré theorem.* Given n random events $A_1, \ldots, A_n$, the probability of their union is given by

$$\begin{aligned}\Pr\left[\bigcup_{i=1}^{n} A_i\right] &= \sum_{i=1}^{n} \Pr[A_i] - \sum_{j=2}^{n}\sum_{i=1}^{j-1} \Pr\left[A_i \bigcap A_j\right] \\ &\quad + \sum_{j=3}^{n}\sum_{k=2}^{j-1}\sum_{i=1}^{k-1} \Pr\left[A_i \bigcap A_j \bigcap A_k\right] - \cdots \\ &\quad + (-1)^n \Pr\left[\bigcap_{i=1}^{n} A_i\right].\end{aligned}$$

  If the events are *mutually exclusive*, i.e., $\forall i, j \quad A_i \cap A_j = \varnothing$, then

$$\Pr\left[\bigcup_{i=1}^{n} A_i\right] = \sum_{i=1}^{n} \Pr[A_i].$$

- *Conditional probability.* Given the random events A and B, the conditional probability of A for observed B is defined by

$$\Pr[A \mid B] \triangleq \frac{\Pr[A \cap B]}{\Pr[B]}.$$

  Two events A and B are defined as *independent* if $\Pr[A \mid B] = \Pr[A]$, or $\Pr[A \cap B] = \Pr[A]\Pr[B]$. Given n random events $A_1, \ldots, A_n$, the probability of their intersection is given by

$$\begin{aligned}\Pr\left[\bigcap_{i=1}^{n} A_i\right] &= \Pr\left[A_1 \mid \bigcap_{i=2}^{n} A_i\right] \Pr\left[A_2 \mid \bigcap_{i=3}^{n} A_i\right] \cdots \\ &\quad \Pr[A_{n-1} \mid A_n] \Pr[A_n].\end{aligned}$$

- *Total probability theorem.* Given n mutually exclusive events $A_1, \ldots, A_n$, whose probabilities sum to unity, then

$$\Pr[B] = \Pr[B \mid A_1] \Pr[A_1] + \cdots + \Pr[B \mid A_n] \Pr[A_n],$$

where B is an arbitrary event, and $\Pr[B \mid A_i]$ is the conditional probability of B assuming A_i.

- *Bayes theorem.* For the same settings, the Bayes theorem gives the conditional probability

$$\Pr[A_i \mid B] = \frac{\Pr[B \mid A_i] \Pr[A_i]}{\sum_{k=1}^{n} \Pr[B \mid A_k] \Pr[A_k]}.$$

D.2 Expectation, Variance, and Covariance

For scalar continuous random variables X and Y with joint probability density $f(x, y)$, marginals and conditionals are refined as

$$f(x) = \int_y f(x, y) \, \mathrm{d}y, \qquad f(y) = \int_x f(x, y) \, \mathrm{d}x$$

and

$$f(x \mid y) = f(x, y) / f(y), \qquad f(y \mid x) = f(x, y) / f(x),$$

respectively. The statistical characteristics up to second order of X and Y are:

- *Expectation.* It can be interpreted as the center of gravity of random variables:

$$E[X] = \int_x x f(x) \, \mathrm{d}x, \qquad E[Y] = \int_y y f(y) \, \mathrm{d}y$$

on X and Y axes, respectively, or

$$E[XY] = \int_{xy} xy f(x, y) \, \mathrm{d}x \, \mathrm{d}y$$

on X, Y plan. If X and Y are independent, $E[XY] = E[X] E[Y]$.

- *Variance*: It can be interpreted as the inertia about the centers of gravity $E[X]$ and $E[Y]$:

$$Var[X] = \int_x \{x - E[X]\}^2 f(x) \, \mathrm{d}x, \quad Var[Y] = \int_y \{y - E[Y]\}^2 f(y) \, \mathrm{d}y.$$

- *Covariance*:

$$\begin{aligned} Cov[X, Y] &= \int_{x,y} \{x - E[X]\} \{y - E[Y]\} f(x, y) \, \mathrm{d}x\mathrm{d}y \\ &= E[XY] - E[X] E[Y] \end{aligned}$$

and *correlation*:

$$Cor[X, Y] = \frac{Cov[X, Y]}{\sqrt{Var[X] Var[Y]}}.$$

D.3 Conditional Expectation and Variance

- *Conditional expectation.* It is defined as

$$E[X \mid y] = \int_x x f(x \mid y)\,\mathrm{d}x, \qquad E[Y \mid x] = \int_y y f(y \mid x)\,\mathrm{d}y,$$

  and these are functions of y and x, respectively [604]. It follows that

$$E[X] = E_y E[X \mid y], \qquad E[Y] = E_x E[Y \mid x].$$

  In these expressions, $E[X \mid y]$ and $E[Y \mid x]$ are considered as random variables and subscripts in E_x or E_y mean that expectation is taken with respect to x or y by using their respective marginals. The two last expressions are also known as *total expectations.*

- *Conditional variance.* It is defined as

$$Var[X \mid y] = \int_x \{x - E[X \mid y]\}^2 f(x \mid y)\,\mathrm{d}x$$

  and

$$Var[Y \mid x] = \int_y \{y - E[Y \mid x]\}^2 f(y \mid x)\,\mathrm{d}y,$$

  and they are functions of y and x, respectively [604]. It follows that

$$Var[X] = Var_y E[X \mid y] + E_y Var[X \mid y]$$

  and

$$Var[Y] = Var_x E[Y \mid x] + E_x Var[Y \mid x]$$

  As above, $Var[X \mid y]$ and $Var[Y \mid x]$ are considered as random variables and subscripts in Var_x or Var_y mean that variance is taken with respect to x or y using their respective marginals. The two last expressions are also known as *total variances.*

D.4 Generating Functions

Generating functions are coming into widespread use as methodological tools [385]. They may be used to obtain numerical summary measures of probability distributions in an analytical form by computing its moments and cumulants. For the nonnegative integer-valued random variable $X(t)$:

- The *probability generating function* $\mathcal{P}(s,t)$ is defined as

$$\mathcal{P}(s,t) = \sum s^x p_x(t),$$

  where s is a "dummy variable" such that $|s| < 1$. It follows that one could obtain any probability, say $p_i(t)$, by differentiating $\mathcal{P}(s,t)$ with respect to s; specifically,

$$p_i(t) = \mathcal{P}^{(i)}(0,t),$$

  where $\mathcal{P}^{(i)}(0,t)$ denotes the ith derivative with respect to s evaluated at $s = 0$.

- The *moment generating function* $\mathcal{M}(\theta, t)$ is defined as

$$\mathcal{M}(\theta, t) = \sum \exp(\theta x) p_x(t),$$

where θ is a "dummy variable." Clearly, using the previous relation one has $\mathcal{M}(\theta, t) = \mathcal{P}(\exp(\theta), t)$. If $\mathcal{M}(\theta, t)$ is expressed as the power series

$$\mathcal{M}(\theta, t) = \sum_{i \geq 0} \frac{\mu_i(t)\, \theta^i}{i!},$$

the coefficients $\mu_i(t)$ in this series expansion are the ith moments of $X(t)$, which are usually defined as $\mu_i(t) = \sum x^i p_x(t)$ with $\mu_0 = 1$. It follows that the ith moment may be obtained from the moment generating function as

$$\mu_i(t) = \mathcal{M}^{(i)}(0, t).$$

- The *cumulant generating function* $\mathcal{K}(\theta, t)$ is defined as

$$\mathcal{K}(\theta, t) = \log \mathcal{M}(\theta, t),$$

with power series expansion

$$\mathcal{K}(\theta, t) = \sum_{i \geq 0} \frac{\kappa_i(t)\, \theta^i}{i!}.$$

This equation formally defines a cumulant $\kappa_i(t)$ as a coefficient in the series expansion of $\mathcal{K}(\theta, t)$. It too is easily found from its generating function as

$$\kappa_i(t) = \mathcal{K}^{(i)}(0, t).$$

The first three cumulants may be obtained as

$$\begin{aligned} \kappa_1(t) &= \mu_1(t), \\ \kappa_2(t) &= \mu_2(t) - \mu_1^2(t), \\ \kappa_3(t) &= \mu_3(t) - 3\mu_1(t)\,\mu_2(t) + 2\mu_1^3(t), \end{aligned}$$

which give the mean, variance, and skewness functions for $X(t)$ from the $\mu_i(t)$ moment functions.

Appendix E

Convolution in Probability Theory

A convolution is an integral that expresses the amount of overlap of one function g as it is shifted over another function f. It therefore "blends" one function with another. The convolution is sometimes also known by its German name, *Faltung* (*folding*). Abstractly, a convolution is defined as a product of functions f and g that are objects in the algebra of Schwartz functions in $\mathbb{R}^n$. Convolution of two functions $f(z)$ and $g(z)$ over a finite range $[0, t]$ is given by

$$f * g\,(t) \triangleq \int_0^t f\,(\tau - t)\, g\,(\tau)\, \mathrm{d}\tau,$$

where the symbol $f * g$ denotes convolution of f and g.

There is also a definition of the convolution that arises in probability theory and is given by

$$F * G\,(t) = \int_0^t F\,(t - z)\, \mathrm{d}G\,(z)\,,$$

where

$$\int_0^t F\,(t - z)\, \mathrm{d}G\,(z)$$

is a Stieltjes integral.

The Stieltjes integral is a generalization of the Riemann integral. Let $f(z)$ and $h\,(z)$ be real-valued bounded functions defined on a closed interval $[a, b]$. Take a partition of the interval $a = z_1 < z_2 < \cdots < z_{n-1} < z_n = b$ and consider the Riemann sum

$$\sum_{i=1}^{n-1} f\,(\xi_i)\,[h\,(z_{i+1}) - h\,(z_i)]$$

with $\xi_i \in [z_i, z_{i+1}]$. If the sum tends to a fixed number $\mathcal{I}$ as $\max\,(z_{i+1} - z_i) \to 0$, then $\mathcal{I}$ is called the Stieltjes integral, or sometimes the Riemann–Stieltjes integral. The Stieltjes integral of f with respect to h is denoted by $\int f(z)\mathrm{d}h(z)$. If f and h have a common point of discontinuity, then the integral does not exist. However, if f is continuous and h' is Riemann integrable over the specified interval, then

$$\int f(z)\,\mathrm{d}h(z) = \int f(z)\,h'(z)\,\mathrm{d}z.$$

For enumeration of many of the Stieltjes integral properties, cf. [605] (p.105). In the following, we present some useful convolution relationships:

- $f * g(t) = g * f(t) \triangleq \int_0^t f(\tau - t)\,g(\tau)\mathrm{d}\tau = \int_0^t g(\tau - t)\,f(\tau)\mathrm{d}\tau$
- $f(t) * [k_1 g(t) + k_2 h(t)] = k_1 f * g(t) + k_2 f * h(t)$
- $f * g(t)|_{t=0} = 0$
- $\int_0^t f * g(\tau)\mathrm{d}\tau = f(t) * \int_0^t g(\tau)\mathrm{d}\tau = g(t) * \int_0^t f(\tau)\mathrm{d}\tau$
- $\int_0^\infty f * g(\tau)\mathrm{d}\tau = \left[\int_0^\infty f(\tau)\,\mathrm{d}\tau\right]\left[\int_0^\infty g(\tau)\,\mathrm{d}\tau\right]$
- $\frac{\mathrm{d}}{\mathrm{d}t}[f * g(t)] = f(t)\,g(0) + f * \frac{\mathrm{d}g(t)}{\mathrm{d}t} = g(t)\,f(0) + g * \frac{\mathrm{d}f(t)}{\mathrm{d}t}$
- $k * f(t) = k\int_0^t f(\tau)\mathrm{d}\tau$
- $\frac{\mathrm{d}}{\mathrm{d}t}[k * f(t)] = kf(t)$
- $\delta * f(t) = f(t)$

where k is a scalar constant and $\delta(t)$ is the Dirac delta function.

Appendix F

Laplace Transform

The Laplace transform $\widetilde{f}(s)$ of the function $f(t)$ of the non negative variable t is defined by

$$\widetilde{f}(s) \triangleq L\{f(t)\} = \int_0^\infty \exp(-st)\, f(t)\, \mathrm{d}t.$$

This transform is widely used to formulate semi-Markov stochastic models, where t and $f(t)$ are the random variable and its probability density function, respectively. In Table F.1, we briefly report some Laplace transform pairs.

Table F.1: Some Laplace transform properties and pairs of functions used as probability density functions for semi-Markov modeling.

$f(t)$	$\widetilde{f}(s)$
$\exp(-\alpha t)\, f(t)$	$\widetilde{f}(s+\alpha)$
$f(t-\alpha)\, u(t-\alpha)$	$\exp(-\alpha s)\, \widetilde{f}(s)$
$f(kt)$	$\widetilde{f}(s/k)/k$
$f_1 * f_2(t)$	$\widetilde{f}_1(s)\, \widetilde{f}_2(s)$
$\mathrm{Exp}(\kappa)$	$\kappa/(s+\kappa)$ with $s > -\kappa$
$\mathrm{Erl}(\lambda, \nu)$	$[1+(s/\lambda)]^{-\nu}$ with $s > -\lambda$
$\mathrm{Chi}(\nu)$	$(1+2s)^{-\nu/2}$ with $s > -\frac{1}{2}$
$\mathrm{Gam}(\lambda, \mu)$	$[1+(s/\lambda)]^{-\mu}$ with $s > -\lambda$
$\mathrm{Rec}(\alpha, \beta)$	$\{\exp(-\alpha s) - \exp[-(\alpha+\beta)s]\}/(\beta s)$

The probability density functions $\mathrm{Exp}(\kappa)$, $\mathrm{Erl}(\lambda, \nu)$, $\mathrm{Gam}(\lambda, \mu)$, and $\mathrm{Rec}(\alpha, \beta)$ are defined in Tables 9.1 and 9.2, and $\mathrm{Chi}(\nu)$ is the χ^2 distribution with ν degrees of freedom. After modeling in frequency s-space, the solution in time t-space must be obtained by inverse Laplace transform. Nevertheless, given the complexity of the obtained model, the inverse transform may be rarely obtained from the above table. Usually, the numerical inverse Laplace transform is used [353, 360].

Appendix G

Estimation

Since this monograph is devoted only to the conception of mathematical models, the inverse problem of estimation is not fully detailed. Nevertheless, estimating parameters of the models is crucial for verification and applications. Any parameter in a deterministic model can be sensibly estimated from time-series data only by embedding the model in a statistical framework. It is usually performed by assuming that instead of exact measurements on concentration, we have these values blurred by observation errors that are independent and normally distributed. The parameters in the deterministic formulation are estimated by nonlinear least-squares or maximum likelihood methods.

Let us point out the meaning of estimation in the models with heterogeneous particles. The situation is qualitatively different from that described above because the constants have been replaced by random variables. Now, by fitting the model to the observed data, we obtain estimates of the parameters of the distribution of the random variables. Moreover, the weighting scheme in the nonlinear regression is not the same as in the deterministic case; we need to take into account the process uncertainty and the measurement error components blurring the observations [304,340]. Special computational methods in nonlinear regression are available (unconditional and conditional generalized least squares, etc.) [372], and the classical maximum likelihood approach is also possible based on the multinomial distribution of particles in the compartments.

Appendix H

Theorem on Continuous Functions

Lemma 13 *Let $h(z)$ be a derivable function of z over $[a,b]$ satisfying $h(z) \neq 0$ for all $z \in]a,b[$, $h(a) = h(b) = 0$, and $h'(a) < 0$. Then, for all $z \in]a,b[$, $h(z) < 0$ and $h'(b) \geq 0$.*

When the derivative $h'(a)$ is approximated by the quotient difference, we have

$$h'(a) \approx \frac{h(z) - h(a)}{z - a} = \frac{h(z)}{z - a} < 0,$$

and therefore, using continuity, $h(z) < 0$ for $z \in]a, a+\Delta a]$. Since $h(z)$ is continuous over $]a,b[$ and $h(z) \neq 0$, $h(z)$ preserves its sign for $z \in]a,b[$; consequently, $h(z) < 0$ for $z \in]a,b[$. Conversely, for all $z \in]a,b[$, we have

$$\frac{h(z) - h(b)}{z - b} = \frac{h(z)}{z - b} > 0$$

and

$$h'(b) = \lim_{z \to b} \frac{h(z) - h(b)}{z - b} \geq 0$$

■

Proposition 14 *Let $f(z)$ and $g(z)$ be derivable functions of z over $[a,b]$ satisfying: $g(z)$ is a monotone increasing function over $[a,b]$, $f(z) \neq g(z)$ for all $z \in]a,b[$, $f(a) = g(a)$ and $f(b) = g(b)$, and $f'(a) < 0$. Then $f(z) < g(z)$ and $f'(b) \geq 0$.*

Let $h(z) = f(z) - g(z)$. Since $f'(a) < 0$ and $g'(a) \geq 0$ (g is monotone increasing), $h'(a) = f'(a) - g'(a) < 0$. Because of $f(z) \neq g(z)$, $h(z) \neq 0$ for all $z \in]a,b[$. According to the previous lemma, it follows that $h(z) < 0$, i.e., $f(z) < g(z)$ for all $z \in]a,b[$, and $h'(b) = f'(b) - g'(b) \geq 0$. Since $g'(b) \geq 0$ (g is monotone increasing), we have also $f'(b) \geq 0$. ■

A similar proof may be delineated for the dual proposition:

Proposition 15 *Let $f(z)$ and $g(z)$ be derivable functions of z over $[a, b]$ satisfying: $g(z)$ is a monotone increasing function over $[a, b]$, $f(z) \neq g(z)$ for all $z \in]a, b[$, $f(a) = g(a)$ and $f(b) = g(b)$, and $f'(a) > 0$. Then $f(z) > g(z)$ and $f'(b) \leq 0$.*

From the last two propositions, we can state the following result:

Theorem 16 *Let $f(z)$ and $g(z)$ be derivable functions of z over the interval I and let $g(z)$ be a monotone increasing function. Let also $a_1 < a_2 < \cdots < a_n$ be n reals over I satisfying: $f(a_i) = g(a_i)$ for $i = 1, \ldots, n$ and $f(z) \neq g(z)$ for all $z \in]a_i, a_{i+1}[$ with $i = 1, \ldots, n-1$. Then $f(a_i)g(a_i) \leq 0$ for $i = 1, \ldots, n$. In other words, the derivatives on two successive intersection points between two continuous functions, one of which is monotone, have opposite signs.*

Simply, apply the previous propositions to the segment $[a_i, a_{i+1}]$. ■

Appendix I

List of Symbols

The symbols in the following tables are classified in several lists according to their significance and form: symbols associated with functions and distributions (Table I.1), time-dependent variables (Table I.2), random variables (Table I.3), constants and parameters (Tables I.4, I.5, I.6), and Greek symbols (Table I.7).

In order to respect the initial writing in the literature of symbols, sometimes but in a different place the same symbol has been used for more than one purpose. For example, $s(t)$ denotes the substrate variable in Chapters 8 and 9, whereas it refers to the neutrophil myelocytes in Chapter 11. In such cases, we systematically report as reference for each use the number of the corresponding chapter. For random variables, a pair of symbols is used with the same character in uppercase and lowercase form to denote the name of a random variable and an element of that variable, respectively. For instance, A denotes the random variable "age" and a a given age. Underscored lowercase characters and bold uppercase denote vectors and matrices, respectively, e.g., $\underline{y}$ and $\mathbf{H}$. Usually, Greek letters κ, λ, μ, ν stand for the parameters of statistical distributions, and α, β, and γ are used as unspecified constants or parameters.

Table I.1: Functions and distributions.

SYMBOL	LABEL
$B(t)$	Brownian motion
$\xi(t)$	Gaussian white noise, Chapter 5
$\varphi(t)$	Fraction of dose dissolved, Chapters 5, 6
$c_B(y)$	Binding curve
$c_F(y)$	Feedback curve
$\delta(\cdot)$	Dirac delta function
$f(a)$	Density function
$F(a)$	Distribution function
$\Phi(\cdot)$	Feedback control function
$\phi(\cdot)$	Dimensionless feedback function
$g(\cdot)$	Functional form
$I(\cdot)$	Intensity function
$J_0(\cdot)$	Zero-order Bessel function
$\mathcal{K}(\cdot,\cdot)$	Cumulant generating function
$\mathcal{M}(\cdot,\cdot)$	Moment generating function
$\mathcal{P}(\cdot,\cdot)$	Probability generating function
$\mathcal{S}(a)$	Survival function
$\mathcal{T}(\cdot)$	Transducer function
$r(t)$	Input function
$\rho(t)$	Dimensionless input function
$u(\cdot)$	Heaviside step function
$\Gamma(\cdot)$	Gamma function
$\psi(t)$	History function, Chapter 11
$\mathrm{Bin}(\cdot,\cdot)$	Binomial distribution
$\mathrm{Chi}(\cdot)$	χ^2 distribution
$\mathrm{Erl}(\cdot,\cdot)$	Erlang distribution
$\mathrm{Exp}(\cdot,\cdot)$	Exponential distribution
$\mathrm{Gam}(\cdot,\cdot)$	Gamma distribution
$\mathrm{Rec}(\cdot,\cdot)$	Rectangular (uniform) distribution
$\mathrm{Wei}(\cdot,\cdot)$	Weibull distribution
$E[\cdot]$	Expectation
$Var[\cdot]$	Variance
$Cov[\cdot,\cdot]$	Covariance
$Cor[\cdot,\cdot]$	Correlation

Table I.2: Time-dependent variables.

SYMBOL	LABEL
$c(t)$	Drug concentration, Chapters 10, 11
$e(t)$	Enzyme, Chapters 8, 9
	Cytokines, Chapter 11
$E(t)$	Pharmacological effect, Chapters 10, 11
$s(t)$	Substrate, Chapters 8, 9
	Neutrophil myelocytes, Chapter 11
$v(t)$	Substrate–enzyme complex, Chapter 8
	Drug–receptor complex, Chapters 10, 11
$w(t)$	Product of enzymatic reaction, Chapter 8
	Blood neutrophils, Chapter 11
$x(\tau), y(\tau), z(\tau)$	Dimensionless state variables
$y(t)$	State variable

Table I.3: Random variables.

SYMBOL	LABEL	DIMENSION
a, A	Age, Chapter 9	Time
$c(t), C(t)$	Concentration, Chapter 9	Mass$\times$Volume^{-1}
$n(t), N(t)$	no. of particles, Chapter 9	
$q(t), Q(t)$	Amount, quantity, Chapter 9	Mass
t, T	Time, Chapter 5	Time
θ, Θ	Characteristic parameter, Chapter 8	

Table I.4: Constants, parameters (part 1). [apu] denotes arbitrary pharmacological units.

SYMBOL	LABEL	DIMENSION
$\mathcal{A}$	Area	Area
A_n	Absorption number	
AUC	Area under curve	Mass×Volume^{-1}×Time
$B_{..}$	Coefficients in a sum of exponentials	Mass×Volume^{-1}
$b_{.}$	Exponents in a sum of exponentials	Time^{-1}
$\mathbf{B}_{.}, \underline{b}_{.}$	Parameters in pseudocompartments	
c_s	Solubility	Mass×Volume^{-1}
c_0	Initial concentration	Mass×Volume^{-1}
$c_{\max}$	Peak drug concentration	Mass×Volume^{-1}
CL	Clearance	Volume×Time^{-1}
CV	Coefficient of variation	
d_t	Topological dimension	
d_f	Fractal dimension	
d_c	Capacity dimension	
d_e	Embedding dimension	
d_s	Spectral dimension	
d_w	Random walk dimension	
d_o	Cover dimension	
d^*	Index of nonlinear coherence	
$\mathcal{D}$	Diffusion coefficient	Area×Time^{-1}
$\mathcal{D}'$	Modified diffusion coefficient	Area×Time^{-1}
$\mathcal{D}_\gamma$	Fractional diffusion coefficient	Area×Time^{-1}
D	Dispersion coefficient	Area×Time^{-1}
D_n	Dissolution no.	
e_0	Initial enzyme amount, Chapters 8, 9	Mass
	Initial cytokine amount, Chapter 11	Mass
E_0	Baseline in $E_{\max}$ model	[apu]
$E_{\max}$	Maximum pharmacological effect	[apu]
Ec_{50}	Concentration at half $E_{\max}$	Mass×Volume^{-1}
f_1	Difference factor	
f_2	Similarity factor	
f_u	Drug unbound fraction	
f_{un}	Fraction of unionized species	
F_a	Fraction of dose absorbed	
$h_{..}$	Hazard rates	Time^{-1}
$\mathbf{H}$	Transfer intensity matrix	
$I_{\max}$	Maximum inhibition rate	
Ic_{50}	Concentration at half $I_{\max}$	Mass×Volume^{-1}
J	Net flux	Mass×Area^{-1}×Time^{-1}

Table I.5: Constants, parameters (part 2). [apu] denotes arbitrary pharmacological units.

Symbol	Label	Dimension
k	First-order rate const. (generic)	Time^{-1}
$k_{\circ}$	Reference rate const.	Time^{-1}
k_{0}	Case-II relaxation const.	Mass×Area^{-1}×Time^{-1}
k_{+1}	Forward enzyme reaction rate const.	Mass^{-1}×Time^{-1}
k_{-1}	Backward enzyme reaction rate const.	Time^{-1}
k_{+2}	Enzymatic product formation rate const.	Time^{-1}
k_{2}	Pharmacological proportionality const.	[apu]×Mass^{-1}×Volume
k_{a}	Macroscopic absorption rate const.	Time^{-1}
k_{a}'	Microscopic absorption rate const.	Time^{-1}
k_{d}	Dissolution rate const.	Time^{-1}
$k_{d,eff}$	Effective dissolution rate const.	Time^{-1}
k_{c}	Controlled dissolution rate const., Chapter 6	Time^{-1}
	Effect–compartment rate const., Chapter 10	Time^{-1}
k_{s}	Surface area dissolution rate const.	
k_{D}	Dissociation const.	Mass
k_{i}, k_{o}	Input (orders 0 and 1), output rate const.	
k_{e}, k_{s}, k_{w}	Rate const. in hemopoiesis	
k_{y}	Effect–compartment rate const.	Time^{-1}
k_{M}	Michaelis–Menten const.	Mass
$k_{..}$	Fractional flow rates	Time^{-1}
$\mathbf{K}$	Matrix of fractional flow rates	
L	Height of cylinder	Length
m	no. of objects (except particles): samples, compartments, individuals, vessels, administrations, sites	
$m_{\circ}$	no. of reaction channels	
n_{0}	Initial number of particles	
$n(t)$	no. of visited sites, Chapter 2	
	no. of remained particles, Chapter 4	
$\widetilde{n}(t)$	no. of escaped particles	
N_{leak}	no. of leak sites	
N_{tot}	Total no. of sites	
N_{villi}	no. of villi	
p	Probability	
p_{a}	Probability for absorption by villi	
p_{f}	Forward probability to the output	
p_{c}	Critical probability	

Table I.6: Constants, parameters (part 3).

SYMBOL	LABEL	DIMENSION
P	Permeability	Length×Time^{-1}
P_{app}	Apparent permeability	Length×Time^{-1}
P_{eff}	Effective permeability	Length×Time^{-1}
P_c	Partition coefficient	
$\mathbf{P}$	State probability matrix	
q_∞	Maximum cumulative amount	Mass
Q	Volumetric flow rate	Volume×Time^{-1}
R	Radius of tube	Length
$R_{..}$	Transfer rate	Mass×Time^{-1}
R^2	Coefficient of determination	
$R_{\max}$	Maximum biotransformation rate	Mass×Time^{-1}
r_0	Total number of receptors	
R_i	Reference drug dissolved at i	Mass
$S_{\max}$	Maximum stimulation rate	
Sc_{50}	Concentration at half $S_{\max}$	Mass×Volume^{-1}
t	Time	Time
τ	Dimensionless time	
t°	Time delay	Time
τ°	Dimensionless time delay	
$t_\circ$	Time reference	Time
t_0	Initial time	Time
t_{sim}	Maximum simulation time	Time
t_{diff}	Diffusion time	Time
t_{reac}	Reaction time	Time
$t_{\max}$	Time to $c_{\max}$	Time
$t_{..}$	Observation times	Time
T	Infusion duration	Time
T_E	Infusion ending time	Time
T_S	Infusion starting time	Time
T_{si}	Small-intestinal transit time	Time
T_i	Test drug dissolved at i	Mass
u_i	Unit geometric vector in direction i	
v	Velocity	Length×Time^{-1}
V	Volume of distribution	Volume
V_c	Central compartment volume	Volume
V_y	Effect–compartment volume	Volume
$V_{\max}$	Maximum transport rate	Mass×Time^{-1}
$y_{..}$	Observations	Mass×Volume^{-1}
z	Spatial coordinates	Length

Table I.7: Greek symbols. [apu] denotes arbitrary pharmacological units.

SYMBOL	LABEL	DIMENSION
α, β, γ	Constants, parameters (usually)	
δ	Thickness, elementary distance	Length
Δ	Finite difference symbol	
ε	Intrinsic efficacy, Chapter 10	[apu]$\times$Mass$^{-1}\times$Volume
	Bias factor, Chapter 6	
	Subinterval length, Chapter 9	Length
$\varepsilon_{..}$	Prediction error	
ζ	Eigenvalue	
η	Eigenvector	
θ	Characteristic parameter, Chapters 1, 3, 9, 11	
	Dose–solubility ratio, Chapters 5, 6	
$\kappa, \lambda, \mu, \nu$	Distribution parameters (usually)	
$\kappa_{.}$	Cumulants	
$\mu_{.}$	Moments	
ξ	Characteristic multiplier, Chapter 3	
	Correlation length, Chapter 1	
ϱ	Density of the drug	Mass$\times$Volume^{-1}
ρ	Radius of particle	Length
ρ_c	Critical radius	Length
ρ_0	Initial radius	Length
σ^2	Variance, Chapters 5, 8	
$\varphi_{..}$	Changes in population size by the reaction, Chapter 9	
$\psi_{..}$	no. of particles implied in the reaction, Chapter 9	
$\underline{\Psi}$	Vector of population parameters	
ω	Resolution, Chapter 1	
$\omega_{..}$	Transition probabilities, Chapter 9	

Bibliography

[1] Kac, M., Some mathematical models in science, *Science*, Vol. 166, 1969, pp. 695–699.

[2] Mandelbrot, B., *Fractals: Form, Chance, and Dimension*, W.H. Freeman, San Francisco, 1977.

[3] Parker, T. and Chua, L., *Practical Numerical Algorithms for Chaotic Systems*, Springer-Verlag, New York, 1989.

[4] Bassingthwaighte, J., Liebovitch, L., and West, B., *Fractal Physiology*, Methods in Physiology Series, Oxford University Press, New York, 1994.

[5] Flory, P., Molecular size distribution in three dimensional polymers. I. Gelation, *Journal of American Chemical Society*, Vol. 63, No. 11, 1941, pp. 3083–3090.

[6] Flory, P., Molecular size distribution in three dimensional polymers. II. Trifunctional branching units, *Journal of American Chemical Society*, Vol. 63, No. 11, 1941, pp. 3091–3096.

[7] Flory, P., Molecular size distribution in three dimensional polymers. III. Tetrafunctional branching units, *Journal of American Chemical Society*, Vol. 63, No. 11, 1941, pp. 3096–3100.

[8] Stockmayer, W., Theory of molecular size distribution and gel formation in branched-chain polymers, *Journal of Chemical Physics*, Vol. 11, 1943, pp. 45–55.

[9] Kopelman, R., Rate processes on fractals: Theory, simulations, and experiments, *Journal of Statistical Physics*, Vol. 42, No. 1/2, 1986, pp. 185–200.

[10] Crank, J., *The Mathematics of Diffusion*, Oxford University Press, 2nd ed., 1975.

[11] Ovchinnikov, A. and Zeldovich, Y., Role of density fluctuations in bimolecular reaction kinetics, *Chemical Physics*, Vol. 28, 1978, pp. 215–218.

[12] Toussaint, D. and Wilczek, F., Particle-antiparticle annihilation in diffusive motion, *Journal of Chemical Physics*, Vol. 78, No. 5, 1983, pp. 2642–2647.

[13] Lindenberg, K., West, B., and Kopelman, R., Steady-state segregation in diffusion limited reactions, *Physical Review Letters*, Vol. 60, No. 18, 1988, pp. 1777–1780.

[14] Wenzel, C., Lehre Von der Verwandtschaft der Körper, *Dresden*, Vol. 8, 1777.

[15] Farin, D. and Avnir, D., Reactive fractal surfaces, *Journal of Physical Chemistry*, Vol. 91, No. 22, 1987, pp. 5517–5521.

[16] Kopelman, R., Fractal reaction kinetics, *Science*, Vol. 241, 1988, pp. 1620–1626.

[17] Clement, E., Kopelman, R., and Sander, L., The diffusion-limited reaction A+B –>0 on a fractal substrate, *Journal of Statistical Physics*, Vol. 65, No. 5-6, 1991, pp. 919–924.

[18] Klymko, P. and Kopelman, R., Fractal reaction-kinetics: Exciton fusion on clusters, *Journal of Physical Chemistry*, Vol. 87, No. 23, 1983, pp. 4565–4567.

[19] Alexander, S. and Orbach, R., Density of states on fractals: Fractons, *Journal of Physical Letters*, Vol. 43, 1982, pp. L625–631.

[20] Zumofen, G., Blumen, A., and Klafter, J., Concentration fluctuations in reaction kinetics, *Journal of Chemical Physics*, Vol. 82, No. 7, 1985, pp. 3198–3206.

[21] Newhouse, J. and Kopelman, R., Steady-state chemical kinetics on surface clusters and islands: Segregation of reactants, *Journal of Physical Chemistry*, Vol. 92, No. 6, 1988, pp. 1538–1541.

[22] Jiang, Z. and Ebner, C., Simulation study of reaction fronts, *Physical Review A*, Vol. 42, No. 12, 1990, pp. 7483–7486.

[23] DeGennes, P., Kinetics of diffusion-controlled processes in dense polymer systems I. Nonentangled regimes, *Journal of Chemical Physics*, Vol. 76, No. 6, 1982, pp. 3316–3321.

[24] DeGennes, P., Kinetics of diffusion-controlled processes in dense polymer systems II. Effects of entanglements, *Journal of Chemical Physics*, Vol. 76, No. 6, 1982, pp. 3322–3326.

[25] Savageau, M., Michaelis-Menten mechanism reconsidered: Implication of fractal kinetics, *Journal of Theoretical Biology*, Vol. 176, 1995, pp. 115–124.

[26] Srere, P., Jones, M., and Mathews, C., *Structural and Organizational Aspects of Metabolic Regulation*, Wiley-Liss, 1990.

[27] Savageau, M., Development of fractal kinetic theory for enzyme-catalysed reactions and implications for the design of biochemical pathways, *Biosystems*, Vol. 47, 1998, pp. 9–36.

[28] Schroeder, M., *Fractals, Chaos, Power Laws: Minutes from an Infinite Paradise*, W.H. Freeman, New York, 1991.

[29] Sokolov, I., Klafter, J., and Blumen, A., Fractional kinetics, *Physics Today*, Vol. 55, 2002, pp. 48–54.

[30] Berding, C., On the heterogeneity of reaction-diffusion generated patterns, *Bulletin of Mathematical Biology*, Vol. 49, No. 2, 1987, pp. 233–252.

[31] Glass, L. and Mackey, M., *From Clock to Chaos. The Rhythms of Life*, Princeton University Press, New Jersey, 1988.

[32] Tsonis, A., *Chaos: From Theory to Applications*, Kluwer Academic Publishers, 1992.

[33] Dokoumetzidis, A., Iliadis, A., and Macheras, P., Nonlinear dynamics and chaos theory: Concepts and applications relevant to pharmacodynamics, *Pharmaceutical Research*, Vol. 18, No. 4, 2001, pp. 415–426.

[34] Farmer, J., Ott, E., and Yorke, J., The dimension of chaotic attractors, *Physica D*, Vol. 7, No. 1-3, 1983, pp. 153–180.

[35] Packard, N., Crutchfield, J., Farmer, J., and Shaw, R., Geometry from a time series, *Physical Review Letters*, Vol. 45, No. 9, 1980, pp. 712–716.

[36] Froehling, H., Crutchfield, J., Farmer, D., Packard, N., and Shaw, R., On determining the dimension of chaotic flows, *Physica D*, Vol. 3, No. 3, 1981, pp. 605–617.

[37] Nayfeh, A. and Balachandran, B., *Applied Nonlinear Dynamics: Analytical, Computational, and Experimental Methods*, John Wiley, New York, 1995.

[38] Baake, E., Baake, M., Bock, H., and Briggs, K., Fitting ordinary differential equations to chaotic data, *Physical Review A*, Vol. 45, No. 8, 1992, pp. 5524–5529.

[39] Grassberger, P. and Procaccia, I., Characterization of strange attractors, *Physical Review Letters*, Vol. 50, No. 5, 1983, pp. 346–349.

[40] Thompson, J. and Stewart, H., *Nonlinear Dymanics and Chaos*, John Wiley, New York, 1986.

[41] Albano, A., Mess, A., DeGuzman, G., and Rapp, P., Data requirements for reliable estimation of correlation dimensions, *Chaos in Biological Systems*, edited by H. Degn, A. Holden, and L. Olsen, Plenum Press, New York, 1987, pp. 207–220.

[42] Kaplan, D. and Glass, L., Direct test for determinism in a time series, *Physical Review Letters*, Vol. 68, No. 4, 1992, pp. 427–430.

[43] Kaplan, D. and Glass, L., *Understanding Nonlinear Dynamics*, Textbooks in Mathematical Sciences, Springer-Verlag, Berlin, 1995.

[44] Lorenz, E., Deterministic nonperiodic flow, *Journal of the Atmospheric Sciences*, Vol. 20, No. 2, 1963, pp. 130–148.

[45] Shinbrot, T., Grebogi, C., Ott, E., and Yorke, J., Using small perturbations to control chaos, *Nature*, Vol. 363, 1993, pp. 411–417.

[46] Kapitaniak, T., *Controlling Chaos: Theoretical and Practical Methods in Non-linear Dynamics*, Academic Press, London, 1996.

[47] Mackey, M. and Glass, L., Oscillation and chaos in physiological control systems, *Science*, Vol. 197, No. 4300, 1977, pp. 287–289.

[48] Glass, L. and Mackey, M., Pathological conditions resulting from instabilities in physiological control systems, *Annals of the New York Academy of Sciences*, Vol. 316, 1979, pp. 214–235.

[49] Mackey, M. and Milton, J., Dynamical diseases, *Annals of the New York Academy of Sciences*, Vol. 504, 1987, pp. 16–32.

[50] Skinner, J., Low-dimensional chaos in biological systems, *Biotechnology*, Vol. 12, No. 6, 1994, pp. 596–600.

[51] Dighe, S., Development of dissolution tests for immediate release oral dosage forms, *Bio-lnternational 2, Bioavailabilily, Bioequivalence and Pharmacokinetic Studies*, edited by H. Blume and K. K. Midha, Medpharm, Stuttgart, 1995, pp. 247–255.

[52] Fagerholm, U. and Lennernas, H., Experimental estimation of the effective unstirred water layer thickness in the human jejunum, and its importance in oral drug absorption, *European Journal of Pharmaceutical Sciences*, Vol. 3, No. 5, 1995, pp. 247–253.

[53] Dressman, J. and Fleisher, D., Mixing tank model for predicting dissolution rate control of oral absorption, *Journal of Pharmaceutical Sciences*, Vol. 75, No. 2, 1986, pp. 109–116.

[54] Sinko, P., Leesman, G. D., and Amidon, G. L., Predicting fraction dose absorbed in humans using a macroscopic mass balance approach, *Pharmacological Research*, Vol. 8, No. 8, 1991, pp. 979–988.

[55] Oh, D., Curl, R., and Amidon, G., Estimating the fraction dose absorbed from suspensions of poorly soluble compounds in humans: A mathematical model, *Pharmaceutical Research*, Vol. 10, No. 2, 1993, pp. 264–270.

[56] Higuchi, T., Rate of release of medicaments from ointment bases containing drugs in suspension, *Journal of Pharmaceutical Sciences*, Vol. 50, 1961, pp. 874–875.

[57] Higuchi, T., Mechanism of sustained-action medication. Theoretical analysis of rate of release of solid drugs dispersed in solid satrices, *Journal of Pharmaceutical Sciences*, Vol. 52, 1963, pp. 1145–1149.

[58] Lapidus, H. and Lordi, N., Some factors affecting the release of a water-soluble drug from a compressed hydrophilic matrix, *Journal of Pharmaceutical Sciences*, Vol. 55, No. 8, 1966, pp. 840–843.

[59] Lapidus, H. and Lordi, N., Drug release from compressed hydrophilic matrices, *Journal of Pharmaceutical Sciences*, Vol. 57, No. 8, 1968, pp. 1292–1301.

[60] Siepmann, J. and Peppas, N., Modeling of drug release from delivery systems based on hydroxypropyl methylcellulose (HPMC), *Advanced Drug Delivery Reviews*, Vol. 48, No. 2-3, 2001, pp. 139–157.

[61] Wang, T., Kwei, T., and Frisch, H., Diffusion in glassy polymers. III, *Journal of Polymer Science*, Vol. 7, No. 12, 1969, pp. 2019–2028.

[62] Enscore, D., Hopfenberg, H., and Stannett, V., Effect of particle size on the mechanism controlling n-hexane sorption in glassy polystyrene microspheres, *Polymer*, Vol. 18, No. 8, 1977, pp. 793–800.

[63] Frisch, H., Sorption and transport in glassy polymers - A review, *Polymer Engineering and Science*, Vol. 20, 1980, pp. 2–13.

[64] Peppas, N., Analysis of Fickian and non-Fickian drug release from polymers, *Pharmaceutica Acta Helvetiae*, Vol. 60, No. 4, 1985, pp. 110–111.

[65] Ritger, P. and Peppas, N., A simple equation for description of solute release I. Fickian and non-Fickian release from non-swellable devices in the form of slabs, spheres, cylinders or discs, *Journal of Controlled Release*, Vol. 5, No. 1, 1987, pp. 23–36.

[66] Ritger, P. and Peppas, N., A simple equation for description of solute release II. Fickian and anomalous release from swellable devices, *Journal of Controlled Release*, Vol. 5, No. 1, 1987, pp. 37–42.

[67] Kosmidis, K., Rinaki, E., Argyrakis, P., and Macheras, P., Analysis of Case II drug transport with radial and axial release from cylinders, *International Journal of Pharmaceutics*, Vol. 254, No. 2, 2003, pp. 183–188.

[68] Peppas, N. and Korsmeyer, R., Dynamically swelling hydrogels in controlled applications, *Hydrogels in Medicine and Pharmacy, Vol. 3. Properties and Applications*, edited by N. Peppas, CRC Press, Boca Raton, 1987, pp. 109–136.

[69] Rinaki, E., Valsami, G., and Macheras, P., The power law can describe the 'entire' drug release curve from HPMC-based matrix tablets: A hypothesis, *International Journal of Pharmaceutics*, Vol. 255, No. 1-2, 2003, pp. 199–207.

[70] Fu, J., Hagemeir, C., and Moyer, D., A unified mathematical model for diffusion from drug-polymer composite tablets, *Journal of Biomedical Materials Research*, Vol. 10, No. 5, 1976, pp. 743–758.

[71] Gao, P. and Fagerness, P., Diffusion in HPMC gels. I. Determination of drug and water diffusivity by pulsed-field-gradient spin-echo NMR, *Pharmaceutical Research*, Vol. 12, No. 7, 1995, pp. 955–964.

[72] Gao, P., Nixon, P., and Skoug, J., Diffusion in HPMC gels. II. Prediction of drug release rates from hydrophilic matrix extended-release dosage forms, *Pharmaceutical Research*, Vol. 12, No. 7, 1995, pp. 965–971.

[73] Narasimhan, B. and Peppas, N., Disentanglement and reptation during dissolution of rubbery polymers, *Journal of Polymer Science*, Vol. 34, No. 5, 1996, pp. 947–961.

[74] Narasimhan, B. and Peppas, N., On the importance of chain reptation in models of dissolution of glassy polymers, *Macromolecules*, Vol. 29, No. 9, 1996, pp. 3283 – 3291.

[75] Narasimhan, B. and Peppas, N., Molecular analysis of drug delivery systems controlled by dissolution of the polymer carrier, *Journal of Pharmaceutical Sciences*, Vol. 86, No. 3, 1997, pp. 297–304.

[76] Siepmann, J., Kranz, H., Bodmeier, R., and Peppas, N., HPMC-matrices for controlled drug delivery: A new model combining diffusion, swelling, and dissolution mechanisms and predicting the release kinetics, *Pharmaceutical Research*, Vol. 16, No. 11, 1999, pp. 1748–1756.

[77] Siepmann, J., Podual, K., Sriwongjanya, M., Peppas, N., and Bodmeier, R., A new model describing the swelling and drug release kinetics from hydroxypropyl methylcellulose tablets, *Journal of Pharmaceutical Sciences*, Vol. 88, No. 1, 1999, pp. 65–72.

[78] Siepmann, J., Kranz, H., Peppas, N., and Bodmeier, R., Calculation of the required size and shape of hydroxypropyl methylcellulose matrices to achieve desired drug release profiles, *International Journal of Pharmaceutics*, Vol. 201, No. 2, 2000, pp. 151–164.

[79] Siepmann, J. and Peppas, N., Hydrophilic matrices for controlled drug delivery: An improved mathematical model to predict the resulting drug release kinetics (the 'sequential layer' model), *Pharmaceutical Research*, Vol. 17, No. 10, 2000, pp. 1290–1298.

[80] Streubel, A., Siepmann, J., Peppas, N., and Bodmeier, R., Bimodal drug release achieved with multi-layer matrix tablets: transport mechanisms and device design, *Journal of Controlled Release*, Vol. 69, No. 3, 2000, pp. 455–468.

[81] Siepmann, J., Streubel, A., and Peppas, N., Understanding and predicting drug delivery from hydrophilic matrix tablets using the 'sequential layer' model, *Pharmaceutical Research*, Vol. 19, No. 3, 2002, pp. 306–314.

[82] Kosmidis, K., Argyrakis, P., and Macheras, P., A reappraisal of drug release laws using Monte Carlo simulations: The prevalence of the Weibull function, *Pharmaceutical Research*, Vol. 20, No. 7, 2003, pp. 988–995.

[83] Bonny, J. and Leuenberger, H., Matrix type controlled release systems: I. Effect of percolation on drug dissolution kinetics, *Pharmaceutica Acta Helvetiae*, Vol. 66, No. 5-6, 1991, pp. 160–164.

[84] Bunde, A., Havlin, S., Nossal, R., Stanley, H., and Weiss, G., On controlled diffusion-limited drug release from a leaky matrix, *Journal of Chemical Physics*, Vol. 83, No. 11, 1985, pp. 5909–5913.

[85] Pham, A. and Lee, P., Probing the mechanisms of drug release from hydroxypropylmethyl cellulose matrices, *Pharmaceutical Research*, Vol. 11, No. 10, 1994, pp. 1379–1384.

[86] Bettini, R., Catellani, P., Santi, P., Massimo, G., Peppas, N., and Colombo, P., Translocation of drug particles in HPMC matrix gel layer: Effect of drug solubility and influence on release rate, *Journal of Controlled Release*, Vol. 70, No. 3, 2001, pp. 383–391.

[87] Kosmidis, K., Argyrakis, P., and Macheras, P., Fractal kinetics in drug release from finite fractal matrices, *Journal of Chemical Physics*, Vol. 119, No. 12, 2003, pp. 6373–6377.

[88] Weiss, G., *Aspects and Applications of the Random Walk. Random Materials and Processes*, North-Holland, Amsterdam, 1994.

[89] Papadopoulou, V., Kosmidis, K., Vlachou, M., and Macheras, P., On the use of the Weibull function for the discernment of drug release mechanisms, *International Journal of Pharmaceutics*, In press.

[90] Rinaki, E., Dokoumetzidis, A., and Macheras, P., The mean dissolution time depends on the dose/solubility ratio, *Pharmaceutical Research*, Vol. 20, No. 3, 2003, pp. 406–408.

[91] Siepmann, J., Faisant, N., and Benoit, J., A new mathematical model quantifying drug release from bioerodible microparticles using Monte Carlo simulations, *Pharmaceutical Research*, Vol. 19, No. 12, 2002, pp. 1885–1893.

[92] Siepmann, J., Faisant, N., Akiki, J., Richard, J., and Benoit, J., Effect of the size of biodegradable microparticles on drug release: Experiment and theory, *Journal of Controlled Release*, Vol. 96, No. 1, 2004, pp. 123–134.

[93] Mazer, N., Pharmacokinetic and pharmacodynamic aspects of polypeptide delivery, *Journal of Controlled Release*, Vol. 11, No. 1-3, 1990, pp. 343–356.

[94] Heller, J., Modulated release from drug delivery devices, *Critical Reviews in Therapeutic Drug Carrier Systems*, Vol. 10, No. 3, 1993, pp. 253–305.

[95] Siegel, R. and Ziaie, B., Biosensing and drug delivery at the microscale, *Advanced Drug Delivery Reviews*, Vol. 56, No. 2, 2004, pp. 121–123.

[96] Misra, G. and Siegel, R., Ionizable drugs and pH oscillators: Buffering effects, *Journal of Pharmaceutical Sciences*, Vol. 91, No. 9, 2002, pp. 2003–2015.

[97] Leroux, J. and Siegel, R., Autonomous gel/enzyme oscillator fueled by glucose: Preliminary evidence for oscillations, *Chaos*, Vol. 9, No. 2, 1999, pp. 267–275.

[98] Giannos, S., Dinh, S., and Berner, B., Temporally controlled drug delivery systems: Coupling of pH oscillators with membrane diffusion, *Journal of Pharmaceutical Sciences*, Vol. 84, No. 5, 1995, pp. 539–543.

[99] Misra, G. and Siegel, R., Multipulse drug permeation across a membrane driven by a chemical pH-oscillator, *Journal of Controlled Release*, Vol. 79, No. 1-3, 2002, pp. 293–297.

[100] Misra, G. and Siegel, R., New mode of drug delivery: Long term autonomous rhythmic hormone release across a hydrogel membrane, *Journal of Controlled Release*, Vol. 81, No. 1-2, 2002, pp. 1–6.

[101] Li, B. and Siegel, R., Global analysis of a model pulsing drug delivery oscillator based on chemomechanical feedback with hysteresis, *Chaos*, Vol. 10, No. 3, 2000, pp. 682–690.

[102] Levich, V., *Physicochemical Hydrodynamics*, Prentice Hall, New Jersey, 1962.

[103] Noyes, A. and Whitney, W., The rate of solution of solid substances in their own solutions, *Journal of American Chemical Society*, Vol. 19, 1897, pp. 930–934.

[104] Nernst, W. and Brunner, E., Theorie der Reaktionsgeschwindigkeit in heterogen Systemen, *Zeitschrift für Physikalische Chemie*, Vol. 47, 1904, pp. 52–110.

[105] Hixson, A. and Crowell, J., Dependence of reaction velocity upon surface and agitation. I. Theoretical considerations, *Industrial and Engineering Chemistry Research*, Vol. 51, 1931, pp. 923–931.

[106] Avnir, D., Farin, D., and Pfeifer, P., Surface geometric irregularity of particulate materials, *Journal of Colloid and Interface Science*, Vol. 103, 1985, pp. 112–123.

[107] Li, T. and Park, K., Fractal analysis of pharmaceutical particles by atomic force microscopy, *Pharmacological Research*, Vol. 15, No. 8, 1998, pp. 1222–1232.

[108] Schroder, M. and Kleinebudde, P., Structure of disintegrating pellets with regard to fractal geometry, *Pharmacological Research*, Vol. 12, No. 11, 1995, pp. 1694–1700.

[109] Tromelin, A., Hautbout, G., and Pourcelot, Y., Application of fractal geometry to dissolution kinetic study of a sweetener excipient, *International Journal of Pharmaceutics*, Vol. 224, No. 1-2, 2001, pp. 131–140.

[110] Cavallari, C., Abertini, B., Gonzalez-Rodriguez, M., and Rodriguez, L., Improved dissolution behavior of steam-granulated piroxicam, *European Journal of Pharmaceutics and Biopharmaceutics*, Vol. 54, No. 1, 2002, pp. 65–73.

[111] Fini, A., Holgado, M., Rodriguez, L., and Cavallari, C., Ultrasound-compacted indomethacin/polyvinylpyrrolidone systems: Effect of compaction process on particle morphology and dissolution behavior, *Journal of Pharmaceutical Sciences*, Vol. 91, No. 8, 2002, pp. 1880–1890.

[112] Pettit, F. and Bowie, J., Protein surface roughness and small molecular binding sites, *Journal of Molecular Biology*, Vol. 285, No. 4, 1999, pp. 1377–1382.

[113] Farin, D. and Avnir, D., Use of fractal geometry to determine effects of surface morphology on drug dissolution, *Journal of Pharmaceutical Sciences*, Vol. 81, No. 1, 1992, pp. 54–57.

[114] Farin, D. and Avnir, D., The reaction dimension in catalysis on dispersed metals, *Journal of American Chemical Society*, Vol. 110, 1988, pp. 2039–2045.

[115] Valsami, G. and Macheras, P., Determination of fractal reaction dimension in dissolution studies, *European Journal of Pharmaceutical Sciences*, Vol. 3, No. 3, 1995, pp. 163–169.

[116] Weibull, W., A statistical distribution function of wide applicability, *Journal of Applied Mechanics*, Vol. 18, 1951, pp. 293–297.

[117] Langenbucher, F., Linearization of dissolution rate curves by the Weibull distribution, *Journal of Pharmaceutics and Pharmacology*, Vol. 24, No. 12, 1972, pp. 979–981.

[118] Djordjevic, A. and Mendas, I., Method for modelling in vitro dissolution profiles of drugs using gamma distribution, *European Journal of Pharmaceutics and Biopharmaceutics*, Vol. 44, No. 2, 1997, pp. 215–217.

[119] Smoluchowski, M., Versuch Einer Mathematischen Theorie der Koagulationskinetik Kolloider Lösungen, *Zeitschrift für Physikalische Chemie*, Vol. 92, 1917, pp. 129–168.

[120] Elkoshi, Z., On the variability of dissolution data, *Pharmaceutical Research*, Vol. 14, No. 10, 1997, pp. 1355–1362.

[121] Dokoumetzidis, A. and Macheras, P., On the heterogeneity of drug dissolution and release, *Pharmaceutical Research*, Vol. 17, No. 2, 2000, pp. 108–112.

[122] Lansky, P. and Weiss, M., Does the dose-solubility ratio affect the mean dissolution time of drugs? *Pharmaceutical Research*, Vol. 16, No. 9, 1999, pp. 1470–1476.

[123] Dokoumetzidis, A. and Macheras, P., A population growth model of dissolution, *Pharmaceutical Research*, Vol. 14, No. 9, 1997, pp. 1122–1126.

[124] May, R., Simple mathematical models with very complicated dynamics, *Nature*, Vol. 261, No. 5560, 1976, pp. 459–467.

[125] Wiggins, S., *Introduction to Applied Nonlinear Dynamical Systems and Chaos*, Springer-Verlag, New York, 3rd ed., 1990.

[126] Shah, V., Noory, A., Noory, C., McCullough, B., Clarke, S., Everett, R., Naviasky, H., Srinivasan, B., Fortman, D., and Skelly, J., In vitro dissolution of sparingly water-soluble drug dosage forms, *International Journal of Pharmaceutics*, Vol. 125, 1995, pp. 99–106.

[127] Yamamoto, K., Nakano, M., Arita, T., Takayama, Y., and Nakai, Y., Dissolution behavior and bioavailability of phenytoin from a ground mixture with microcrystalline cellulose, *Journal of Pharmaceutical Sciences*, Vol. 65, No. 10, 1976, pp. 1484–1488.

[128] Suzuki, H. and Sunada, H., Influence of water-soluble polymers on the dissolution of nifedipine solid dispersions with combined carriers, *Chemical Pharmaceutical Bulletin*, Vol. 46, No. 3, 1998, pp. 482–487.

[129] Valsami, G., Dokoumetzidis, A., and Macheras, P., Modeling of supersaturated dissolution data, *International Journal of Pharmaceutics*, Vol. 181, No. 2, 1999, pp. 153–157.

[130] Lansky, P. and Weiss, M., Modeling heterogeneity of particles and random effects in drug dissolution, *Pharmaceutical Research*, Vol. 18, No. 7, 2001, pp. 1061–1067.

[131] Lansky, P. and Weiss, M., Classification of dissolution profiles in terms of fractional dissolution rate and a novel measure of heterogeneity, *Journal of Pharmaceutical Sciences*, Vol. 92, No. 8, 2003, pp. 1632–1647.

[132] Lansky, P., Lanska, V., and Weiss, M., A stochastic differential equation model for drug dissolution and its parameters, *Journal of Controlled Release*, Vol. 100, No. 2, 2004, pp. 267–274.

[133] Lansky, P. and Weiss, M., Role of heterogeneity in deterministic models of drug dissolution and their statistical characteristics, *Biosystems*, Vol. 71, No. 1-2, 2003, pp. 123–131.

[134] Tsong, Y., Hammerstrom, T., Sathe, P., and Shah, V., Statistical assessment of mean differences between two dissolution data sets, *Drug Information Journal*, Vol. 30, 1996, pp. 1105–1112.

[135] Chow, S. and Ki, F., Statistical comparison between dissolution profiles of drug products, *Journal of Biopharmaceutical Statistics*, Vol. 7, No. 2, 1997, pp. 241–258.

[136] Costa, P. and Lobo, J., Modeling and comparison of dissolution profiles, *European Journal of Pharmaceutical Sciences*, Vol. 13, 2001, pp. 123–133.

[137] FDA, Guidance for Industry. Dissolution testing of immediate release solid oral dosage forms, Tech. rep., Center for Drug Evaluation and Research, 1997.

[138] FDA, Guidance for Industry. Waiver of in vivo bioavailability and bioequivalence studies for immediate release solid oral dosage forms based on a biopharmaceutics classification system, Tech. rep., Center for Drug Evaluation and Research, 2000.

[139] EAEMP, Note for guidance on the investigation of bioavailability and bioequivalence, Tech. rep., Committee for Proprietary Medicinal Products, 2000.

[140] Liu, J., Ma, M., and Chow, S., Statistical evaluation of similarity factor f2 as a criterion for assessment of similarity between dissolution profiles, *Drug Information Journal*, Vol. 31, 1997, pp. 1255–1271.

[141] Shah, V., Tsong, Y., Sathe, P., and Liu, J., In vitro dissolution profile comparison-statistics and analysis of the similarity factor, f2, *Pharmaceutical Research*, Vol. 15, No. 6, 1998, pp. 889–896.

[142] Vertzoni, M., Symillides, M., Iliadis, A., Nicolaides, E., and Reppas, C., Comparison of simulated cumulative drug versus time data sets with indices, *European Journal of Pharmaceutics and Biopharmaceutics*, Vol. 56, No. 3, 2003, pp. 421–428.

[143] Jacobs, M., Some aspects of cell permeability to weak electrolytes, *Cold Spring Harbor Symp. Quant. Biol.*, Vol. 8, 1940, pp. 30–39.

[144] Shore, P., Brodie, B., and Hogben, C., The gastric secretion of drugs: A pH partition hypothesis, *Journal of Pharmacology and Experimental Therapeutics*, Vol. 119, No. 3, 1957, pp. 361–369.

[145] Schanker, L., Shore, P., Brodie, B., and Hogben, C., Absorption of drugs from the stomach. I. The rat, *Journal of Pharmacology and Experimental Therapeutics*, Vol. 120, No. 4, 1957, pp. 528–539.

[146] Hogben, C., Schanker, L., Tocco, D., and Brodie, B., Absorption of drugs from the stomach. II. The human, *Journal of Pharmacology and Experimental Therapeutics*, Vol. 120, No. 4, 1957, pp. 540–545.

[147] Schanker, L., Tocco, D., Brodie, B., and Hogben, C., Absorption of drugs from the rat small intestine, *Journal of Pharmacology and Experimental Therapeutics*, Vol. 123, No. 1, 1958, pp. 81–88.

[148] Hogben, C., Tocco, D., Brodie, B., and Schanker, L., On the mechanism of intestinal absorption of drugs, *Journal of Pharmacology and Experimental Therapeutics*, Vol. 125, No. 4, 1959, pp. 275–282.

[149] Schanker, L., On the mechanism of absorption of drugs from the gastrointestinal tract, *Journal of Medicinal and Pharmaceutical Chemistry*, Vol. 2, 1960, pp. 343–359.

[150] Suzuki, A., Higuchi, W., and Ho, N., Theoretical model studies of drug absorption and transport in the gastrointestinal tract. I, *Journal of Pharmaceutical Sciences*, Vol. 59, No. 5, 1970, pp. 644–651.

[151] Suzuki, A., Higuchi, W., and Ho, N., Theoretical model studies of drug absorption and transport in the gastrointestinal tract. II, *Journal of Pharmaceutical Sciences*, Vol. 59, No. 5, 1970, pp. 651–659.

[152] Winne, D., Shift of pH-absorption curves, *Journal of Pharmacokinetics and Biopharmaceutics*, Vol. 5, No. 1, 1977, pp. 53–94.

[153] Amidon, G., Lennernas, H., Shah, V., and Crison, J., A theoretical basis for a biopharmaceutic drug classification: The correlation of in vitro drug product dissolution and in vivo bioavailability, *Pharmaceutical Research*, Vol. 12, No. 3, 1995, pp. 413–420.

[154] Testa, B., Carrupt, P., Gaillard, P., Billois, F., and Weber, P., Lipophilicity in molecular modeling, *Pharmaceutical Research*, Vol. 13, No. 3, 1996, pp. 335–343.

[155] VanDeWaterbeemd, H. and Testa, B., The parametrization of lipophilicity and other structural properties in drug design, *Advances in Drug Research*, edited by B. Testa, Vol. 16, Academic Press, London, 1988, pp. 87–227.

[156] Dressman, J., Amidon, G., and Fleisher, D., Absorption potential: Estimating the fraction absorbed for orally administered compounds, *Journal of Pharmaceutical Sciences*, Vol. 74, No. 5, 1985, pp. 588–589.

[157] Boxenbaum, H., Absorption potential and its variants, *Pharmaceutical Research*, Vol. 16, No. 12, 1999, pp. 1893.

[158] Macheras, P. and Symillides, M., Toward a quantitative approach for the prediction of the fraction of dose absorbed using the absorption potential concept, *Biopharmaceutics Drug Disposition*, Vol. 10, No. 1, 1989, pp. 43–53.

[159] Balon, K., Riebesehl, B., and Muller, B., Drug liposome partitioning as a tool for the prediction of human passive intestinal absorption, *Pharmaceutical Research*, Vol. 16, No. 6, 1999, pp. 882–888.

[160] Sanghvi, T., Ni, N., and Yalkowsky, S., A simple modified absorption potential, *Pharmaceutical Research*, Vol. 18, No. 12, 2001, pp. 1794–1796.

[161] Sanghvi, T., Ni, N., Mayersohn, M., and Yalkowsky, S., Predicting passive intestinal absorption using a single parameter, *Quantitative Structure Activity Relationships and Combinatorial Science*, Vol. 22, No. 2, 2003, pp. 247–257.

[162] Johnson, K. and Swindell, A., Guidance in the setting of drug particle size specifications to minimize variability in absorption, *Pharmaceutical Research*, Vol. 13, No. 12, 1996, pp. 1795–1798.

[163] Komiya, I., Park, J., Kamani, A., Ho, N., and Higuchi, W., Quantitative mechanistic studies in simultaneous fluid flow and intestinal absorption using steroids as model solutes, *International Journal of Pharmaceutics*, Vol. 4, No. 3, 1980, pp. 249–262.

[164] Ho, N., Raub, T., Burton, P., Barsuhn, C., Adson, A., Audus, K., and Borchardt, R., Quantitative approaches to delineate transport mechanisms in cell culture monolayers, *Transport Processes in Pharmaceutical Systems*, edited by G. Amidon and P. Lee, Marcel Dekker, New York, 2000, pp. 219–316.

[165] Hilgendorf, C., Spahn-Langguth, H., Regardh, C., Lipka, E., Amidon, G., and Langguth, P., Caco-2 versus Caco-2/HT29-MTX co-cultured cell lines: Permeabilities via diffusion, inside- and outside-directed carrier-mediated transport, *Journal of Pharmaceutical Sciences*, Vol. 89, No. 1, 2000, pp. 63–75.

[166] Irvine, J., Takahashi, L., Lockhart, K., Cheong, J., Tolan, J., Selick, H., and Grove, J., MDCK (Madin-Darby canine kidney) cells: A tool for membrane permeability screening, *Journal of Pharmaceutical Sciences*, Vol. 88, No. 1, 1999, pp. 28–33.

[167] Zhu, C., Jiang, L., Chen, T., and Hwang, K., A comparative study of artificial membrane permeability assay for high throughput profiling of drug absorption potential, *European Journal of Medicinal Chemistry*, Vol. 37, No. 5, 2002, pp. 399–407.

[168] Bergstrom, C., Strafford, M., Lazorova, L., Avdeef, A., Luthman, K., and Artursson, P., Absorption classification of oral drugs based on molecular surface properties, *Journal of Medicinal Chemistry*, Vol. 46, No. 4, 2003, pp. 558–570.

[169] Artursson, P., Palm, K., and Luthman, K., Caco-2 monolayers in experimental and theoretical predictions of drug transport, *Advanced Drug Delivery Reviews*, Vol. 46, No. 1-3, 2001, pp. 27–43.

[170] Sun, D., Lennernas, H., Welage, L., Barnett, J., Landowski, C., Foster, D., Fleisher, D., Lee, K., and Amidon, G., Comparison of human duodenum and Caco-2 gene expression profiles for 12,000 gene sequences tags and correlation with permeability of 26 drugs, *Pharmaceutical Research*, Vol. 19, No. 10, 2002, pp. 1400–1416.

[171] Dressman, J., Fleisher, D., and Amidon, G., Physicochemical model for dose-dependent drug absorption, *Journal of Pharmaceutical Sciences*, Vol. 73, No. 9, 1984, pp. 1274–1279.

[172] Macheras, P., Reppas, C., and Dressman, J., Estimate of volume/flow ratio of gastrointestinal fluids in humans using pharmacokinetic data, *Pharmaceutical Research*, Vol. 7, No. 5, 1990, pp. 518–522.

[173] Yu, L., Crison, J., and Amidon, G., Compartmental transit and dispersion model analysis of small intestinal transit flow in humans, *International Journal of Pharmaceutics*, Vol. 140, 1996, pp. 111–118.

[174] Yu, L. and Amidon, G., Characterization of small intestinal transit time distribution in humans, *International Journal of Pharmaceutics*, Vol. 171, No. 2, 1998, pp. 157–163.

[175] Yu, L. and Amidon, G., Saturable small intestinal drug absorption in humans: Modeling and interpretation of cefatrizine data, *European Journal of Pharmaceutics and Biopharmaceutics*, Vol. 45, No. 2, 1998, pp. 199–203.

[176] Agoram, B., Woltosz, W., and Bolger, M., Predicting the impact of physiological and biochemical processes on oral drug bioavailability, *Advanced Drug Delivery Reviews*, Vol. 50 S1, 2001, pp. S41–67.

[177] Ni, P., Ho, N., Fox, J., Leuenberger, H., and Higuchi, W., Theoretical model studies of intestinal drug absorption V. Non-steady-state fluid flow and absorption, *International Journal of Pharmaceutics*, Vol. 5, No. 1, 1980, pp. 33–47.

[178] Ho, N., Park, J., Ni, P., and Higuchi, W., Advancing quantitative and mechanistic approaches in interfacing gastrointestinal drug absorption studies in animals and humans, *Animal Models for Oral Drug Delivery. In Situ and In Vivo Approaches*, edited by W. Crouthamel and A. Sarapu, American Pharmaceutical Association, Washington, 1983, pp. 27–106.

[179] Unice, K. and Logan, B., The insignificant role of hydrodynamic dispersion on bacterial transport, *Journal of Environmental Engineering*, Vol. 126, No. 6, 2000, pp. 491–500.

[180] Willmann, S., Schmitt, W., Keldenich, J., and Dressman, J., A physiologic model for simulating gastrointestinal flow and drug absorption in rats, *Pharmaceutical Research*, Vol. 20, No. 11, 2003, pp. 1766–1771.

[181] Willmann, S., Schmitt, W., Keldenich, J., Lippert, J., and Dressman, J., A physiological model for the estimation of the fraction dose absorbed in humans, *Journal of Medicinal Chemistry*, Vol. 47, No. 16, 2004, pp. 4022–4031.

[182] Dokoumetzidis, A. and Macheras, P., A dispersion-convection model for the study of the gastrointestinal drug absorption, *AAPS PharmSci*, Vol. 5, No. 4, 2003, pp. R6086.

[183] Agutter, P., Malone, P., and Wheatley, D., Intracellular transport mechanisms: A critique of diffusion theory, *Journal of Theoretical Biology*, Vol. 176, No. 2, 1995, pp. 261–272.

[184] Wheatley, D., Difusion theory in biology: Its validity and relevance, *Journal of Biological Education*, Vol. 27, 1993, pp. 181–188.

[185] Macheras, P. and Argyrakis, P., Gastrointestinal drug absorption: Is it time to consider heterogeneity as well as homogeneity? *Pharmaceutical Research*, Vol. 14, 1997, pp. 842–847.

[186] Davenport, H., *Physiology of the Digestive Tract*, Year Book Medical, Chicago, 5th ed., 1982.

[187] Davis, S., Hardy, J., and Fara, J., Transit of pharmaceutical dosage forms through the small intestine, *Gut*, Vol. 27, No. 8, 1986, pp. 886–892.

[188] Digenis, G. and Sandefer, E., Gamma scintigraphy and neutron activation techniques in the in vivo assessment of orally administered dosage forms, *Critical Reviews in Therapeutic Drug Carrier Systems*, Vol. 7, No. 4, 1991, pp. 309–345.

[189] Davis, S., Stockwell, A., Taylor, M., Hardy, J., Whalley, D., Wilson, C., Bechgaard, H., and Christensen, F., The effect of density on the gastric emptying of single- and multiple-unit dosage forms, *Pharmaceutical Research*, Vol. 3, No. 4, 1986, pp. 208–213.

[190] Khosla, R., Feely, L., and Davis, S., Gastrointestinal transit of non-disintegrating tablets in fed subjects, *International Journal of Pharmaceutics*, Vol. 53, No. 2, 1989, pp. 107–117.

[191] Wilding, I., Hardy, J., Maccari, M., Ravelli, V., and Davis, S., Scintigraphic and pharmacokinetic assessment of a multiparticulate sustained release formulation of diltiazem, *International Journal of Pharmaceutics*, Vol. 76, No. 1-2, 1991, pp. 133–143.

[192] Spiller, R., Brown, M., and Phillips, S., Emptying of the terminal ileum in intact humans. Influence of meal residue and ileal motility, *Gastroenterology*, Vol. 92, No. 3, 1987, pp. 724–729.

[193] Krevsky, B., Malmud, L., D'Ercole, F., Maurer, A., and Fisher, R., Colonic transit scintigraphy. A physiologic approach to the quantitative measurement of colonic transit in humans, *Gastroenterology*, Vol. 91, No. 5, 1986, pp. 1102–1112.

[194] Adkin, D., Davis, S., Sparrow, R., and Wilding, I., Colonic transit of different sized tablets in healthy subjects, *Journal of Controlled Release*, Vol. 23, No. 2, 1993, pp. 147–156.

[195] Scher, H., Shlesinger, M., and Bendler, J., Time-scale invariance in transport and relaxation, *Physics Today*, Vol. 44, No. 1, 1991, pp. 26–34.

[196] Katori, N., Aoyagi, N., and Terao, T., Estimation of agitation intensity in the GI tract in humans and dogs based on in vitro/in vivo correlation, *Pharmaceutical Research*, Vol. 12, No. 2, 1995, pp. 237–243.

[197] Shameem, M., Katori, N., Aoyagi, N., and Kojima, S., Oral solid controlled release dosage forms: Role of GI-mechanical destructive forces and colonic release in drug absorption under fasted and fed conditions in humans, *Pharmaceutical Research*, Vol. 12, No. 7, 1995, pp. 1049–1054.

[198] Boxenbaum, H., Jodhka, G., Ferguson, A., Riegelman, S., and MacGregor, T., The influence of bacterial gut hydrolysis on the fate of orally administered isonicotinuric acid in man, *Journal of Pharmacokinetics and Biopharmaceutics*, Vol. 2, No. 3, 1974, pp. 211–237.

[199] Sawamoto, T., Haruta, S., Kurosaki, Y., Higaki, K., and Kimura, T., Prediction of the plasma concentration profiles of orally administered drugs in rats on the basis of gastrointestinal transit kinetics and absorbability, *Journal of Pharmaceutics and Pharmacology*, Vol. 49, No. 4, 1997, pp. 450–457.

[200] Nittmann, J., Daccord, G., and Stanley, H., Fractal growth of viscous fingers: A quantitative characterization of a fluid instability phenomenon, *Nature*, Vol. 314, 1985, pp. 141–144.

[201] VanDamme, H., Flow and interfacial instabilities in Newtonian and colloidal fluids (or the birth, life and death of a fractal), *The Fractal Approach to Heterogeneous Chemistry*, edited by D. Avnir, Wiley, Chishester, 1989, pp. 199–226.

[202] Bhaskar, K., Garik, P., Turner, B., Bradley, J., Bansil, R., Stanley, H., and LaMont, J., Viscous fingering of HCl through gastric mucin, *Nature*, Vol. 360, No. 6403, 1992, pp. 458–461.

[203] Holm, L. and Flemstrom, G., Microscopy of acid transport at the gastric surface in vivo, *Journal of Internal Medicine. Supplement*, Vol. 732, 1990, pp. 91–95.

[204] Daccord, G. and Lenormand, R., Fractal patterns from chemical dissolution, *Nature*, Vol. 325, 1987, pp. 41–43.

[205] Daccord, G., Chemical dissolution of a porous medium by a reactive fluid, *Physical Review Letters*, Vol. 58, No. 5, 1987, pp. 479–482.

[206] Veng-Pedersen, P. and Modi, N., Optimal extravascular dosing intervals, *Journal of Pharmacokinetics and Biopharmaceutics*, Vol. 19, No. 4, 1991, pp. 405–412.

[207] Charter, M. and Gull, S., Maximum entropy and its application to the calculation of drug absorption rates, *Journal of Pharmacokinetics and Biopharmaceutics*, Vol. 15, No. 6, 1987, pp. 645–655.

[208] Verotta, D., Two constrained deconvolution methods using spline functions, *Journal of Pharmacokinetics and Biopharmaceutics*, Vol. 21, No. 5, 1993, pp. 609–636.

[209] Cutler, D., Numerical deconvolution by least squares: Use of prescribed input functions, *Journal of Pharmacokinetics and Biopharmaceutics*, Vol. 6, No. 3, 1978, pp. 227–241.

[210] Cutler, D., Numerical deconvolution by least squares: Use of polynomials to represent the input function, *Journal of Pharmacokinetics and Biopharmaceutics*, Vol. 6, No. 3, 1978, pp. 243–263.

[211] Veng-Pedersen, P., An algorithm and computer program for deconvolution in linear pharmacokinetics, *Journal of Pharmacokinetics and Biopharmaceutics*, Vol. 8, No. 5, 1980, pp. 463–481.

[212] Higaki, K., Yamashita, S., and Amidon, G., Time-dependent oral absorption models, *Journal of Pharmacokinetics and Pharmacodynamics*, Vol. 28, No. 2, 2001, pp. 109–128.

[213] Macheras, P., Argyrakis, P., and Polymilis, C., Fractal geometry, fractal kinetics and chaos en route to biopharmaceutical sciences, *European Journal of Drug Metabolism and Pharmacokinetics*, Vol. 21, No. 2, 1996, pp. 77–86.

[214] Tsang, Y., Pop, R., Gordon, P., Hems, J., and Spino, M., High variability in drug pharmacokinetics complicates determination of bioequivalence: Experience with verapamil, *Pharmaceutical Research*, Vol. 13, No. 6, 1996, pp. 846–850.

[215] Tozer, T., Bois, F., Hauck, W., Chen, M., and Williams, R., Absorption rate vs. exposure: Which is more useful for bioequivalence testing? *Pharmaceutical Research*, Vol. 13, No. 3, 1996, pp. 453–456.

[216] Kalampokis, A., Argyrakis, P., and Macheras, P., Heterogeneous tube model for the study of small intestinal transit flow, *Pharmaceutical Research*, Vol. 16, No. 1, 1999, pp. 87–91.

[217] Kalampokis, A., Argyrakis, P., and Macheras, P., A heterogeneous tube model of intestinal drug absorption based on probabilistic concepts, *Pharmaceutical Research*, Vol. 16, No. 11, 1999, pp. 1764–1769.

[218] Lipinski, C., Lombardo, F., Dominy, B., and Feeney, P., Experimental and computational approaches to estimate solubility and permeability in drug discovery and development settings, *Advanced Drug Delivery Reviews*, Vol. 46, No. 1-3, 2001, pp. 3–26.

[219] Stenberg, P., Bergstrom, C., Luthman, K., and Artursson, P., Theoretical predictions of drug absorption in drug discovery and development, *Clinical Pharmacokinetics*, Vol. 41, No. 11, 2002, pp. 877–899.

[220] Zhao, Y., Le, J., Abraham, M., Hersey, A., Eddershaw, P., Luscombe, C., Butina, D., Beck, G., Sherborne, B., Cooper, I., and Platts, J., Evaluation of human intestinal absorption data and subsequent derivation of a quantitative structure-activity relationship (QSAR) with the Abraham descriptors, *Journal of Pharmaceutical Sciences*, Vol. 90, No. 6, 2001, pp. 749–784.

[221] Zhao, Y., Abraham, M., Le, J., Hersey, A., Luscombe, C., Beck, G., Sherborne, B., and Cooper, I., Rate-limited steps of human oral absorption and QSAR studies, *Pharmaceutical Research*, Vol. 19, No. 10, 2002, pp. 1446–1457.

[222] Turner, J., Maddalena, D., and Agatonovic-Kustrin, S., Bioavailability prediction based on molecular structure for a diverse series of drugs, *Pharmaceutical Research*, Vol. 21, No. 1, 2004, pp. 68–82.

[223] FDA/CDER, Guidance for Industry. Waiver of in vivo bioavailability studies for immediate release solid oral dosageforms based on a biopharmaceutics classification system, Tech. rep., Center for Drug Evaluation and Research, 2000.

[224] Rinaki, E., Valsami, G., and Macheras, P., Quantitative biopharmaceutics classification system: The central role of dose/solubility ratio, *Pharmaceutical Research*, Vol. 20, No. 12, 2003, pp. 1917–1925.

[225] Yu, L., Amidon, G., Polli, J., Zhao, H., Mehta, M., Conner, D., Shah, V., Lesko, L., Chen, M., Lee, V., and Hussain, A., Biopharmaceutics classification system: The scientific basis for biowaiver extensions, *Pharmaceutical Research*, Vol. 19, No. 7, 2002, pp. 921–925.

[226] Polli, J., Yu, L., Cook, J., Amidon, G., Borchardt, R., Burnside, B., Burton, P., Chen, M., Conner, D., Faustino, P., Hawi, A., Hussain, A., Joshi, H., Kwei, G., Lee, V., Lesko, L., Lipper, R., Loper, A., Nerurkar, S., Polli, J., Sanvordeker, D., Taneja, R., Uppoor, R., Vattikonda, C., Wilding, I., and Zhang, G., Summary workshop report: biopharmaceutics classification system - Implementation challenges and extension opportunities, *Journal of Pharmaceutical Sciences*, Vol. 93, No. 6, 2004, pp. 1375–1381.

[227] Yazdanian, M., Briggs, K., Jankovsky, C., and Hawi, A., The 'high solubility' definition of the current FDA Guidance on Biopharmaceutical Classification System may be too strict for acidic drugs, *Pharmaceutical Research*, Vol. 21, No. 2, 2004, pp. 293–299.

[228] Yu, L., Carlin, A., Amidon, G., and Hussain, A., Feasibility studies of utilizing disk intrinsic dissolution rate to classify drugs, *International Journal of Pharmaceutics*, Vol. 270, No. 1-2, 2004, pp. 221–227.

[229] Ginski, M., Taneja, R., and Polli, J., Prediction of dissolution-absorption relationships from a continuous dissolution/Caco-2 system, *AAPS PharmSci*, Vol. 1, No. 2, 1999, pp. E3.

[230] Rinaki, E., Dokoumetzidis, A., Valsami, G., and Macheras, P., Identification of biowaivers among Class II drugs: Theoretical justification and practical examples, *Pharmaceutical Research*, Vol. 21, No. 9, 2004, pp. 1567–1572.

[231] Davies, N. and Anderson, K., Clinical pharmacokinetics of diclofenac. Therapeutic insights and pitfalls, *Clinical Pharmacokinetics*, Vol. 33, No. 3, 1997, pp. 184–213.

[232] Blume, H. and Schug, B., The biopharmaceutics classification system (BCS): Class III drugs - better candidates for BA/BE waiver? *European Journal of Pharmaceutical Sciences*, Vol. 9, No. 2, 1999, pp. 117–121.

[233] Wu, C. and Benet, L., Predicting drug disposition via application of BCS: Transport/absorption/elimination interplay and development of a biopharmaceutics drug disposition classification system, *Pharmaceutical Research*, Vol. 22, No. 1, 2005, pp. 11–23.

[234] VanRossum, J. and DeBie, J., Chaos and illusion, *Trends in Pharmacological Sciences*, Vol. 12, No. 10, 1991, pp. 379–383.

[235] Dokoumetzidis, A. and Macheras, P., Investigation of absorption kinetics by the phase plane method, *Pharmaceutical Research*, Vol. 15, No. 8, 1998, pp. 1262–1269.

[236] Dokoumetzidis, A., Iliadis, A., and Macheras, P., An alternative method for the estimation of the terminal slope when a few data points are available, *Journal of Pharmaceutical Sciences*, Vol. 88, No. 5, 1999, pp. 557–560.

[237] Tothfalusi, L., Endrenyi, L., and Midha, K., Scaling or wider bioequivalence limits for highly variable drugs and for the special case of C(max), *International Journal of Clinical Pharmacology and Therapeutics*, Vol. 41, No. 5, 2003, pp. 217–225.

[238] Tothfalusi, L. and Endrenyi, L., Limits for the scaled average bioequivalence of highly variable drugs and drug products, *Pharmaceutical Research*, Vol. 20, No. 3, 2003, pp. 382–389.

[239] Karalis, V., Symillides, M., and Macheras, P., Novel scaled average bioequivalence limits based on GMR and variability considerations, *Pharmaceutical Research*, Vol. 21, No. 10, 2004, pp. 1933–1942.

[240] Wang, Z., Zhenya, H., and Chen, J., Chaotic behavior of gastric migrating myoelectrical complex, *IEEE Transactions on Biomedical Engineering*, Vol. 51, No. 8, 2004, pp. 1401–1406.

[241] Marathe, P., Sandefer, E., Kollia, G., Greene, D., Barbhaiya, R., Lipper, R., Page, R., Doll, W., Ryo, U., and Digenis, G., In vivo evaluation of the absorption and gastrointestinal transit of avitriptan in fed and fasted subjects using gamma scintigraphy, *Journal of Pharmacokinetics and Biopharmaceutics*, Vol. 26, No. 1, 1998, pp. 1–29.

[242] Richalet, J., Rault, A., and Pouliquen, R., *Identification Des Processus Par la Méthode Du Modèle*, Gordon and Breach, Paris, 1971.

[243] Balant, L. and Gex-Fabry, M., Modelling during drug development, *European Journal of Pharmaceutics and Biopharmaceutics*, Vol. 50, No. 1, 2000, pp. 13–26.

[244] Wise, M., Negative power functions of time in pharmacokinetics and their implications, *Journal of Pharmacokinetics and Biopharmaceutics*, Vol. 13, No. 3, 1985, pp. 309–346.

[245] Weiss, M., Use of gamma distributed residence times in pharmacokinetics, *European Journal of Clinical Pharmacology*, Vol. 25, 1983, pp. 695–702.

[246] Weiss, M., A note on the role of generalized inverse Gaussian distributions of circulatory transit times in pharmacokinetics, *Journal of Mathematical Biology*, Vol. 20, 1984, pp. 95–102.

[247] Beard, D. and Bassingthwaighte, J., Power-law kinetics of tracer washout from physiological systems, *Annals of Biomedical Engineering*, Vol. 26, 1998, pp. 775–779.

[248] Norwich, K., Noncompartment models of whole-body clearance of tracers: A review, *Annals of Biomedical Engineering*, Vol. 25, 1997, pp. 421–439.

[249] Macheras, P., Carrier-mediated transport can obey fractal kinetics, *Pharmaceutical Research*, Vol. 12, No. 4, 1995, pp. 541–548.

[250] Havlin, S. and Ben-Avraham, D., Diffusion in disordered media, *Advances in Physics*, Vol. 51, No. 1, 2002, pp. 187–292.

[251] Mandelbrot, B., *The Fractal Geometry of Nature*, W.H. Freeman, New York, 1983.

[252] VanBeek, J., Roger, S., and Bassingthwaighte, J., Regional myocardial flow heterogeneity explained with fractal networks, *American Journal of Physiology*, Vol. 257, No. 5(2), 1989, pp. H1670–1680.

[253] VanDerGraff, K. and Fox, S., *Concepts of Human Anatomy and Physiology*, McGraw Hill, New York, 5th ed., 2000.

[254] Suwa, N., Niwa, T., Fukasawa, H., and Sasaki, Y., Estimation of intravascular blood pressure gradient by mathematical analysis of arterial casts, *Tohoku Journal of Experimental Medicine*, Vol. 79, 1963, pp. 168–198.

[255] Suwa, N. and Takahashi, T., *Morphological and Morphometrical Analysis of Circulation in Hypertension and Ischemic Kidney*, Urban und Schwarzenberg, München, 1971.

[256] Macheras, P., A fractal approach to heterogeneous drug distribution: Calcium pharmacokinetics, *Pharmaceutical Research*, Vol. 13, No. 5, 1996, pp. 663–670.

[257] Gerlowski, L. and Jain, R., Physiologically based pharmacokinetic modeling: Principles and applications, *Journal of Pharmaceutical Sciences*, Vol. 72, No. 10, 1983, pp. 1103–1127.

[258] Wiltshire, H., Sutton, B., Heeps, G., Betty, A., Angus, D., Harris, S., Worth, E., and Welker, H., Metabolism of the calcium antagonist, mibefradil (POSICOR, Ro 40-5967). Part III. Comparative pharmacokinetics of mibefradil and its major metabolites in rat, marmoset, cynomolgus monkey and man, *Xenobiotica*, Vol. 27, No. 6, 1997, pp. 557–571.

[259] Campra, J. and Reynolds, T., The hepatic circulation, *In The Liver: Biology and Pathobiology*, edited by I. Arias, H. Popper, D. Schachter, and D. Shafritz, Raven Press, New York, 1982, pp. 627–645.

[260] Glenny, R. and Robertson, H., Fractal properties of pulmonary blood flow: Characterization of spatial heterogeneity, *Journal of Applied Physiology*, Vol. 69, No. 2, 1990, pp. 532–545.

[261] Glenny, R. and Robertson, H., Fractal modeling of pulmonary blood flow heterogeneity, *Journal of Applied Physiology*, Vol. 70, No. 3, 1991, pp. 1024–1030.

[262] Lin, J. and Lu, A., Role of pharmacokinetics and metabolism in drug discovery and development, *Pharmacological Reviews*, Vol. 49, No. 4, 1997, pp. 403–449.

[263] Houston, J. B. and Carlile, D., Prediction of hepatic clearance from microsoms, hepatocytes, and liver slices, *Drug Metabolism Reviews*, Vol. 29, No. 4, 1997, pp. 891–922.

[264] Obach, R., Prediction of human clearance of twenty-nine drugs from hepatic microsomal intrinsic clearance data: An examination of in vitro half-life approach and nonspecific binding to microsomes, *Drug Metabolism and Disposition*, Vol. 27, No. 11, 1999, pp. 1350–1359.

[265] Ito, K. and Houston, J., Comparison of the use of liver models for predicting drug clearance using in vitro kinetic data from hepatic microsomes and isolated hepatocytes, *Pharmaceutical Research*, Vol. 21, No. 5, 2004, pp. 785–792.

[266] Rowland, M., Benet, L., and Graham, G., Clearance concepts in pharmacokinetics, *Journal of Pharmacokinetics and Biopharmaceutics*, Vol. 1, No. 2, 1973, pp. 123–136.

[267] Winkler, K., Keiding, S., and Tygstrup, N., Clearance as a quantitative measure of liver function, *The Liver: Quantitative Aspects of Structure and Function*, edited by P. Paumgartner and R. Preisig, Karger, Basle, 1973, pp. 144–155.

[268] Roberts, M., Donaldson, J., and Rowland, M., Models of hepatic elimination: Comparison of stochastic models to describe residence time distributions and to predict the influence of drug distribution, enzyme heterogeneity, and systemic recycling on hepatic elimination, *Journal of Pharmacokinetics and Biopharmaceutics*, Vol. 16, No. 1, 1988, pp. 41–83.

[269] Roberts, M. and Rowland, M., A dispersion model of hepatic elimination: 1. Formulation of the model and bolus concentrations, *Journal of Pharmacokinetics and Biopharmaceutics*, Vol. 14, No. 3, 1986, pp. 227–260.

[270] Anissimov, Y., Bracken, A., and Roberts, M., Interconnected-tubes model of hepatic elimination, *Journal of Theoretical Biology*, Vol. 188, No. 1, 1997, pp. 89–101.

[271] Anissimov, Y., Bracken, A., and Roberts, M., Interconnected-tubes model of hepatic elimination: Steady-state considerations, *Journal of Theoretical Biology*, Vol. 199, No. 4, 1999, pp. 435–447.

[272] Tirona, R., Schwab, A., Geng, W., and Pang, K., Hepatic clearance models: Comparison of the dispersion and Goresky models in outflow profiles from multiple indicator dilution rat liver studies, *Drug Metabolism and Disposition*, Vol. 26, No. 5, 1998, pp. 465–475.

[273] Niro, R., Byers, J., Fournier, R., and Bachmann, K., Application of a convective-dispersion model to predict in vivo hepatic clearance from in vitro measurements utilizing cryopreserved human hepatocytes, *Current Drug Metabolism*, Vol. 4, No. 5, 2003, pp. 357–369.

[274] Ridgway, D., Tuszynski, J., and Tam, Y., Reassessing models of hepatic extraction, *Journal of Biological Physics*, Vol. 29, No. 1, 2003, pp. 1–21.

[275] Javanaud, C., The application of a fractal model to the scattering of ultrasound in biological media, *Journal of the Acoustical Society of America*, Vol. 86, No. 2, 1989, pp. 493–496.

[276] Liang, E. and Derendorf, H., Pitfalls in pharmacokinetic multicompartment analysis, *Journal of Pharmacokinetics and Biopharmaceutics*, Vol. 26, No. 2, 1998, pp. 247–260.

[277] Berry, H., Monte Carlo simulations of enzyme reactions in two dimensions: Fractal kinetics and spatial segregation, *Biophysical Journal*, Vol. 83, No. 4, 2002, pp. 1891–1901.

[278] Fuite, J., Marsh, R., and Tuszynski, J., Fractal pharmacokinetics of the drug mibefradil in the liver, *Physical Review. E, Statistical, Nonlinear, and Soft Matter Physics*, Vol. 66, No. 2, 2002, pp. 021904.

[279] Skerjanec, A., Tawfik, S., and Tam, Y., Nonlinear pharmacokinetics of mibefradil in the dog, *Journal of Pharmaceutical Sciences*, Vol. 85, No. 2, 1996, pp. 189–192.

[280] Welker, H., Single- and multiple-dose mibefradil pharmacokinetics in normal and hypertensive subjects, *Journal of Pharmacy and Pharmacology*, Vol. 50, No. 9, 1998, pp. 983–987.

[281] Kosmidis, K., Karalis, V., Argyrakis, P., and Macheras, P., Michaelis-Menten kinetics under spatially constrained conditions: Application to mibefradil pharmacokinetics, *Biophysical Journal*, Vol. 87, No. 3, 2004, pp. 1498–1506.

[282] Sun, H. and Charef, A., Fractal system - A time domain approach, *Annals of Biomedical Engineering*, Vol. 18, 1990, pp. 597–621.

[283] Suki, B., Barabasi, A., Hantos, Z., Petak, F., and Stanley, H., Avalanches and power-law behaviour in lung inflation, *Nature*, Vol. 368, 1994, pp. 615–618.

[284] Maddox, J., Long-range correlations within DNA, *Nature*, Vol. 358, 1992, pp. 103.

[285] Peng, C., Buldyrev, S., Goldberger, A., Havlin, S., Sciortino, F., Simons, M., and Stanley, H., Long-range correlations in nucleotide sequences, *Nature*, Vol. 356, 1992, pp. 168–170.

[286] Mandelbrot, B. and Wallis, J., Some long-run properties of geophysical records, *Water Resources Research*, Vol. 5, 1969, pp. 321–340.

[287] Tang, C. and Bak, P., Critical exponents and scaling relations for self organized critical phenomena, *Physical Review Letters*, Vol. 60, 1988, pp. 2347–2350.

[288] Rhodes, C. and Anderson, R., Power laws governing epidemics in isolated populations, *Nature*, Vol. 381, 1996, pp. 600–602.

[289] Mantegna, R. and Stanley, H., Scaling behaviour in the dynamics of an economic index, *Nature*, Vol. 376, 1995, pp. 46–49.

[290] West, B. and Shlesinger, M., On the ubiquity of 1/f noise, *International Journal of Modern Physics B*, Vol. 3, No. 6, 1989, pp. 795–819.

[291] Stanley, M., Amaral, L., Buldyrev, S., Havlin, S., Leschhorn, H., Maass, P., Salinger, M., and Stanley, H., Can statistical physics contribute to the science of economics? *Fractals*, Vol. 4, No. 3, 1996, pp. 415–425.

[292] Bak, P., Tang, C., and Wiesenfeld, K., Self-organized criticality: An explanation of 1/f noise, *Physical Review Letters*, Vol. 59, 1987, pp. 381–384.

[293] Bak, P., Tang, C., and Wiesenfeld, K., Self-organized criticality, *Physical Review A*, Vol. 38, 1988, pp. 364–374.

[294] Havlin, S., Molecular diffusion and reactions, *The Fractal Approach to Heterogeneous Chemistry*, edited by D. Avnir, Wiley, Chishester, 1989, pp. 251–270.

[295] Giona, M., First-order reaction-diffusion kinetics in complex fractal media, *Chemical Engineering Science*, Vol. 47, No. 6, 1992, pp. 1503–1515.

[296] Koo, Y. and Kopelman, R., Space- and time-resolved diffusion-limited binary reaction kinetics in capillaries. Experimental observation of segregation, anomalous exponents and depletion zone, *Journal of Statistical Physics*, Vol. 65, 1991, pp. 893–918.

[297] Kopelman, R., Lin, A., and Argyrakis, P., Non-classical kinetics and reactant segregation in d-dimensional tubular spaces, *Physics Letters A*, Vol. 232, No. 1-2, 1997, pp. 34–40.

[298] Wise, M., Osborn, S., Anderson, J., and Tomlinson, R., A stochastic model for turnover of radiocalcium based on the observed power laws, *Mathematical Biosciences*, Vol. 2, 1968, pp. 199–224.

[299] Wise, M., Interpreting both short- and long-term power laws in physiological clearance curves, *Mathematical Biosciences*, Vol. 20, 1974, pp. 327–337.

[300] Marcus, A., Power laws in compartmental analysis. Part I: A unified stochastic model, *Mathematical Biosciences*, Vol. 23, 1975, pp. 337–350.

[301] Matis, J. and Wehrly, T., Generalized stochastic compartmental models with Erlang transit times, *Journal of Pharmacokinetics and Biopharmaceutics*, Vol. 18, No. 6, 1990, pp. 589–607.

[302] Eckerman, K., Leggett, R., and Williams, L., An elementary method for solving compartmental models with time-dependent coefficients, *Radiation Protection Dosimetry*, Vol. 41, 1992, pp. 257–263.

[303] Farris, F., Dedrick, R., Allen, P., and Smith, J., Physiological model for the pharmacokinetics of methyl mercury in the growing rat, *Toxicology and Applied Pharmacology*, Vol. 119, No. 1, 1993, pp. 74–90.

[304] Matis, J. and Wehrly, T., Stochastic models of compartmental systems, *Biometrics*, Vol. 35, 1979, pp. 199–220.

[305] Matis, J., An introduction to stochastic compartmental models in pharmacokinetics, *Pharmacokinetics-Mathematical and Statistical Approaches in Metabolism and Distribution of Chemicals and Drugs*, edited by A. Pecile and A. Rescigno, Plenum Press, New York, 1988, pp. 113–128.

[306] Matis, J. and Wehrly, T., A general approach to non-Markovian compartment models, *Journal of Pharmacokinetics and Biopharmaceutics*, Vol. 26, No. 4, 1998, pp. 437–456.

[307] Wagner, J., *Fundamentals of Clinical Pharmacokinetics*, Drug Intelligence, Hamilton, 2nd ed., 1979.

[308] Riggs, D., *The Mathematical Approach to Physiological Problems*, The MIT Press, Cambridge, 1963.

[309] Press, W., Teukolsky, S., Vetterling, W., and Flannery, B., *Numerical Recipies in C. The Art of Scientific Computing*, Cambridge University Press, Victoria, 2nd ed., 1992.

[310] Gibaldi, M. and Perrier, D., *Pharmacokinetics*, Vol. 15 of *Drugs and the Pharmaceutical Science*, Marcel Dekker, 2nd ed., 1982.

[311] Matis, J., Wehrly, T., and Metzler, C., On some stochastic formulations and related statistical moments of pharmacokinetic models, *Journal of Pharmacokinetics and Biopharmaceutics*, Vol. 11, No. 1, 1983, pp. 77–92.

[312] Soong, T., Pharmacokinetics with uncertainties in rate constants, *Mathematical Biosciences*, Vol. 12, 1971, pp. 235–243.

[313] Soong, T. and Dowdee, J., Pharmacokinetics with uncertainties in rate constants, III: The inverse problem, *Mathematical Biosciences*, Vol. 19, 1974, pp. 343–353.

[314] Tsokos, J. and Tsokos, C., Statistical modeling of pharmacokinetic systems, *Journal of Dynamic Systems, Measurement, and Control*, Vol. 98, 1976, pp. 37–43.

[315] Michaelis, L. and Menten, M., Die Kinetik der Invertinwirkung, *Biochemische Zeitschrift*, Vol. 49, 1913, pp. 333–369.

[316] Segel, L., Simplification and scaling, *SIAM Review*, Vol. 14, No. 4, 1972, pp. 547–571.

[317] Fung, Y., *Biomechanics: Circulation*, Springer-Verlag, New York, 1997.

[318] Brown, J. and West, G., *Scaling in Biology*, Oxford University Press, New York, 2000.

[319] Lefevre, J., Teleonomical optimization of a fractal model of the pulmonary arterial bed, *Journal of Theoretical Biology*, Vol. 102, No. 2, 1983, pp. 225–248.

[320] Horsfield, K., Diameters, generations, and orders of branches in the bronchial tree, *Journal of Applied Physiology*, Vol. 68, No. 2, 1990, pp. 457–461.

[321] Zamir, M., On fractal properties of arterial trees, *Journal of Theoretical Biology*, Vol. 197, No. 4, 1999, pp. 517–526.

[322] West, G., Brown, J., and Enquist, B., A general model for the origin of allometric scaling laws in biology, *Science*, Vol. 276, 1997, pp. 122–126.

[323] Wagner, J., *Pharmacokinetics for the Pharmaceutical Scientist*, Technomic, Lancaster, 1993.

[324] Weiss, M. and Föster, W., Pharmacokinetic model based on circulatory transport, *European Journal of Clinical Pharmacology*, Vol. 16, 1979, pp. 287–293.

[325] Vicini, P., Bonadonna, R., Lehtovirta, M., Groop, L., and Cobelli, C., Estimation of blood flow heterogeneity in human skeletal muscle using intravascular tracer data: Importance for modeling transcapillary exchange, *Annals of Biomedical Engineering*, Vol. 26, 1998, pp. 764–774.

[326] Beard, D. and Bassingthwaighte, J., Advection and diffusion of substances in biological tissues with complex vascular networks, *Annals of Biomedical Engineering*, Vol. 28, No. 3, 2000, pp. 253–268.

[327] Oliver, R., Jones, A., and Rowland, M., A whole-body physiologically based pharmacokinetic model incorporating dispersion concepts: Short and long time characteristics, *Journal of Pharmacokinetics and Biopharmaceutics*, Vol. 28, No. 1, 2001, pp. 27–55.

[328] Dokoumetzidis, A. and Macheras, P., A model for transport and dispersion in the circulatory system based on the vascular fractal tree, *Annals of Biomedical Engineering*, Vol. 31, No. 3, 2003, pp. 284–293.

[329] Edwards, D., A general theory of the macrotransport of nondepositing particles in the lung by convective dispersion, *Journal of Aerosol Science*, Vol. 25, No. 3, 1994, pp. 543–565.

[330] Huang, W., Yen, R., McLaurine, M., and Bledsoe, G., Morphometry of the human pulmonary vasculature, *Journal of Applied Physiology*, Vol. 81, No. 5, 1996, pp. 2123–2133.

[331] Niemann, C., Henthorn, T., Krejcie, T., Shanks, C., Enders-Klein, C., and Avram, M., Indocyanine green kinetics characterize blood volume and flow distribution and their alteration by propranolol, *Clinical Pharmacology and Therapeutics*, Vol. 67, No. 4, 2000, pp. 342–350.

[332] Karalis, V., Dokoumetzidis, A., and Macheras, P., A physiologically based approach for the estimation of recirculatory parameters, *Journal of Pharmacology and Experimental Therapeutics*, Vol. 308, No. 1, 2004, pp. 198–205.

[333] Hashimoto, M. and Watanabe, G., Simultaneous measurement of effective hepatic blood flow and systemic circulation, *Hepatogastroenterology*, Vol. 47, No. 36, 2000, pp. 1669–1674.

[334] Rescigno, A. and Segre, G., *Drug and Tracer Kinetics*, Blaisdell, 1966.

[335] Jacquez, J., *Compartmental Analysis in Biology and Medicine*, BioMedware, Ann Arbor, 3rd ed., 1996.

[336] Wise, M. and Borsboom, G., Two exceptional sets of physiological clearance curves and their mathematical form: Test cases? *Bulletin of Mathematical Biology*, Vol. 51, 1989, pp. 579–596.

[337] Rescigno, A., The rise and fall of compartmental analysis, *Pharmacological Research*, Vol. 44, No. 4, 2001, pp. 335–340.

[338] Weiss, M., The anomalous pharmacokinetics of amiodarone explained by nonexponential tissue trapping, *Journal of Pharmacokinetics and Biopharmaceutics*, Vol. 27, No. 4, 1999, pp. 383–396.

[339] Dokoumetzidis, A., *Contribution on the study of heterogeneity of biopharmaceutical, pharmacokinetic and pharmacodynamic processes*, Ph.D. thesis, National University of Athens, 2002.

[340] Matis, J. and Hartley, H., Stochastic compartmental analysis: Model and least squares estimation from time series data, *Biometrics*, Vol. 27, 1971, pp. 77–102.

[341] Matis, J., Gamma time-dependency in Blaxter's compartmental model, *Biometrics*, Vol. 28, 1972, pp. 597–602.

[342] Takur, A., Rescigno, A., and Schafer, D., On the stochastic theory of compartments. II. Multi-compartment systems, *Bulletin of Mathematical Biophysics*, Vol. 35, 1973, pp. 263–271.

[343] Purdue, P., Stochastic theory of compartments, *Bulletin of Mathematical Biology*, Vol. 36, 1974, pp. 305–309.

[344] Purdue, P., Stochastic theory of compartments: One and two compartment systems, *Bulletin of Mathematical Biology*, Vol. 36, 1974, pp. 577–587.

[345] Epperson, J. and Matis, J., On the distribution of the general irreversible n-compartmental model having time-dependent transition probabilities, *Bulletin of Mathematical Biology*, Vol. 41, 1979, pp. 737–749.

[346] Chiang, C., *Introduction to Stochastic Processes and Their Applications*, Krieger, New York, 1980.

[347] Mehata, K. and Selvam, D., A stochastic model for the n-compartment irreversible system, *Bulletin of Mathematical Biology*, Vol. 43, No. 5, 1981, pp. 549–561.

[348] Weiss, M., Theorems on log-convex disposition curves in drug and tracer kinetics, *Journal of Theoretical Biology*, Vol. 116, 1985, pp. 355–368.

[349] Weiss, M., Generalizations in linear pharmacokinetics using properties of certain classes of residence time distributions. I. Log-convex drug disposition curves, *Journal of Pharmacokinetics and Biopharmaceutics*, Vol. 14, No. 6, 1986, pp. 635–657.

[350] Weiss, M., Generalizations in linear pharmacokinetics using properties of certain classes of residence time distributions. II. Log-concave concentration-time curves following oral administration, *Journal of Pharmacokinetics and Biopharmaceutics*, Vol. 15, No. 1, 1987, pp. 57–74.

[351] Agrafiotis, G., On the stochastic theory of compartments: A semi-Markov approach for the analysis of the k-compartmental systems, *Bulletin of Mathematical Biology*, Vol. 44, No. 6, 1982, pp. 809–817.

[352] Ljung, L., *System Identification: Theory for the User*, PTR Prentice Hall Information and System Sciences Series, Prentice Hall, New Jersey, 2nd ed., 1999.

[353] Schalla, M. and Weiss, M., Pharmacokinetic curve fitting using numerical inverse Laplace transformation, *European Journal of Pharmaceutical Sciences*, Vol. 7, 1999, pp. 305–309.

[354] Cutler, D., A linear recirculation model for drug disposition, *Journal of Pharmacokinetics and Biopharmaceutics*, Vol. 7, No. 1, 1979, pp. 101–116.

[355] Yamaoka, K., Kanba, M., Toyoda, Y., Yano, Y., and Nakagawa, T., Analysis of enterohepatic circulation of ceficime in rat by fast inverse Laplace transform (FILT), *Journal of Pharmacokinetics and Biopharmaceutics*, Vol. 18, No. 6, 1990, pp. 545–559.

[356] Gallo, J., Lam, F., and Perrier, D., Area method for the estimation of partition coefficients for physiological pharmacokinetic models, *Journal of Pharmacokinetics and Biopharmaceutics*, Vol. 15, No. 3, 1987, pp. 271–280.

[357] Veng-Pedersen, P., Linear and nonlinear system approaches in pharmacokinetics: How much do they offer. I. General considerations, *Journal of Pharmacokinetics and Biopharmaceutics*, Vol. 16, No. 4, 1988, pp. 413–472.

[358] Weiss, M., Moments of physiological transit time distributions and the time course of drug disposition in the body, *Journal of Mathematical Biology*, Vol. 15, 1982, pp. 305–318.

[359] Pipes, L. and Harvill, L., *Applied Mathematics for Engineers and Scientists*, Pure and Applied Mathematics, McGraw Hill, Kogakusha, 3rd ed., 1970.

[360] Duffy, D., On the numerical inversion of Laplace transforms: Comparison of three new methods on characteristic problems from applications, *ACM Transactions on Mathematical Software*, Vol. 19, No. 3, 1993, pp. 333–359.

[361] Durisova, M., Dedik, L., and Balan, M., Building a structured model of a complex pharmacokinetic system with time delays, *Bulletin of Mathematical Biology*, Vol. 57, No. 6, 1995, pp. 787–808.

[362] Purdue, P., Stochastic theory of compartments: An open, two-compartment, reversible system with independent Poisson arrivals, *Bulletin of Mathematical Biology*, Vol. 37, 1975, pp. 269–275.

[363] Neuts, M., *Matrix-Geometric Solutions in Stochastic Models: An Algorithmic Approach*, Johns Hopkins University Press, Baltimore, 1981.

[364] Faddy, M., Compartmental models with phase-type residence-time distributions, *Applied Stochastic Models and Data Analysis*, Vol. 6, 1990, pp. 121–127.

[365] Faddy, M., A structured compartmental model for drug kinetics, *Biometrics*, Vol. 49, No. 1, 1993, pp. 243–248.

[366] Johnson, M. and Taaffe, M., Matching moments to phase distributions: Mixtures of Erlang distributions of common order, *Communications in Statistics - Stochastic Models*, Vol. 5, 1989, pp. 711–743.

[367] Ramaswami, V. and Latouche, G., An experimental evaluation of the matrix-geometric method for the G1/PH/1 queue, *Communications in Statistics - Stochastic Models*, Vol. 5, 1989, pp. 629–667.

[368] Karalis, V., Claret, L., Iliadis, A., and Macheras, P., Fractal volume of drug distribution: It scales proportionally to body mass, *Pharmaceutical Research*, Vol. 18, No. 7, 2001, pp. 1056–1060.

[369] Matis, J. and Wehrly, T., On the use of residence time moments in the statistical analysis of age-department stochastic compartmental systems, *Mathematics in Biology and Medicine*, edited by V. Capasso, E. Grosso, and S. Paveri-Fontana, Springer-Verlag, New York, 1985, pp. 386–398.

[370] McInnis, B., El-Asfouri, S., and Kapadia, S., On stochastic compartmental modeling, *Bulletin of Mathematical Biology*, Vol. 41, 1979, pp. 611–613.

[371] Rescigno, A. and Matis, J., On the relevance of stochastic compartmental models to pharmacokinetic systems, *Bulletin of Mathematical Biology*, Vol. 43, No. 2, 1981, pp. 245–247.

[372] Seber, G. and Wild, C., *Nonlinear Regression Analysis*, Wiley Series in Probability and Mathematical Statistics, John Wiley, New York, 1989.

[373] Kodell, R. and Matis, J., Estimating the rate constants in a two-compartment stochastic model, *Biometrics*, Vol. 32, No. 2, 1976, pp. 377–390.

[374] Piotrovskii, V., Pharmacokinetic stochastic model with Weibull-distributed residence times of drug molecules in the body, *European Journal of Clinical Pharmacology*, Vol. 32, 1987, pp. 515–523.

[375] Matis, J. and Tolley, H., Compartmental models with multiple sources of stochastic variability: The one compartment, time invariant hazard rate case, *Bulletin of Mathematical Biology*, Vol. 41, 1979, pp. 491–515.

[376] Matis, J. and Tolley, H., On the stochastic modeling of tracer kinetics, *Federation Proceedings*, Vol. 39, 1980, pp. 104–109.

[377] Rao, C., *Linear Statistical Inference and Its Applications*, General Probability and Mathematical Statistics, John Wiley, New York, 2nd ed., 2002.

[378] Matis, J., Wehrly, T., and Ellis, W., Some generalized stochastic compartment models for digesta flow, *Biometrics*, Vol. 45, 1989, pp. 703–720.

[379] Wing, G., *A Primer on Integral Equations of the First Kind: The Problem of Deconvolution and Unfolding*, Society for Industrial and Applied Mathematics, Philadelphia, 1991.

[380] Charter, M. and Gull, S., Maximum entropy and drug absorption, *Journal of Pharmacokinetics and Biopharmaceutics*, Vol. 19, No. 5, 1991, pp. 497–520.

[381] Gillespie, D., Exact stochastic simulation of coupled chemical reactions, *Journal of Physical Chemistry*, Vol. 81, No. 25, 1977, pp. 2340–2361.

[382] BenAvraham, D. and Havlin, S., *Diffusion and Reactions in Fractals and Disordered Systems*, Cambridge University Press, Cambridge, 2000.

[383] McQuarrie, D., Stochastic approach to chemical kinetics, *Journal of Applied Probability*, Vol. 4, 1967, pp. 413–478.

[384] Gardenas, M. and Matis, J., On the time-dependent reversible stochastic compartmental model. II. A class of n-compartment systems, *Bulletin of Mathematical Biology*, Vol. 37, 1975, pp. 555–564.

[385] Matis, J. and Kiffe, T., *Stochastic Population Models. A Compartmental Perspective*, Vol. 145 of *Lecture Notes in Statistics*, Springer-Verlag, New York, 2000.

[386] Renshaw, E., Saddlepoint approximations for stochastic processes with truncated cumulant generating functions, *Journal of Mathematical Applications in Medicine and Biology*, Vol. 15, 1998, pp. 1–12.

[387] Bailey, N., *Elements of Stochastic Processes with Applications to the Natural Sciences*, General Applied Probability and Statistics, John Wiley, New York, 1964.

[388] Gillespie, D., A general method for numerically simulating the stochastic time evolution of coupled chemical reactions, *Journal of Computational Physics*, Vol. 22, 1976, pp. 403–434.

[389] Gibson, M. and Bruck, J., Efficient exact stochastic simulation of chemical systems with many species and many channels, *Journal of Physical Chemistry*, Vol. 104, No. 9, 2000, pp. 1876–1889.

[390] Gillespie, D., Approximate accelerated stochastic simulation of chemically reacting systems, *Journal of Chemical Physics*, Vol. 115, No. 4, 2001, pp. 1716–1733.

[391] Stundzia, A. and Lumsden, C., Stochastic simulation of coupled reaction-diffusion processes, *Journal of Computational Physics*, Vol. 127, No. 1, 1996, pp. 196–207.

[392] Oppenheim, I., Shuler, K., and Weiss, G., *Stochastic Processes in Chemical Physics: The Master Equation*, MIT Press, Cambridge, 1977.

[393] Matheson, I., Walls, D., and Gardiner, C., Stochastic models of first-order nonequilibrium phase transitions in chemical reactions, *Journal of Statistical Physics*, Vol. 12, 1975, pp. 21–34.

[394] Liebovitch, L., Fischbarg, J., and Koniarek, J., Ion channel kinetics: A model based on fractal scaling rather than multistate Markov processes, *Mathematical Biosciences*, Vol. 84, No. 1, 1987, pp. 37–68.

[395] Shooman, M., *Probabilistic Reliability: An Engineering Approach*, McGraw-Hill Electrical and Electronic Engineering Series, McGraw-Hill, New York, 1968.

[396] Mann, N., Schafer, R., and Singpurwalla, N., *Methods for Statistical Analysis of Reliability and Life Data*, John Wiley, New York, 1974.

[397] Billinton, R. and Allan, R., *Reliability Evaluation of Engineering Systems. Concepts and Techniques*, Plenum Press, New York, 2nd ed., 1992.

[398] Kalbfleisch, J. and Prentice, R., *The Statistical Analysis of Failure Time Data*, Wiley Series in Probability and Statistics, John Wiley, New Jersey, 2nd ed., 2002.

[399] Singer, K., Application of the theory of stochastic processes to the study of irreproducible chemical reactions and nucleation processes, *Journal of the Royal Statistical Society*, Vol. 15, 1953, pp. 92–106.

[400] McAdams, H. and Arkin, A., Stochastic mechanisms in gene expression, *Proceedings of the National Academy of Sciences USA*, Vol. 94, 1997, pp. 814–819.

[401] Lee, J., Hulse, J., and Colburn, W., Surrogate biochemical markers: Precise measurement for strategic drug and biologics development, *Journal of Clinical Pharmacology*, Vol. 35, No. 5, 1995, pp. 464–470.

[402] McCullagh, P., Regression models for ordinal data, *Journal of the Royal Statistical Society*, Vol. 42, No. 2, 1980, pp. 109–142.

[403] Black, J., Receptors as pharmaceutical targets, *Textbook of Receptor Pharmacology*, edited by J. Foreman and T. Johansen, CRC Press, New York, 2nd ed., 2002, pp. 271–278.

[404] Gieschke, R., Reigner, B., and Steimer, J., Exploring clinical study design by computer simulation based on pharmacokinetic/pharmacodynamic modelling, *International Journal of Clinical Pharmacology and Therapeutics*, Vol. 55, No. 10, 1997, pp. 469–474.

[405] Derendorf, H. and Meibohm, B., Modeling of pharmacokinetic/pharmacodynamic (PK/PD) relationships: Concepts and perspectives, *Pharmaceutical Research*, Vol. 16, No. 2, 1999, pp. 176–185.

[406] Hill, A., The mode of action of nicotine and curari, determined by the form of the contraction curve and the method of temperature coefficients, *Journal of Physiology*, Vol. 39, 1909, pp. 361–373.

[407] Clark, A., *Applied Pharmacology*, Churchill, London, 5th ed., 1933.

[408] Clark, A., *General Pharmacology*, Vol. 4 of *Handbook of Experimental Pharmacology Series*, Springer-Verlag, Berlin, 1937.

[409] Black, J. and Leff, P., Operational models of pharmacological agonism, *Proceedings of the Royal Society of London. Series B. Biological sciences*, Vol. 220, No. 1219, 1983, pp. 141–162.

[410] Kenakin, T., Principles: Receptor theory in pharmacology, *Trends in Pharmacological Sciences*, Vol. 25, No. 4, 2004, pp. 186–192.

[411] Visser, S., Wolters, F., Gubbens-Stibbe, J., Tukker, E., VanDer-Graaf, P., Peletier, L., and Danhof, M., Mechanism-based pharmacokinetic/pharmacodynamic modeling of the electroencephalogram effects of GABAA receptor modulators: in vitro-in vivo correlations, *Journal of Pharmacology and Experimental Therapeutics*, Vol. 304, No. 1, 2003, pp. 88–101.

[412] Levy, G., Relationship between elimination rate of drugs and rate of decline of their pharmacological effects, *Journal of Pharmaceutical Sciences*, Vol. 53, 1964, pp. 342–343.

[413] Levy, G., Kinetics of pharmacologic effects, *Clinical Pharmacology and Therapeutics*, Vol. 7, 1966, pp. 362–372.

[414] Wagner, J., Aghajanian, G., and Bing, O., Correlation of performance test scores with 'tissue concentration' of lysergic acid diethylamide in human subjects, *Clinical Pharmacology and Therapeutics*, Vol. 9, No. 5, 1968, pp. 635–638.

[415] Segre, G., Kinetics of interaction between drugs and biological systems, *Il Farmaco*, Vol. 23, 1968, pp. 907–918.

[416] Sheiner, L., Stanski, D., Vozeh, S., Miller, R., and Ham, J., Simultaneous modeling of pharmacokinetics and pharmacodynamics: Application to d-tubocurarine, *Clinical Pharmacology and Therapeutics*, Vol. 25, No. 3, 1979, pp. 358–371.

[417] Holford, N. and Sheiner, L., Kinetics of pharmacologic response, *Pharmacology and Therapeutics*, Vol. 16, 1982, pp. 143–166.

[418] Racine-Poon, A., Botta, L., Chang, T., Davis, F., Gygax, D., Liou, R., Rohane, P., Staehelin, T., VanSteijn, A., and Frank, W., Efficacy, pharmacodynamics, and pharmacokinetics of CGP 51901, an anti-immunoglobulin E chimeric monoclonal antibody, in patients with seasonal allergic rhinitis, *Clinical Pharmacology and Therapeutics*, Vol. 62, No. 6, 1997, pp. 675–690.

[419] D'Hollander, A. and Delcroix, C., An analytical pharmacodynamic model for nondepolarizing neuromuscular blocking agents, *Journal of Pharmacokinetics and Biopharmaceutics*, Vol. 9, No. 1, 1981, pp. 27–40.

[420] Perry, S. and Inturrisi, C., Analgesia and morphine disposition in burn patients, *Journal of Burn Care and Rehabilitation*, Vol. 4, No. 4, 1983, pp. 276–279.

[421] Inturrisi, C., Colburn, W., Kaiko, R., Houde, R., and Foley, K., Pharmacokinetics and pharmacodynamics of methadone in patients with chronic pain, *Clinical Pharmacology and Therapeutics*, Vol. 41, No. 4, 1987, pp. 392–401.

[422] Scott, J., Cooke, J., and Stanski, D., Electroencephalographic quantitation of opioid effect: Comparative pharmacodynamics of fentanyl and sufentanil, *Anesthesiology*, Vol. 74, No. 1, 1991, pp. 34–42.

[423] Stanski, D., Hudson, R., Homer, T., Saidman, L., and Meathe, E., Pharmacodynamic modeling of thiopental anesthesia, *Journal of Pharmacokinetics and Biopharmaceutics*, Vol. 12, No. 2, 1984, pp. 223–240.

[424] Homer, T. and Stanski, D., The effect of increasing age on thiopental disposition and anesthetic requirement, *Anesthesiology*, Vol. 62, No. 6, 1985, pp. 714–724.

[425] Dingemanse, J., Haussler, J., Hering, W., Ihmsen, H., Albrecht, S., Zell, M., Schwilden, H., and Schuttler, J., Pharmacokinetic-pharmacodynamic modelling of the EEG effects of Ro 48-6791, a new short-acting benzodiazepine, in young and elderly subjects, *British Journal of Anaesthesia*, Vol. 79, No. 5, 1997, pp. 567–574.

[426] Salazar, D., Much, D., Nichola, P., Seibold, J., Shindler, D., and Slugg, P., A pharmacokinetic-pharmacodynamic model of d-sotalol Q-Tc prolongation during intravenous administration to healthy subjects, *Journal of Clinical Pharmacology*, Vol. 37, No. 9, 1997, pp. 799–809.

[427] Derendorf, H. and Hochhaus, G., *Handbook of Pharmacokinetic/Pharmacodynamic Correlation*, CRC, 1995.

[428] Iliadis, A. and Cano, J., A new pharmacodynamic modelling approach based on receptors theory, *3th European Congress of Biopharmaceutics and Pharmacokinetics*, edited by J. Aiache and J. Hirtz, Vol. 3, Imprimerie de l'Université, Clermont-Ferrand, Freiburg, 1987, pp. 65–74.

[429] Shimada, S., Nakajima, Y., Yamamoto, K., Sawada, Y., and Iga, T., Comparative pharmacodynamics of eight calcium channel blocking agents in Japanese essential hypertensive patients, *Biological and Pharmaceutical Bulletin*, Vol. 19, No. 3, 1996, pp. 430–437.

[430] Karlsson, M., Molnar, V., Bergh, J., Freijs, A., and Larsson, R., A general model for time-dissociated pharmacokinetic-pharmacodynamic relationships exemplified by paclitaxel myelosuppression, *Clinical Pharmacology and Therapeutics*, Vol. 63, 1998, pp. 11–25.

[431] Dayneka, N., Garg, V., and Jusko, W., Comparison of four basic models of indirect pharmacodynamic response, *Journal of Pharmacokinetics and Biopharmaceutics*, Vol. 21, No. 4, 1993, pp. 457–478.

[432] Ariens, E., Affinity and intrinsic activity in the theory of competitive inhibition. I. Problems and theory, *Archives Internationales de Pharmacodynamie et de Therapie*, Vol. 99, No. 1, 1954, pp. 32–49.

[433] Nagashima, R., O'Reilly, R., and Levy, G., Kinetics of pharmacologic effects in man: The anticoagulant action of warfarin, *Clinical Pharmacology and Therapeutics*, Vol. 10, 1969, pp. 22–35.

[434] Jusko, W. and Ko, H., Physiologic indirect response models characterize diverse types of pharmacodynamic effects, *Clinical Pharmacology and Therapeutics*, Vol. 56, No. 4, 1994, pp. 406–419.

[435] Mager, D., Wyska, E., and Jusko, W., Diversity of mechanism-based pharmacodynamic models, *Drug Metabolism and Disposition*, Vol. 31, No. 5, 2003, pp. 510–518.

[436] Sharma, A., Ebling, W., and Jusko, W., Precursor-dependent indirect pharmacodynamic response model for tolerance and rebound phenomena, *Journal of Pharmaceutical Sciences*, Vol. 87, No. 12, 1998, pp. 1577–1584.

[437] Lew, K., Ludwig, E., Milad, M., Donovan, K., Middleton, E., Ferry, J., and Jusko, W., Gender-based effects on methylprednisolone pharmacokinetics and pharmacodynamics, *Clinical Pharmacology and Therapeutics*, Vol. 54, No. 4, 1993, pp. 402–414.

[438] Chakraborty, A., Krzyzanski, W., and Jusko, W., Mathematical modeling of circadian cortisol concentrations using indirect response models: Comparison of several methods, *Journal of Pharmacokinetics and Biopharmaceutics*, Vol. 27, No. 1, 1999, pp. 23–43.

[439] Li, H., Meno-Tetang, G., Chiba, K., Arima, N., Heining, P., and Jusko, W., Pharmacokinetics and cell trafficking dynamics of 2-Amino-2-[2-(4-octylphenyl)ethyl]propane-1,3-diol Hydrochloride (FTY720) in cynomolgus monkeys after single oral and intravenous doses, *Journal of Pharmacology and Experimental Therapeutics*, Vol. 301, No. 2, 2002, pp. 519–526.

[440] Wakelkamp, M., Alvan, G., and Paintaud, G., The time of maximum effect for model selection in pharmacokinetic-pharmacodynamic analysis applied to frusemide, *British Journal of Clinical Pharmacology*, Vol. 45, No. 1, 1998, pp. 63–70.

[441] Trzeciakowski, J., Analysis of stimulus-response chains using nonlinear dynamics, *Journal of Pharmacological and Toxicological Methods*, Vol. 36, No. 2, 1996, pp. 103–121.

[442] Yamamoto, K., Abe, M., Katashima, M., Yamada, Y., Sawada, Y., and Iga, T., Pharmacodynamic analysis of antiplatelet effect of aspirin in the literature. Modeling based on inhibition of cyclooxygenase in the platelet and the vessel wall endothelium, *Japan Journal of Hospital Pharmacy*, Vol. 22, 1996, pp. 133–141.

[443] Katashima, M., Yamamoto, K., Tokuma, Y., Hata, T., Sawada, Y., and Iga, T., Pharmacokinetic and pharmacodynamic study of a new nonsteroidal 5 alpha-reductase inhibitor, 4-[3-[3-[Bis(4-isobutylphenyl)methylamino]benzoyl]-1H-indol-1-yl]-butyric acid, in rats, *Journal of Pharmacology and Experimental Therapeutics*, Vol. 284, No. 3, 1998, pp. 914–920.

[444] Abelo, A., Eriksson, U., Karlsson, M., Larsson, H., and Gabrielsson, J., A turnover model of irreversible inhibition of gastric acid secretion by omeprazole in the dog, *Journal of Pharmacology and Experimental Therapeutics*, Vol. 295, No. 2, 2000, pp. 662–669.

[445] Mouton, J., Vinks, A., and Punt, N., Pharmacokinetic-pharmacodynamic modeling of activity of ceftazidime during continuous and intermittent infusion, *Antimicrobial Agents and Chemotherapy*, Vol. 41, No. 4, 1997, pp. 733–738.

[446] Jusko, W., Pharmacodynamics of chemotherapeutic effects: Dose-time-response relationships for phase-nonspecific agents, *Journal of Pharmaceutical Sciences*, Vol. 60, No. 6, 1971, pp. 892–895.

[447] Zhi, J., Nightingale, C., and Quintiliani, R., Microbial pharmacodynamics of piperacillin in neutropenic mice of systematic infection due to Pseudomonas aeruginosa, *Journal of Pharmacokinetics and Biopharmaceutics*, Vol. 16, No. 4, 1988, pp. 355–375.

[448] Meibohm, B., Hochhaus, G., and Derendorf, H., Time dependency of the pharmacologic response to glucocorticoids, *Clinical Pharmacology and Therapeutics*, Vol. 61, No. 2, 1997, pp. 155 (PI–71).

[449] Bauer, J. and Fung, H., Pharmacodynamic models of nitroglycerin-induced hemodynamic tolerance in experimental heart failure, *Pharmaceutical Research*, Vol. 11, No. 6, 1994, pp. 816–823.

[450] Kenakin, T., *Pharmacological Analysis of Drug-Receptor Interaction*, Lippincott-Raven, New York, 1997.

[451] VanDenMeiracker, A., ManIn'tVeld, A., Boomsma, F., Fischberg, D., Molinoff, P., and Schalekamp, M., Hemodynamic and beta-adrenergic receptor adaptations during long-term beta-adrenoceptor blockade. Studies with acebutolol, atenolol, pindolol, and propranolol in hypertensive patients, *Circulation*, Vol. 80, No. 4, 1989, pp. 903–914.

[452] Sheiner, L. and Verotta, D., Further notes on physiologic indirect response models, *Clinical Pharmacology and Therapeutics*, Vol. 58, No. 2, 1995, pp. 238–240.

[453] Tallarida, R., Further characterization of a control model for ligand-receptor interaction: Phase plane geometry, stability, and oscillation, *Annals of Biomedical Engineering*, Vol. 18, 1990, pp. 671–684.

[454] Paalzow, L., Integrated pharmacokinetic-dynamic modeling of drugs acting on the CNS, *Drug Metabolism Reviews*, Vol. 15, 1984, pp. 383–400.

[455] Perez-Urizar, J., Granados-Soto, V., Flores-Murrieta, F., and Castaneda-Hernandez, G., Pharmacokinetic-pharmacodynamic modeling: Why? *Archives of Medical Research*, Vol. 31, 2000, pp. 539–545.

[456] Gabrielsson, J. and Weiner, D., Methodology for pharmacokinetic/pharmacodynamic data analysis, *Pharmaceutical Science and Technology Today*, Vol. 2, No. 6, 1999, pp. 244–252.

[457] Meibohm, B. and Derendorf, H., Pharmacokinetic/pharmacodynamic studies in drug product development, *Journal of Pharmaceutical Sciences*, Vol. 91, No. 1, 2002, pp. 18–31.

[458] Steimer, J., Vozeh, S., Racine-Poon, A., Holford, N., and O'Neil, R., The population approach: Rationale, methods, applications in clinical pharmacology and drug development, *Pharmacokinetics of Drugs, Handbook of Experimental Pharmacology*, edited by P. Welling and L. Balant, Springer-Verlag, Berlin, 1994, pp. 405–451.

[459] Sheiner, L., Rosenberg, B., and Melmon, K., Modelling of individual pharmacokinetics for computer-aided drug dosage, *Computers and Biomedical Research*, Vol. 4, 1972, pp. 441–459.

[460] Sheiner, L., Rosenberg, B., and Marathe, V., Estimation of population characteristics of pharmacokinetic parameters from routine clinical data, *Journal of Pharmacokinetics and Biopharmaceutics*, Vol. 5, No. 5, 1977, pp. 445–479.

[461] Aarons, L., Balant, L., Danhof, M., Gex-Fabry, M., Gundert-Remy, U., Karlsson, M., Mentré, F., Morselli, P., Rombout, F., Rowland, M., Steimer, J., and Vozeh, S., *The Population Approach: Measuring and*

Managing Variability in Response, Concentration and Dose, COST B1 Medicine, Commission of European Communities, Luxembourg, 1997.

[462] Beal, S. and Sheiner, L., The NONMEM system, *American Statistician*, Vol. 34, No. 2, 1980, pp. 118–119.

[463] Lindstrom, M. and Bates, D., Nonlinear mixed effects models for repeated measures data, *Biometrics*, Vol. 46, No. 3, 1990, pp. 673–687.

[464] Lunn, D., Best, N., Thomas, A., Wakefield, J., and Spiegelhalter, D., Bayesian analysis of population PK/PD models: General concepts and software, *Journal of Pharmacokinetics and Pharmacodynamics*, Vol. 29, No. 3, 2002, pp. 271–307.

[465] Davidian, M. and Gallant, R., The nonlinear mixed effect model with a smooth random effects density, *Biometrika*, Vol. 80, 1993, pp. 475–488.

[466] Mallet, A., A maximum likelihood estimation method for random coefficient regression models, *Biometrika*, Vol. 73, No. 3, 1986, pp. 645–656.

[467] Schumitzky, A., Nonparametric EM algorithms for estimating prior distributions, *Applied Mathematics and Computation*, Vol. 45, No. 2, 1991, pp. 143–157.

[468] Aarons, L., Software for population pharmacokinetics and pharmacodynamics, *Clinical Pharmacokinetics*, Vol. 36, No. 4, 1999, pp. 255–264.

[469] Gurrin, L., Moss, T., Sloboda, D., Hazelton, M., Challis, J., and Newnham, J., Using WinBUGS to fit nonlinear mixed models with an application to pharmacokinetic modelling of insulin response to glucose challenge in sheep exposed antenatally to glucocorticoids, *Journal of Biopharmaceutical Statistics*, Vol. 13, No. 1, 2003, pp. 117–139.

[470] Mu, S. and Ludden, T., Estimation of population pharmacokinetic parameters in the presence of non-compliance, *Journal of Pharmacokinetics and Pharmacodynamics*, Vol. 30, No. 1, 2003, pp. 53–81.

[471] Duffull, S., Kirkpatrick, C., Green, B., and Holford, N., Analysis of population pharmacokinetic data using NONMEM and WinBUGS, *Journal of Biopharmaceutical Statistics*, Vol. 15, No. 1, 2005, pp. 53–73.

[472] Milton, J., VanDerHeiden, U., Longtin, A., and Mackey, M., Complex dynamics and noise in simple neural networks with delayed mixed feedback, *Biomedica Biochimica Acta*, Vol. 49, No. 8-9, 1990, pp. 697–707.

[473] Tallarida, R., Control and oscillation in ligand receptor interactions according to the law of mass action, *Life Sciences*, Vol. 46, No. 22, 1990, pp. 1559–1568.

[474] Tallarida, R., On stability and control of ligand-receptor interactions according to the mass action law: A theoretical model of chaos, *Drug Development Research*, Vol. 19, 1990, pp. 257–274.

[475] Tallarida, R. and Freeman, K., Control and stability of ligand receptors interaction in the presence of competitive compound, *Life Sciences*, Vol. 48, No. 6, 1991, pp. PL19–PL24.

[476] Tallarida, R. and Freeman, K., Chaos and control in mass-action binding of endogeneous compounds, *Annals of Biomedical Engineering*, Vol. 22, 1994, pp. 153–161.

[477] Freeman, K. and Tallarida, R., A quantitative study of dopamine control in the rat striatum, *Journal of Pharmacology and Experimental Therapeutics*, Vol. 268, No. 2, 1994, pp. 629–638.

[478] Sun, Y. and Jusko, W., Role of baseline parameters in determining indirect pharmacodynamic responses, *Journal of Pharmaceutical Sciences*, Vol. 88, No. 10, 1999, pp. 987–990.

[479] Mackey, M. and AnDerHeiden, U., The dynamics of recurrent inhibition, *Journal of Mathematical Biology*, Vol. 19, No. 2, 1984, pp. 211–225.

[480] Klausen, T., The feed-back regulation of erythropoietin production in healthy humans, *Danish Medical Bulletin*, Vol. 45, No. 4, 1998, pp. 345–53.

[481] Boggs, D., Athens, J., Cartwright, G., and Wintrobe, M., Leukokinetic studies. IX. Experimental evaluation of a model of granulopoiesis, *Journal of Clinical Investigation*, Vol. 44, 1965, pp. 643–656.

[482] Rubinow, S. and Lebowitz, J., A mathematical model of neutrophil production and control in normal man, *Journal of Mathematical Biology*, Vol. 1, 1975, pp. 187–225.

[483] Lange, R., Cyclic hematopoiesis: Human cyclic neutropenia, *Experimental Hematology*, Vol. 11, No. 6, 1983, pp. 435–451.

[484] Cohen, T. and Cooney, D., Cyclic thrombocytopenia. Case report and review of literature, *Scandinavian Journal of Haematology*, Vol. 12, No. 1, 1974, pp. 9–17.

[485] Mackey, M., Unified hypothesis for the origin of aplastic anemia and periodic hematopoiesis, *Blood*, Vol. 51, No. 5, 1978, pp. 941–956.

[486] Mackey, M., Periodic auto-immune hemolytic anemia: An induced dynamical disease, *Bulletin of Mathematical Biology*, Vol. 41, No. 6, 1979, pp. 829–834.

[487] Dunn, C., Cyclic hematopoiesis: the biomathematics, *Experimental Hematology*, Vol. 11, No. 9, 1983, pp. 779–791.

[488] Lieschke, G., Grail, D., Hodgson, G., Metcalf, D., Stanley, E., Cheers, C., Fowler, K., Basu, S., Zhan, Y., and Dunn, A., Mice lacking granulocyte colony-stimulating factor have chronic neutropenia, granulocyte and macrophage progenitor cell deficiency, and impaired neutrophil mobilization, *Blood*, Vol. 84, No. 6, 1994, pp. 1737–1746.

[489] Price, T., Chatta, G., and Dale, D., Effect of recombinant granulocyte colony-stimulating factor on neutrophil kinetics in normal young and elderly humans, *Blood*, Vol. 88, No. 1, 1996, pp. 335–340.

[490] Haurie, C., Dale, D., and Mackey, M., Cyclical neutropenia and other periodic hematological disorders: A review of mechanisms and mathematical models, *Blood*, Vol. 92, No. 8, 1998, pp. 2629–2640.

[491] Degn, H., Holden, A., and Olsen, L., *Chaos in Biological Systems*, Plenum Press, New York, 1987.

[492] Bai-Lin, H., *Chaos II*, World Scientific, Singapore, 1990.

[493] Quesenberry, P. and Levitt, L., Hematopoietic stem cells (second of three parts), *New England Journal of Medicine*, Vol. 301, No. 15, 1979, pp. 819–823.

[494] Terashi, K., Oka, M., Ohdo, S., Furukubo, T., Ikeda, C., Fukuda, M., Soda, H., Higuchi, S., and Kohno, S., Close association between clearance of recombinant human granulocyte colony-stimulating factor (G-CSF) and G-CSF receptor on neutrophils in cancer patients, *Antimicrobial Agents and Chemotherapy*, Vol. 43, No. 1, 1999, pp. 21–24.

[495] Wheldon, T., Mathematical models of oscillatory blood cell production, *Mathematical Biosciences*, Vol. 24, No. 3-4, 1975, pp. 289–305.

[496] Kazarinoff, N. and VanDenDriessche, P., Control of oscillations in hematopoiesis, *Science*, Vol. 203, No. 4387, 1979, pp. 1348–1349.

[497] Hearn, T., Haurie, C., and Mackey, M., Cyclical neutropenia and the peripheral control of white blood cell production, *Journal of Theoretical Biology*, Vol. 192, No. 2, 1998, pp. 167–181.

[498] Haurie, C., Dale, D., Rudnicki, R., and Mackey, M., Modeling complex neutrophil dynamics in the grey collie, *Journal of Theoretical Biology*, Vol. 204, No. 4, 2000, pp. 505–519.

[499] Bernard, S., Belair, J., and Mackey, M., Oscillations in cyclical neutropenia: new evidence based on mathematical modeling, *Journal of Theoretical Biology*, Vol. 223, No. 3, 2003, pp. 283–298.

[500] Gu, K., Kharitonov, V., and Chen, J., *Stability of Time-Delay Systems*, Control Engineering, Birkhaüser, Boston, 2003.

[501] Karlsson, M., Port, R., Ratain, M., and Sheiner, L., A population model for the leukopenic effect of etoposide, *Clinical Pharmacology and Therapeutics*, Vol. 57, 1995, pp. 325–334.

[502] Minami, H., Sasaki, Y., Saijo, N., Ohtsu, T., Fujii, H., Igarashi, T., and Itoh, K., Indirect-response model for the time course of leukopenia with anticancer drugs, *Clinical Pharmacology and Therapeutics*, Vol. 64, No. 5, 1998, pp. 511–521.

[503] Fokas, A., JB, J. K., and Clarkson, B., Mathematical model of granulocytopoiesis and chronic myelogenous leukemia, *Cancer Research*, Vol. 51, No. 8, 1991, pp. 2084–2091.

[504] Iliadis, A. and Barbolosi, D., Optimizing drug regimens in cancer chemotherapy by an efficacy-toxicity mathematical model, *Computers and Biomedical Research*, Vol. 33, No. 3, 2000, pp. 211–226.

[505] Viens, P., Roche, H., Kerbrat, P., Fumoleau, P., Guastalla, J., and Delozier, T., Epirubicin-docetaxel combination in first-line chemotherapy for patients with metastatic breast cancer: Final results of a dose-finding and efficacy study, *American Journal of Clinical Oncology*, Vol. 24, No. 4, 2001, pp. 328–335.

[506] Seymour, A., DeCampos, E., Thatcher, N., DeGreve, J., Cunningham, D., Howell, A., Tueni, E., Bron, D., Steward, W., Berdel, W., Knuth, A., Lorenz, J., Timothy, A., Yver, A., and Richards, M., A single-blind, randomised, vehicle-controlled dose-finding study of recombinant human granulocyte colony-stimulating factor (lenograstim) in patients undergoing chemotherapy for solid cancers and lymphoma, *European Journal of Cancer*, Vol. 31A, No. 13-14, 1995, pp. 2157–2163.

[507] Mackey, M. and AnDerHeiden, U., Dynamic diseases and bifurcations in physiological control systems, *Funktionelle Biologie und Medizin*, Vol. 1, 1982, pp. 156–164.

[508] Gatti, R., Robinson, W., Deinard, A., Nesbit, M., McCullough, J., Ballow, M., and Good, R., Cyclic leukocytosis in chronic myelogenous leukemia: New perspectives on pathogenesis and therapy, *Blood*, Vol. 41, No. 6, 1973, pp. 771–782.

[509] Chikkappa, G., Borner, G., Burlington, H., Chanana, A., Cronkite, E., Ohl, S., Pavelec, M., and Robertson, J., Periodic oscillation of blood leukocytes, platelets, and reticulocytes in a patient with chronic myelocytic leukemia, *Blood*, Vol. 47, No. 6, 1976, pp. 1023–1030.

[510] Goldman, J. and Melo, J., Chronic myeloid leukemia. Advances in biology and new approaches to treatment, *New England Journal of Medicine*, Vol. 349, No. 15, 2003, pp. 1451–1464.

[511] Beresford, C., Time: A biological dimension, *Journal of the Royal College of Physicians of London*, Vol. 22, No. 2, 1988, pp. 94–96.

[512] Reimann, H., Periodic diseases in the aged, *Geriatrics*, Vol. 24, No. 5, 1969, pp. 146–149.

[513] Hellman, L., Nakada, F., Curti, J., Weitzman, E., Kream, J., Roffwarg, H., Ellman, S., Fukushima, D., and Gallagher, T., Cortisol is secreted episodically by normal man, *Journal of Clinical Endocrinology and Metabolism*, Vol. 30, No. 4, 1970, pp. 411–422.

[514] Ilias, I., Vgontzas, A., Provata, A., and Mastorakos, G., Complexity and non-linear description of diurnal cortisol and growth hormone secretory patterns before and after sleep deprivation., *Endocrine Regulations*, Vol. 36, No. 2, 2002, pp. 63–72.

[515] West, B., *Fractal Physiology and Chaos in Medicine*, World Scientific, Singapore, 1990.

[516] Lenbury, Y. and Pacheenburawana, P., Modelling fluctuation phenomena in the plasma cortisol secretion system in normal man, *Biosystems*, Vol. 26, No. 2, 1991, pp. 117–125.

[517] Dokoumetzidis, A., Iliadis, A., and Macheras, P., Nonlinear dynamics in clinical pharmacology: The paradigm of cortisol secretion and suppression, *British Journal of Clinical Pharmacology*, Vol. 54, No. 1, 2002, pp. 21–29.

[518] Kraan, P., Dullaart, R., Pratt, J., Wolthers, B., Drayer, N., and De-Bruin, R., The daily cortisol production reinvestigated in healthy men. The serum and urinary cortisol production rates are not significantly different, *Journal of Clinical Endocrinology and Metabolism*, Vol. 83, No. 4, 1998, pp. 1247–1252.

[519] Murray, J., *Mathematical Biology*, Vol. 19 of *Biomathematics*, Springer-Verlag, Berlin, 2nd ed., 1993.

[520] Goodwin, B., Oscillatory behavior in enzymatic control processes, *Advances in Enzyme Regulation*, Vol. 3, 1965, pp. 425–438.

[521] Leloup, J. and Goldbeter, A., A model for circadian rhythms in drosophila incorporating the formation of a complex between the PER and TIM proteins, *Journal of Biological Rhythms*, Vol. 13, No. 1, 1998, pp. 70–87.

[522] Rohatagi, S., Bye, A., Mackie, A., and Derendorf, H., Mathematical modeling of cortisol circadian rhythm and cortisol suppression, *European Journal of Pharmaceutics and Biopharmaceutics*, Vol. 4, No. 6, 1996, pp. 341–350.

[523] Hairer, E., Norsett, S., and Wanner, G., *Solving Ordinary Differential Equations: Nonstiff Problems*, Springer-Verlag, Berlin, 2nd ed., 1993.

[524] Farmer, J., Chaotic attractors of an infinite-dimensional dynamical system, *Physica D*, Vol. 4, No. 3, 1982, pp. 366–393.

[525] Gupta, S., Ritchie, J., Ellinwood, E., Wiedemann, K., and Holsboer, F., Modeling the pharmacokinetics and pharmacodynamics of dexamethasone in depressed patients, *European Journal of Clinical Pharmacology*, Vol. 43, No. 1, 1992, pp. 51–55.

[526] Meibohm, B., Hochhaus, G., Mollmann, H., Barth, J., Wagner, M., Krieg, M., Stockmann, R., and Derendorf, H., A pharmacokinetic/pharmacodynamic approach to predict the cumulative cortisol suppression of inhaled corticosteroids, *Journal of Pharmacokinetics and Pharmacodynamics*, Vol. 27, No. 2, 1999, pp. 127–147.

[527] Smith, W., Hypothalamic regulation of pituitary secretion of luteinizing hormone. II. Feedback control of gonadotropin secretion, *Bulletin of Mathematical Biology*, Vol. 42, No. 1, 1980, pp. 57–78.

[528] Cartwright, M. and Husain, M., A model for the control of testosterone secretion, *Journal of Theoretical Biology*, Vol. 123, No. 2, 1986, pp. 239–250.

[529] Liu, B. and Deng, G., An improved mathematical model of hormone secretion in the hypothalamo-pituitary-gonadal axis in man, *Journal of Theoretical Biology*, Vol. 150, No. 1, 1991, pp. 51–58.

[530] Das, P., Roy, A., and Das, A., Stability and oscillations of a negative feedback delay model for the control of testosterone secretion, *Biosystems*, Vol. 32, No. 1, 1994, pp. 61–69.

[531] Topp, B., Promislow, K., Devries, G., Miura, R., and Finegood, D., A model of [beta] -cell mass, insulin, and glucose kinetics: Pathways to diabetes, *Journal of Theoretical Biology*, Vol. 206, No. 4, 2000, pp. 605.

[532] Londergan, C. and Peacock-Lopez, E., Dynamic model of hormonal systems coupled by negative feedback, *Biophysical Chemistry*, Vol. 73, No. 1-2, 1998, pp. 85–107.

[533] Fattinger, K., Verotta, D., Porchet, H., Munafo, A., LeCotonnec, J., and Sheiner, L., Modeling a bivariate control system: LH and testosterone response to the GnRH antagonist antide, *American Journal of Physiology*, Vol. 271, No. 4/1, 1996, pp. E775–E787.

[534] Francheteau, P., Steimer, J., Dubray, C., and Lavene, D., Mathematical model for in vivo pharmacodynamics integrating fluctuation of the response: Application to the prolactin suppressant effect of the dupaminomimetic drug DCN 230-922, *Journal of Pharmacokinetics and Biopharmaceutics*, Vol. 19, No. 3, 1991, pp. 287–309.

[535] Lalonde, R., Gaudreault, J., Karhu, D., and Marriott, T., Mixed-effects modeling of the pharmacodynamic response to the calcimimetic agent R-568, *Clinical Pharmacology and Therapeutics*, Vol. 64, No. 1, 1999, pp. 45–49.

[536] Gobburu, J., Agerso, H., Jusko, W., and Ynddal, L., Pharmacokinetic-pharmacodynamic modeling of ipamorelin, a growth hormone releasing peptide, in human volunteers, *Pharmaceutical Research*, Vol. 16, No. 9, 1999, pp. 1412–1416.

[537] Tolic, I., Mosekilde, E., and Sturis, J., Modeling the insulin-glucose feedback system: The significance of pulsatile insulin secretion, *Journal of Theoretical Biology*, Vol. 207, No. 3, 2000, pp. 361–375.

[538] Prank, K., Harms, H., Dammig, M., Brabant, G., Mitschke, F., and Hesch, R., Is there low-dimensional chaos in pulsatile secretion of parathyroid hormone in normal human subjects? *American Journal of Physiology Endocrinology and Matabolism*, Vol. 266, No. 4, 1994, pp. E653–E658.

[539] Papavasiliou, S., Brue, T., Jaquet, P., and Castanas, E., Pattern of prolactin diurnal secretion in normal humans: Evidence for nonlinear dynamics, *Neuroendocrinology*, Vol. 62, No. 5, 1995, pp. 444–453.

[540] Pincus, S., Approximate entropy as a measure of system complexity, *Proceedings of the National Academy of Sciences USA*, Vol. 88, No. 6, 1991, pp. 2297–2301.

[541] Veldhuis, J. and Pincus, S., Orderliness of hormone release patterns: A complementary measure to conventional pulsatile and circadian analyses, *European Journal of Endocrinology*, Vol. 138, No. 2, 1998, pp. 164–169.

[542] Kroll, M., Biological variation of glucose and insulin includes a deterministic chaotic component, *Biosystems*, Vol. 50, No. 3, 1999, pp. 189–201.

[543] Zeisberger, E., Central modulators of thermoregulation, *Journal of Basic and Clinical Physiology and Pharmacology*, Vol. 1, No. 1-4, 1990, pp. 277–289.

[544] Zeisberger, E., Biogenic amines and thermoregulatory changes, *Progress in Brain Research*, Vol. 115, 1998, pp. 159–176.

[545] Schwartz, P., Rosenthal, N., and Wehr, T., Serotonin 1A receptors, melatonin, and the proportional control thermostat in patients with winter depression, *Archives of General Psychiatry*, Vol. 55, No. 10, 1998, pp. 897–903.

[546] Yu, H. and Lewander, T., Pharmacokinetic and pharmacodynamic studies of (R)-8-hydroxy-2-(di-n-propylamino)tetralin in the rat, *European Neuropsychopharmacology*, Vol. 7, No. 3, 1997, pp. 165–172.

[547] Abdel-Fattah, A., Matsumoto, K., El-Hady, K., and Watanabe, H., 5-HT1A and 5-HT2 receptors mediate hypo- and hyperthermic effects of tryptophan in pargyline-pretreated rats, *Pharmacology Biochemistry and Behavior*, Vol. 52, No. 2, 1995, pp. 379–384.

[548] Salmi, P. and Ahlenius, S., Evidence for functional interactions between 5-HT1A and 5-HT2A receptors in rat thermoregulatory mechanisms, *Pharmacology and Toxicology*, Vol. 82, No. 3, 1998, pp. 122–127.

[549] Hjorth, S., Hypothermia in the rat induced by the potent serotoninergic agent 8-OH-DPAT, *Journal of Neural Transmission*, Vol. 61, No. 1-2, 1985, pp. 131–135.

[550] Goodwin, G., DeSouza, R., Green, A., and Heal, D., The pharmacology of the behavioural and hypothermic responses of rats to 8-hydroxy-2-(di-n-propylamino)tetralin (8-OH-DPAT), *Psychopharmacology (Berl)*, Vol. 91, No. 4, 1987, pp. 506–511.

[551] Wakelkamp, M., Alvan, G., Gabrielsson, J., and Paintaud, G., Pharmacodynamic modeling of furosemide tolerance after multiple intravenous administration, *Clinical Pharmacology and Therapeutics*, Vol. 60, No. 1, 1996, pp. 75–88.

[552] Bauer, J., Balthasar, J., and Fung, H., Application of pharmacodynamic modeling for designing time-variant dosing regimens to overcome nitroglycerin tolerance in experimental heart failure, *Pharmaceutical Research*, Vol. 14, No. 9, 1997, pp. 1140–1145.

[553] Werner, J., Modeling homeostatic responses to heat and cold, *Handbook of Physiology*, edited by C. Blatteis and M. Fregly, Vol. Section 4: Adaptation to the Environment, Oxford University Press, Oxford, 1996, pp. 613–626.

[554] Zuideveld, K., Maas, H., Treijtel, N., Hulshof, J., VanDerGraaf, P., Peletier, L., and Danhof, M., A set-point model with oscillatory behavior predicts the time course of 8-OH-DPAT-induced hypothermia, *American Journal of Physiology. Regulatory, Integrative and Comparative Physiology*, Vol. 281, No. 6, 2001, pp. R2059–R2071.

[555] Gabrielsson, J., Jusko, W., and Alari, L., Modeling of dose-response-time data: Four examples of estimating the turnover parameters and generating kinetic functions from response profiles, *Biopharmaceutics Drug Disposition*, Vol. 21, No. 2, 2000, pp. 41–52.

[556] Kitney, R., A nonlinear model for studying oscillations in the blood pressure control system, *Journal of Biomedical Engineering*, Vol. 1, No. 2, 1979, pp. 89–99.

[557] Gordon, C. and Heath, J., Reassessment of the neural control of body temperature: Importance of oscillating neural and motor components, *Comparative Biochemistry and Physiology A*, Vol. 74, No. 3, 1983, pp. 479–489.

[558] Zuideveld, K., Treijtel, N., Maas, H., Gubbens-Stibbe, J., Peletier, L., VanDerGraaf, P., and Danhof, M., A competitive interaction model predicts the effect of WAY-100,635 on the time course of R-(+)-8-hydroxy-2-(di-n-propylamino)tetralin-induced hypothermia, *Journal of Pharmacology and Experimental Therapeutics*, Vol. 300, No. 1, 2002, pp. 330–338.

[559] Zuideveld, K., Rusic-Pavletic, J., Maas, H., Peletier, L., VanDerGraaf, P., and Danhof, M., Pharmacokinetic-pharmacodynamic modeling of buspirone and its metabolite 1-(2-pyrimidinyl)-piperazine in rats, *Journal of Pharmacology and Experimental Therapeutics*, Vol. 303, No. 3, 2002, pp. 1130–1137.

[560] Zuideveld, K., VanDerGraaf, P., Newgreen, D., Thurlow, R., Petty, N., Jordan, P., Peletier, L., and Danhof, M., Mechanism-based pharmacokinetic-pharmacodynamic modeling of 5-HT1A receptor agonists: Estimation of in vivo affinity and intrinsic efficacy on body temperature in rats, *Journal of Pharmacology and Experimental Therapeutics*, Vol. 308, No. 3, 2004, pp. 1012–1020.

[561] Accardo, A., Affinito, M., Carrozzi, M., and Bouquet, F., Use of the fractal dimension for the analysis of electroencephalographic time series, *Biological Cybernetics*, Vol. 77, No. 5, 1997, pp. 339–350.

[562] Silva, C., Pimentel, I., Andrade, A., Foreid, J., and Ducla-Soares, E., Correlation dimension maps of EEG from epileptic absences, *Brain Topography*, Vol. 11, No. 3, 1999, pp. 201–209.

[563] Stam, K., Tavy, D., Jelles, B., Achtereekte, H., Slaets, J., and Keunen, R., Non-linear dynamical analysis of multichannel EEG: Clinical applications in dementia and Parkinson's disease, *Brain Topography*, Vol. 7, No. 2, 1994, pp. 141–150.

[564] Jeong, J., Kim, D., Chae, J., Kim, S., Ko, H., and Paik, I., Nonlinear analysis of the EEG of schizophrenics with optimal embedding dimension, *Medical Engineering and Physics*, Vol. 20, No. 9, 1998, pp. 669–676.

[565] Ehlers, C., Havstad, J., Prichard, D., and Theiler, J., Low doses of ethanol reduce evidence for nonlinear structure in brain activity, *The Journal of Neuroscience*, Vol. 18, No. 18, 1998, pp. 7474–7486.

[566] Mandema, J. and Danhof, M., Electroencephalogram effect measures and relationships between pharmacokinetics and pharmacodynamics of centrally acting drugs, *Clinical Pharmacokinetics*, Vol. 23, No. 3, 1992, pp. 191–215.

[567] Lehnertz, K., Non-linear time series analysis of intracranial EEG recordings in patients with epilepsy - An overview, *International Journal of Psychophysiology*, Vol. 34, No. 1, 1999, pp. 45–52.

[568] Elger, C., Widman, G., Andrzejak, R., Arnhold, J., David, P., and Lehnertz, K., Nonlinear EEG analysis and its potential role in epileptology, *Epilepsia*, Vol. 41 S3, 2000, pp. S34–38.

[569] Roschke, J. and Aldenhoff, J., A nonlinear approach to brain function: Deterministic chaos and sleep EEG, *Sleep*, Vol. 15, No. 2, 1992, pp. 95–101.

[570] Isaacson, R., Varner, J., Baars, J., and DeWied, D., The effects of pregnenolone sulfate and ethylestrenol on retention of a passive avoidance task, *Brain Research*, Vol. 689, No. 1, 1995, pp. 79–84.

[571] DiMascio, M., DiGiovanni, G., DiMatteo, V., and Esposito, E., Decreased chaos of midbrain dopaminergic neurons after serotonin denervation, *Neuroscience*, Vol. 92, No. 1, 1999, pp. 237–243.

[572] Ishizuka, S. and Hayashi, H., Spontaneous epileptiform bursts and long-term potentiation in rat CA3 hippocampal slices induced by chaotic stimulation of mossy fibers, *Brain Research*, Vol. 790, No. 1-2, 1998, pp. 108–114.

[573] Karlsson, M., Schoemaker, R., Kemp, B., Cohen, A., VanGerven, J., Tuk, B., Peck, C., and Danhof, M., A pharmacodynamic Markov mixed-effects model for the effect of temazepam on sleep, *Clinical Pharmacology and Therapeutics*, Vol. 68, No. 2, 2000, pp. 175–188.

[574] Larter, R., Speelman, B., and Worth, R., A coupled ordinary differential equation lattice model for the simulation of epileptic seizures, *Chaos*, Vol. 9, No. 3, 1999, pp. 795–804.

[575] Cleton, A., Mazee, D., Voskuyl, R., and Danhof, M., Rate of change of blood concentrations is a major determinant of the pharmacodynamics of midazolam in rats, *British Journal of Pharmacology*, Vol. 127, No. 1, 1999, pp. 227–235.

[576] Widman, G., Lehnertz, K., Jansen, P., Meyer, W., Burr, W., and Elger, C., A fast general purpose algorithm for the computation of auto- and cross-correlation integrals from single channel data, *Physica D*, Vol. 121, No. 1-2, 1998, pp. 65–74.

[577] Grassberger, P. and Procaccia, I., Measuring the strangeness of strange attractors, *Physica D*, Vol. 9, No. 1-2, 1983, pp. 189–208.

[578] Widman, G., Schreiber, T., Rehberg, B., Hoeft, A., and Elger, C., Quantification of depth of anesthesia by nonlinear time series analysis of brain

electrical activity, *Physical Review. E, Statistical Physics, Plasmas, Fluids, and Related Interdisciplinary Topics*, Vol. 62, No. 4 Pt A, 2000, pp. 4898–4903.

[579] Rampil, I., A primer for EEG signal processing in anesthesia, *Anesthesiology*, Vol. 89, No. 4, 1998, pp. 980–1002.

[580] Bruhn, J., Lehmann, L., Ropcke, H., Bouillon, T., and Hoeft, A., Shannon entropy applied to the measurement of the electroencephalographic effects of desflurane, *Anesthesiology*, Vol. 95, No. 1, 2001, pp. 30–35.

[581] Wagner, C. and Persson, P., Chaos in the cardiovascular system: An update, *Cardiovascular Research*, Vol. 40, No. 2, 1998, pp. 257–264.

[582] Sugihara, G., Allan, W., Sobel, D., and Allan, K., Nonlinear control of heart rate variability in human infants, *Proceedings of the National Academy of Sciences USA*, Vol. 93, No. 6, 1996, pp. 2608–2613.

[583] Karagueuzian, H., Kogan, B., Khan, S., Denton, T., Karplus, W., Mandel, W., and Diamond, G., Induction of cellular chaos during quinidine toxicity. Predictive power of nonlinear dynamic analysis for drug-induced proarrhythmia -A hypothesis, *Journal of Electrocardiology*, Vol. 24 Suppl, 1992, pp. 91–96.

[584] Garfinkel, A., Spano, M., Ditto, W., and Weiss, J., Controlling cardiac chaos, *Science*, Vol. 257, 1992, pp. 1230–1235.

[585] Scheinin, H., Helminen, A., Huhtala, S., Grönroos, P., Bosch, J., Kuusela, T., Kanto, J., and Kaila, T., Spectral analysis of heart rate variability as a quantitative measure of parasympatholytic effect-integrated pharmacokinetics and pharmacodynamics of three anticholinergic drugs, *Therapeutic Drug Monitoring*, Vol. 21, No. 2, 1999, pp. 141–151.

[586] DeBrouwer, S., Edwards, D., and Griffith, T., Simplification of the quasiperiodic route to chaos in agonist-induced vasomotion by iterative circle maps, *American Journal of Physiology*, Vol. 274, No. 4(2), 1998, pp. H1315–1326.

[587] Weiss, J., Garfinkel, A., Karagueuzian, H., Qu, Z., and Chen, P., Chaos and the transition to ventricular fibrillation: A new approach to antiarrhythmic drug evaluation, *Circulation*, Vol. 99, No. 21, 1999, pp. 2819–2826.

[588] Exner, D., Reiffel, J., Epstein, A., Ledingham, R., Reiter, M., Yao, Q., Duff, H., Follmann, D., Schron, E., Greene, H., Carlson, M., Brodsky, M., Akiyama, T., Baessler, C., and Anderson, J., Beta-blocker use and survival in patients with ventricular fibrillation or symptomatic ventricular tachycardia: The Antiarrhythmics Versus Implantable Defibrillators (AVID) trial, *Journal of American College of Cardiology*, Vol. 34, No. 2, 1999, pp. 325–333.

[589] Echt, D., Liebson, P., Mitchell, L., Peters, R., Obias-Manno, D., Barker, A., Arensberg, D., Baker, A., Friedman, L., Greene, H., Juther, M., and Richardson, D., Mortality and morbidity in patients receiving encainide, flecainide, or placebo. The Cardiac Arrhythmia Suppression Trial, *New England Journal of Medicine*, Vol. 324, No. 12, 1991, pp. 781–788.

[590] Starmer, C., Romashko, D., Reddy, R., Zilberter, Y., Starobin, J., Grant, A., and Krinsky, V., Proarrhythmic response to potassium channel blockade. Numerical studies of polymorphic tachyarrhythmias, *Circulation*, Vol. 92, No. 3, 1995, pp. 595–605.

[591] Qu, Z., Weiss, J., and Garfinkel, A., Cardiac electrical restitution properties and stability of reentrant spiral waves: A simulation study, *American Journal of Physiology*, Vol. 276, No. 1(2), 1999, pp. H269–283.

[592] Garfinkel, A., Kim, Y., Voroshilovsky, O., Qu, Z., Kil, J., Lee, M., Karagueuzian, H., Weiss, J., and Chen, P., Preventing ventricular fibrillation by flattening cardiac restitution, *Proceedings of the National Academy of Sciences USA*, Vol. 97, No. 11, 2000, pp. 6061–6066.

[593] Kowey, P., Pharmacological effects of antiarrhythmic drugs. Review and update, *Archives of Internal Medicine*, Vol. 158, No. 4, 1998, pp. 325–332.

[594] Reiffel, J. and Kowey, P., Generic antiarrhythmics are not therapeutically equivalent for the treatment of tachyarrhythmias, *American Journal of Cardiology*, Vol. 85, No. 9, 2000, pp. 1151–1153, A10.

[595] Glass, L. and Malta, C., Chaos in multi-looped negative feedback systems, *Journal of Theoretical Biology*, Vol. 145, No. 2, 1990, pp. 217–223.

[596] Peacocke, A., *An Introduction to the Physical Chemistry of Biological Organization*, Clarendon Press, Oxford, 1983.

[597] Fisher, G., An introduction to chaos theory and some haematological applications, *Comparative Clinical Pathology*, Vol. 3, No. 1, 1992, pp. 43–51.

[598] Cambel, A., *Applied Chaos Theory: A Paradigm for Complexity*, Academic Press, Boston, 1993.

[599] Elbert, T., Ray, W., Kowalik, Z., Skinner, J., Graf, K., and Birbaumer, N., Chaos and physiology: Deterministic chaos in excitable cell assemblies, *Physiological Reviews*, Vol. 74, No. 1, 1994, pp. 1–47.

[600] Ljung, L. and Gunnarsson, S., Adaptation and tracking in system identification: A survey, *Automatica*, Vol. 26, No. 1, 1990, pp. 7–21.

[601] Mohler, R., *Bilinear Control Processes. With Applications to Engineering, Ecology, and Medicine*, Mathematics in Science and Engineering, Academic Press, New York, 1973.

[602] Savageau, M. and Voit, E., Recasting nonlinear differential equations as S-systems: A canonical nonlinear form, *Mathematical Biosciences*, Vol. 87, No. 1, 1987, pp. 83–115.

[603] Irvine, D. and Savageau, M., Efficient solution of nonlinear ordinary differential equations expressed in S-system canonical form, *SIAM Journal of Numerical Analysis*, Vol. 27, 1990, pp. 704–735.

[604] Candy, J., *Signal Processing. The Model-based Approach*, McGraw-Hill Series in Electrical Engineering, McGraw-Hill, Singapore, 2nd ed., 1987.

[605] Dresher, M., *The Mathematics of Games of Strategy: Theory and Applications*, Dover, New York, 1981.

Index

Interdisciplinary Applied Mathematics

1. *Gutzwiller:* Chaos in Classical and Quantum Mechanics
2. *Wiggins:* Chaotic Transport in Dynamical Systems
3. *Joseph/Renardy:* Fundamentals of Two-Fluid Dynamics: Part I: Mathematical Theory and Applications
4. *Joseph/Renardy:* Fundamentals of Two-Fluid Dynamics: Part II: Lubricated Transport, Drops and Miscible Liquids
5. *Seydel:* Practical Bifurcation and Stability Analysis: From Equilibrium to Chaos
6. *Hornung:* Homogenization and Porous Media
7. *Simo/Hughes:* Computational Inelasticity
8. *Keener/Sneyd:* Mathematical Physiology
9. *Han/Reddy:* Plasticity: Mathematical Theory and Numerical Analysis
10. *Sastry:* Nonlinear Systems: Analysis, Stability, and Control
11. *McCarthy:* Geometric Design of Linkages
12. *Winfree:* The Geometry of Biological Time (Second Edition)
13. *Bleistein/Cohen/Stockwell:* Mathematics of Multidimensional Seismic Imaging, Migration, and Inversion
14. *Okubo/Levin:* Diffusion and Ecological Problems: Modern Perspectives (Second Edition)
15. *Logan:* Transport Modeling in Hydrogeochemical Systems
16. *Torquato:* Random Heterogeneous Materials: Microstructure and Macroscopic Properties
17. *Murray:* Mathematical Biology I: An Introduction (Third Edition)
18. *Murray:* Mathematical Biology II: Spatial Models and Biomedical Applications (Third Edition)
19. *Kimmel/Axelrod:* Branching Processes in Biology
20. *Fall/Marland/Wagner/Tyson (Editors):* Computational Cell Biology
21. *Schlick:* Molecular Modeling and Simulation: An Interdisciplinary Guide
22. *Sahimi:* Heterogeneous Materials: Linear Transport and Optical Properties (Vol. I)
23. *Sahimi:* Heterogeneous Materials: Nonlinear and Breakdown Properties and Atomistic Modeling (Vol. II)
24. *Bloch:* Nonholonomic Mechanics and Control
25. *Beuter/Glass/Mackey/Titcombe:* Nonlinear Dynamics in Physiology and Medicine
26. *Ma/Soatto/Kosecka/Sastry:* An Invitation to 3-D Vision
27. *Ewens:* Mathematical Population Genetics (2nd Edition)
28. *Wyatt:* Quantum Dynamics with Trajectories
29. *Karniadakis:* Microflows and Nanoflows
30. *Macheras/Iliadis:* Modeling in Biopharmaceutics, Pharmacokinetics, and Pharmacodynamics: Homogeneous and Heterogeneous Approaches

LaVergne, TN USA
31 August 2009
156583LV00005B/9/P

9 780387 281780